DEARBORN CHEMICAL COMPANY
853
LIBRARY

AF323217

STABILIZATION AND SOLIDIFICATION
OF HAZARDOUS WASTES

STABILIZATION AND SOLIDIFICATION OF HAZARDOUS WASTES

by

Edwin F. Barth, Paul de Percin

U.S. Environmental Protection Agency
Cincinnati, Ohio

M.M. Arozarena, J.L. Zieleniewski, M. Dosani

PEI Associates, Inc.
Cincinnati, Ohio

H.R. Maxey, S.A. Hokanson, C.A. Pryately
T. Whipple, R. Kravitz

Earth Technology Corporation
Alexandria, Virginia

M.J. Cullinane, Jr., L.W. Jones, P.G. Malone

U.S. Army Engineer Waterways Experiment Station
Vicksburg, Mississippi

U.S. Environmental Protection Agency

Cincinnati, Ohio

NOYES DATA CORPORATION

Park Ridge, New Jersey, U.S.A.

1990

Copyright ©1990 by Noyes Data Corporation
Library of Congress Catalog Card Number: 90-39317
ISBN: 0-8155-1245-7
ISSN: 0090-516X
Printed in the United States

Published in the United States of America by
Noyes Data Corporation
Mill Road, Park Ridge, New Jersey 07656

10 9 8 7 6 5 4 3 2 1

Library of Congress Cataloging-in-Publication Data

Stabilization and solidification of hazardous wastes / by Edwin F.
    Barth . . . [et al.] .
         p.   cm. -- (Pollution technology review, ISSN 0090-516X ; no.
    186)
       Includes bibliographical references and index.
       ISBN 0-8155-1245-7 :
       1. Hazardous wastes--Purification. 2. Solidification. I. Barth,
    Edwin F. II. Series.
    TD1060.S73   1990
    628.4'2--dc20   90-39317
                                                         CIP

Foreword

This book provides designers and reviewers of remedial action plans for hazardous waste disposal sites with the information and general guidance necessary to judge the feasibility of stabilization/solidification technology for the control of pollutant migration from hazardous wastes disposed of on land. Stabilization/solidification is an alternative technology that must be identified, analyzed, and evaluated in the feasibility study process.

In general, the book should prove useful to all technical and professional people working in the stabilization/solidification field, including State environmental protection agencies, private industry, commercial treatment and disposal facilities, and environmental consultants. Not only does the book cover a wide range of applications, it has been written to bring the less-experienced professional up to speed quickly while providing detailed information in many of the more technical areas.

Solidification/stabilization technology has been used for over 20 years to treat U.S. industrial waste and more recently for contaminated soils and municipal waste combustion residuals. Definitions for the technology vary depending upon the source. Other terms that have been used are immobilization and fixation. In very general terms, stabilization/solidification, as it relates to managing hazardous waste, refers to a technology where one uses additives or processes to transform the waste into a more manageable form or less toxic form by physically and/or chemically immobilizing the waste constituents.

The book has been written and organized to provide the reader with an informative, yet quick reference-type handbook that can be used by environmental professionals. Part I describes physical tests, chemical testing procedures, technology and screening, and field activities pertaining to stabilization/solidification. Part II, in addition to technology, presents information on reagent costs and typical equipment used in processing. Part III is the summation of an immobilization technology seminar, also covering the multiple aspects of stabilization/solidification technology.

The information in the book is from the following documents:

> *Stabilization/Solidification of CERCLA and RCRA Wastes—Physical Tests, Chemical Testing Procedures, Technology and Screening, and Field Activities,* prepared by Edwin F. Barth and Paul de Percin of the U.S. Environmental Protection Agency; M.M. Arozarena, J.L. Zieleniewski, and M. Dosani of PEI Associates, Inc.; and H.R. Maxey, S.A. Hokanson, C.A. Pryately, T. Whipple, and R. Kravitz of Earth Technology Corporation, for the U.S. Environmental Protection Agency, May 1989.

Handbook for Stabilization/Solidification of Hazardous Wastes, prepared by M.J. Cullinane, Jr., L.W. Jones, and P.G. Malone of the U.S. Army Waterways Experiment Station for the U.S. Environmental Protection Agency, June 1986.

Immobilization Technology Seminar—Speaker Slide Copies and Supporting Information, prepared by the U.S. Environmental Protection Agency, October 1989.

The table of contents is organized in such a way as to serve as a subject index and provides easy access to the information contained in the book.

Advanced composition and production methods developed by Noyes Data Corporation are employed to bring this durably bound book to you in a minimum of time. Special techniques are used to close the gap between "manuscript" and "completed book." In order to keep the price of the book to a reasonable level, it has been partially reproduced by photo-offset directly from the original reports and the cost saving passed on to the reader. Due to this method of publishing, certain portions of the book may be less legible than desired.

NOTICE

Contents and Subject Index

PART III

Part I

The information in Part I is from *Stabilization/Solidification of CERCLA and RCRA Wastes—Physical Tests, Chemical Testing Procedures, Technology and Screening, and Field Activities,* prepared by Edwin F. Barth and Paul de Percin of the U.S. Environmental Protection Agency; M.M. Arozarena, J.L. Zieleniewski, and M. Dosani of PEI Associates, Inc.; and H.R. Maxey, S.A. Hokanson, C.A. Pryately, T. Whipple, and R. Kravitz of Earth Technology Corporation, for the U.S. Environmental Protection Agency, May 1989.

ACKNOWLEDGMENTS

This handbook was developed under Contract No. 68-03-3413, Work Assignment Nos. 0-26 and 1-26, to the U.S. Environmental Protection Agency. The authors were Mr. Michael M. Arozarena, Ms. Jan L. Zieleniewski, and Mr. Majid Dosani with PEI Associates, Inc., and Mr. H. Robert Maxey, P.E.; Ms. Sarah A. Hokanson; Ms. Christine A. Pryately; Mr. Timothy Whipple; and Mr. Robert Kravitz with Earth Technology Corporation. Mr. Maxey and his associates contributed the bulk of the technical input to this handbook, and their dedication to this project is greatly appreciated. The handbook was edited by Ms. Martha H. Phillips of PEI Associates.

The project was conducted under the management supervision of Mr. Clarence A. Clemons. Technical direction and supervision were provided by Mr. Edwin F. Barth and Mr. Paul de Percin.

Development of the handbook involved the participation of the following reviewers, whose comments contributed greatly to its quality:

Dr. Paul L. Bishop	University of Cincinnati
Mr. Jesse R. Connor	Chemical Waste Management, Inc.
Mr. Pierre L. Cote'	Environment Canada, Wastewater Technology Center
Mr. M. John Cullinane, Jr., P.E.	U.S. Army Corps of Engineers, Waterways Experiment Station
Dr. John W. Liskowitz	New Jersey Institute of Technology
Dr. Frank R. Prince	American Petroleum Institute
Dr. R. Soundararajan	RMC Environmental and Analytical Laboratories
Dr. Steven I. Taub, P.E.	CECO Filters, Inc.

We would also like to acknowledge Mr. Kenneth Gutschick, Mr. David Musser, and Mr. Steven Day with the National Lime Association, ENRECO Company, and Geo-Con Company, respectively, for their technical contributions to this handbook. We would also like to acknowledge PEI graphics artist Ms. Joy Wallace, who performed the desktop publishing of this document.

1. Introduction

1.1 Purpose of Handbook

The primary purpose of this handbook is to provide the U.S. Environmental Protection Agency (EPA) regional staff responsible for reviewing CERCLA remedial action plans and RCRA permit applications with a tool for interpreting information on stabilization/solidification treatment. As a practical day-to-day reference guide, it will provide the EPA staff and others with a quick update on particular aspects of stabilization/solidification. It also contains more detailed information on specific subjects and a complete list of recent references.

In general, this handbook should prove useful to all technical and professional people working in the stabilization/solidification field, including State environmental protection agencies, private industry, commercial treatment and disposal facilities, and environmental consultants. Not only does it cover a wide range of applications, it has been written to bring the less-experienced professional up to speed quickly while providing detailed information in many of the more technical areas.

1.2 Terminology and Definitions

In the literature, several terms are used to describe the stabilization/ solidification of hazardous waste, and the meaning of these terms is often not clear to the reader. In this handbook, the meanings of the terms "stabilization" and "solidification" are as defined in the EPA publication "Guide to the Disposal of Chemically Stabilized and Solidified Waste" (Malone et al. 1980). Both stabilization and solidification refer to treatment processes that are designed to accomplish one or more of the following: 1) improve the handling and physical characteristics of the waste, as in the sorption of free liquids; 2) decrease the surface area of the waste mass across which transfer or loss of contaminants can occur; and 3) limit the solubility of any hazardous constituents of the waste, e.g., by pH adjustment or sorption phenomena.

The preliminary benefit of stabilization techniques is that they limit the solubility or mobility of the contaminants with or without changing or improving the physical characteristics of the waste. Stabilization usually involves adding materials that will ensure that the hazardous constituents are maintained in their least mobile or toxic form. The addition of lime or sulfide to a metal hydroxide waste to precipitate the metal ions or the addition of a sorbent to an organic waste are examples of stabilization techniques.

Solidification implies that the beneficial results of treatment are obtained primarily, but not necessarily exclusively, through the production of a solid block of waste material with high structural integrity—a product often referred to as a "monolith." (In many cases, a monolith is not the end product of the stabilization/solidification process; however, after placement, the materials may continue to cure into a facsimile of a monolith.) The monolith can encompass the entire waste disposal site—called a "monofill"—or be as small as the contents of a steel drum. The contaminants do not necessarily interact chemically with reagents. Instead, they are mechanically locked within the solidified matrix. This is known as "microencapsulation." Contaminant loss is limited largely by decreasing the surface area exposed to the environment and/or isolating the contaminants from environmental influences by microencapsulating the waste particles. Wastes can also be "macroencapsulated," that is, bonded to or surrounded by an impervious covering. These techniques are also considered to be stabilization/solidification processes.

The term "fixation" has fallen in and out of favor, but it is widely used in the waste treatment field to mean any of the stabilization/solidification processes just described. "Fixed" wastes are those that have been treated in this manner. The term "immobilization" refers to those techniques that tend to decrease the concentration of toxic compounds by attaching them onto the surfaces of the solid particles.

Both solidification and chemical stabilization are usually included in commercial processes and result in the transformation of liquids or semisolids into environmentally safer forms. For example, a metal-rich sludge would be considered stabilized if it were mixed with a dry sorber such as fly ash or dry soil. If the sorbent and waste were then cemented into an impermeable, monolithic block, benefits would be even greater. In this handbook, the terminology used most often is stabilization/solidification (Cullinane et al. 1986).

1.3 Handbook Overview

This handbook has been written and organized to provide the reader with an informative, yet quick reference-type handbook that can be used by environmental professionals. Section 2 addresses the basis for the stabilization/solidification of hazardous waste and includes a discussion of the Resource Conservation and

Recovery Act (RCRA); the Hazardous and Solid Waste Amendments of 1984 (HSWA); the Comprehensive Environmental Response, Compensation, and Liability Act of 1980 (CERCLA); and the Superfund Amendments Reauthorization Act (SARA) and their requirements with regard to stabilization/ solidification processes. Section 3 presents state-of-the-art stabilization/solidification technologies. Section 4 discusses in depth the physical testing methods used to characterize solid and hazardous wastes before and after stabilization/solidification.

Section 5 addresses chemical testing procedures and includes an overview of leaching mechanisms, leach test methods and applications, factors affecting results, and the selection and interpretation of leach tests. Section 6 provides information on technology screening. Finally, Section 7 discusses the proper application of stabilization/solidification processes and the site conditions that can determine if a particular stabilization/solidification process is appropriate.

2. The Basis for Stabilization/Solidification

Stabilization/solidification is a proven technology for the treatment of hazardous wastes and hazardous waste sites. Technical reasons for the selection of stabilization/solidification as a remediation technology include:

- It improves the handling and physical characteristics of the wastes; e.g., sludges are processed into solids.

- It reduces transfer or loss of contained pollutants by decreasing the surface area.

- It reduces pollutant solubility in the treated waste, generally by chemical changes.

- Techniques for processing sludges are well established; however, they are not set for soils and debris.

- Residues from the treatment of hazardous wastes by physical/chemical, biological, or incineration technologies can be further treated by stabilization/ solidification.

- Because of excavation problems, in situ treatment by stabilization/solidification is the only viable management technique in many cases.

- Alternate hazardous waste treatment and disposal techniques are often economically prohibitive.

Thus, stabilization/solidification has considerable technical merit. Selection of stabilization/solidification as a remediation technology is also supported by recent developments in the environmental regulations. The remainder of this section addresses the Resource Conservation and Recovery Act; the Hazardous and Solid Waste Amendments; the Comprehensive Environmental Response, Compensation, and Liability Act; and the Superfund Amendments as they pertain to the stabilization/solidification of hazardous waste.

2.1 Resource Conservation and Recovery Act (RCRA)

The disposal of hazardous liquid, sludge, or semisolid waste has been a controversial issue since the passage of RCRA in 1976. Prior to RCRA the disposal of liquid waste, other than by underground injection, was regulated under the authority of the Clean Water Act or controlled through State laws. The first liquid-waste-disposal regulations proposed under RCRA in December 1979 brought reactions ranging from suggestions that no restrictions were needed to demands that land disposal of liquids be banned entirely. The EPA believed that bulk liquids, sludge, and semisolid wastes could be placed in a landfill under certain controlled conditions, such as with a secure liner and a system for the collection and removal of leachate. If these measures were not available, EPA required treatment by mixing the waste with materials such as soil, fly ash, or cement kiln dust to stabilize or solidify the waste to ensure that free liquids were no longer present. (Free liquids are defined as liquids that readily separate from the solid portion of a waste under ambient temperature and pressure.) Biodegradable stabilizers were not recommended because their long-term effectiveness was unproven (U.S. EPA 1980a).

2.2 Hazardous and Solid Waste Amendments (HSWA)

During the early 1980's, EPA proposed additional regulations to control the disposal of liquid waste. These included the use of a paint filter test to determine the presence of free liquids in sludges, semisolids, slurries, and other wastes. They were particularly concerned with the disposal of containerized liquid wastes because of possible leachate generation and subsidence of the final landfill cover as a result of container degradation (U.S. EPA 1982). On November 8, 1984, the Hazardous and Solid Waste Amendments (HSWA) to RCRA were signed into law. These amendments significantly expanded both the scope and the requirements of RCRA and specifically addressed the issue of liquid waste disposal. Section 3004(c)(1) of HSWA states:

> "Effective 6 months after the date of enactment of the Hazardous and Solid Waste Amendments of 1984, the placement of bulk or noncontainerized liquid hazardous waste or free liquids contained in hazardous waste (whether or not absorbents have been added) in any landfill is prohibited."

This Congressionally mandated, absolute ban on the placement of bulk liquids in a landfill for any purpose or length of time, regardless of the presence of liners or leachate collection systems, was effective May 8, 1985.

Section 3004(c)(2) of HSWA further requires EPA to promulgate regulations within 15 months that:

> "(A) minimize the disposal of containerized liquid hazardous waste in landfills, and (B) minimize the presence of free liquids in containerized hazardous waste to be disposed of in landfills. Such regulations shall also prohibit the disposal in landfills of liquids that have been absorbed in materials that biodegrade or that release liquids when compressed as might occur during routine landfill operations."

The legislative history of these amendments reveals that Congress considered prohibiting entirely the placement of all liquids, containerized or not, in a landfill, but later reconsidered containerized liquids, particularly those designed to hold small quantities, such as ampules or lab packs (U.S. EPA 1986a).

To comply with Section 3004(c)(1), an owner or operator must first use the Paint Filter Liquids Test (U.S. EPA 1986b) to determine whether a waste is a liquid or contains free liquids. If a sample passes the test, the waste is not subject to the ban. If it does not pass, the waste must be chemically, thermally, or biologically treated prior to landfilling by the application of a technology that does not involve the use of a material that functions primarily as a sorbent (including both absorbents and adsorbents). The EPA believes the purpose of this congressional ban on sorbents is to force the use of treatment technologies that are not reversible, such as chemical stabilization. An absorbent may release the absorbed liquid back into the landfill; chemical stabilization renders liquids permanently unavailable to the environment. The emphasis is on a permanent solution, not a temporary alternative (U.S. EPA 1986c).

The use of a sorbent as part of the chemical stabilization process can make it difficult to determine whether true stabilization has taken place. If there is any doubt, the EPA guidance document "Prohibition on the Disposal of Bulk Liquid Hazardous Waste in Landfills - Statutory Interpretive Guidance" recommends the use of an unconfined compressive strength test as an indirect method for determining the extent to which the waste has been chemically transformed into a solid state. The test should be modeled on ASTM D2166-85, Unconfined Strength of Cohesive Soil. This test is applicable to a wide range of stabilized wastes, regardless of the specific waste type or stabilization process used. A minimum strength of 50 psi is recommended as a measure of adequate bonding. The rationale for selecting this value is an attempt to require a bonding level in excess of that achieved with sorbents. A minimum compressive strength limit of 50 psi should assure that the treated waste has at least as much strength as the soil surrounding the disposal site. In general, the addition of large quantities of stabilization agents resulted in a more solid product with greater strength (U.S. EPA 1986c).

If an owner or operator wishes, it is acceptable to use a different method to prove that a chemical reaction has occurred. Several waste-specific or additive-specific tests are available; however, the data must prove conclusively that chemical stabilization/solidification, not absorption/adsorption, has taken place and that the hazardous constituents are in their least soluble, least toxic state (U.S. EPA 1986c).

Sections 3004(d), (e), and (g) of HSWA prohibit continued land disposal of untreated hazardous waste unless it can be proven there will be "no migration of hazardous constituents...as long as the wastes remain hazardous." Section 3004(m) requires EPA to establish "...levels or methods of treatment, if any, which substantially diminish the toxicity of the waste or substantially reduce the likelihood of migration of hazardous constituents from the waste so that short-term and long-term threats to human health and the environment are minimized."

The Congress set deadlines for developing standards for wastes containing solvents and dioxins, for wastes containing constituents known as the "California list" chemicals, and for all remaining hazardous wastes identified as of November 8, 1984. All the remaining hazardous wastes were ranked into three groups according to their intrinsic hazard and their volume. Based on that ranking, treatment standards are to be developed according to a schedule, with the standards for the highest ranking third being set first (U.S. EPA 1988a).

The key portion of this section of the 1984 HSWA is the mandate for treatment standards for every waste or group of similar wastes. Treatment standards are based on the performance of the best demonstrated available technology (BDAT) to treat the waste. A technology is considered best if its performance is statistically better than that of other technologies. A technology is considered to have been demonstrated for a particular waste if it is in full-scale commercial operation for treatment of that waste or a similar waste. An available technology is one that is commercially available or can be leased. Treatment standards can be established either as a specific technology or as a concentration level based on a BDAT technology. When treatment standards are fixed at a concentration level, the regulated community can use any technology not otherwise prohibited to treat a waste so that it meets the treatment standard. Dilution with aqueous or nonaqueous agents is prohibited either as a partial or as a complete substitute for adequate treatment of a restricted waste (U.S. EPA 1986d). Addition of materials as part of the treatment process, such as fixatives, is not prohibited, however. When treatment standards were first proposed on April 8, 1988, EPA indicated a preference for establishing concentration-based treatment standards rather than treatment standards expressed as specific technologies because this provides greater flexibility in meeting the goal and encourages the development of more efficient and innovative technologies (U.S. EPA 1988a). Compliance is determined by measuring the concentration level of the hazardous constituents in the waste, the residual, or an extract of the waste or treatment residual. When treatment standards are expressed as specific technologies, however, those technologies must be used (U.S. EPA 1988b).

On August 17, 1988, EPA promulgated treatment standards for hazardous wastes listed in 40 CFR 268.10 (the first one-third of the schedule of restricted hazardous wastes). For three of these wastes, F006, K046, and K022 nonwastewaters, the BDAT performance standard is based on stabilization; four other wastes included in this regulation list stabilization as the BDAT for nonwastewater residuals (K001 and K086) or ash residue (K101 and K102) following the initial treatment. On January 11, 1989, EPA proposed treatment standards for hazardous wastes listed in 40 CFR 268.11 (the second one-third of the schedule of restricted hazardous wastes) (U.S. EPA 1989). The EPA also proposed treatment standards for some First Third wastes listed in 40 CFR 268.10 and Third Third wastes listed in 40 CFR 268.12. With regard to these wastes, proposed BDAT performance standards for one Second Third waste (F012) and two First Third wastes (F006 and F019) were based on stabilization. Treatment standards for F006 were previously promulgated in the August 17 regulations; however, at that time, standards for cyanide were reserved. In the January 11 proposal, EPA added proposed treatment standards for total and amenable cyanide in F006 nonwastewaters. Figure 2-1 shows the BDAT treatment standards for these waste codes as published in the <u>Federal Register</u>. In each case, research for the development of the performance standard indicated that full-scale stabilization is widely used throughout the country to bind these waste constituents into a cementitious matrix that immobilizes them, and thereby reduces their leaching potential.

2.3 The Comprehensive Environmental Response, Compensation, and Liability Act (CERCLA)

The Comprehensive Environmental Response, Compensation, and Liability Act (CERCLA) of 1980 was an important landmark in the legislative battle against hazardous waste. It was the first comprehensive Federal law to establish a mechanism of response for the immediate cleanup of hazardous waste contamination from an accidental spill or from chronic degradation (such as from an abandoned waste site). It also required that the National Contingency Plan (NCP) be revised to include a new subpart (known as the National Hazardous Substance Response Plan) that would establish procedures and standards for responding to releases of hazardous substances, pollutants, and contaminants. Promulgated in mid-1982 and revised in late 1985, the Hazardous Substances Response subpart of the NCP first outlined the level of cleanup necessary at a Superfund site and allowed the Federal Government, or a cooperating State government, to "take any appropriate action to abate, minimize, stabilize, mitigate, or eliminate the release..." at any site that posed a potential threat to the public health or the environment (U.S. EPA 1985). Specific techniques mentioned in the NCP for remedial action at hazardous waste sites include solidification for treatment of contaminated soils and sediments.

On December 21, 1988, EPA issued proposed rules that, if promulgated, will revise the NCP. The proposed NCP revisions are intended to implement regulatory changes necessitated by the Superfund Amendments and Reauthorization Act (SARA) as well as to clarify existing NCP language and to reorganize the NCP to coincide more accurately with the sequence of response actions.

2.4 Superfund Amendments and Reauthorization Act (SARA)

The most recent legislation concerned with hazardous waste cleanup and disposal is Public Law 99-499, the Superfund Amendments and Reauthorization Act of 1986 (SARA). The benefits of SARA, which added a new dimension to CERCLA, are twofold: 1) it reaffirms Congress' commitment to take whatever action is necessary to stop further deterioration of the environment through the unwise removal and/or disposal of hazardous waste, and 2) it establishes a strong innovative technology program for the development of new onsite methods of treating hazardous waste.

Section 121 of SARA (Cleanup Standards) strongly recommends remedial actions involving onsite treatment that "... permanently and significantly reduces the volume, toxicity, or mobility of hazardous substances." The actions must assure the protection of human health and the environment, must be cost-effective, and to the extent practicable, also must be in accordance with the NCP. This means that the first step in the selection of an appropriate remedial action is to determine the level of cleanup necessary to protect the environment, and the second step is to choose the most cost-efficient means of achieving that goal. Section 121 further states that "off-site transport and disposal...without such treatment should be the least favored alternative remedial action where practical treatment technologies are available." This strongly encourages the use of the onsite treatment technologies recommended in the RCRA regulations. Once again, emphasis is on a permanent solution, not a temporary alternative.

Figure 2-1. BDAT treatment standards for waste codes based on stabilization.

BDAT TREATMENT STANDARDS METAL FINISHING
SLUDGES SUBCATEGORY FOR F006
[Nonwastewaters]

Constituent	Maximum for any single grab sample	
	Total composition (mg/kg)	TCLP (mg/l)
Cadmium	(1)	0.066
Chromium (total)	(1)	5.2
Lead	(1)	.51
Nickel	(1)	.32
Silver	(1)	.072
Cyanides (total)	110	(1)
Cyanides (amenable)	0.064	(1)

[1] Not applicable.

BDAT TREATMENT STANDARDS FOR K001
(Nonwastewaters)

Constituent	Maximum for any single grab sample	
	Total composition (mg/kg)	TCLP (mg/l)
Naphthalene	0.15	(1)
Pentachlorophenol	.88	(1)
Phenanthrene	.15	(1)
Pyrene	.14	(1)
Toluene	.14	(1)
Xylenes	.16	(1)
Lead	.037	(1)

[1] Not applicable.

BDAT TREATMENT STANDARDS METAL FINISHING
SLUDGES SUBCATEGORY FOR F012 AND F019
[Nonwastewaters]

Constituent	Maximum for any single grab sample	
	Total composition (mg/kg)	TCLP (mg/l)
Cyanides (total)	110	(1)
Cyanides (amenable)	0.064	(1)
Cadmium	(1)	0.066
Chromium	(1)	5.2
Lead	(1)	0.51
Nickel	(1)	0.32
Silver	(1)	0.072

[1] Not applicable.

BDAT TREATMENT STANDARDS FOR K022
[Nonwastewaters]

Constituent	Maximum for any single grab sample	
	Total composition (mg/kg)	TCLP (mg/l)
Acetophenone	19	(1)
Sum of diphenylamine and diphenylnitrosamine	13	(1)
Phenol	12	(1)
Toluene	.034	(1)
Chromium (total)	(1)	5.2
Nickel	(1)	.32

[1] Not applicable.

Figure 2-1. BDAT treatment standards for waste codes based on stabilization. (continued)

BDAT TREATMENT STANDARDS FOR K046
(Nonwastewaters]
[Nonreactive subcategory]

Constituent	Maximum for any single grab sample	
	Total composition (mg/kg)	TCLP (mg/l)
Lead	(1)	0.18

[1] Not applicable.

BDAT TREATMENT STANDARDS FOR K086
(Nonwastewaters]
[Solvent Washes Subcategory]

Constituent	Maximum for any single grab sample	
	Total composition (mg/kg)	TCLP (mg/l)
Acetone	0.37	(1)
bis(2-ethylhexyl) phthalate	.49	(1)
n-Butyl alcohol	.37	(1)
Cyclohexanone	.49	(1)
1,2-Dichlorobenzene	.49	(1)
Ethyl acetate	.37	(1)
Ethyl benzene	.031	(1)
Methanol	.37	(1)
Methylene chloride	.037	(1)
Methyl ethyl ketone	.37	(1)
Methyl isobutyl ketone	.37	(1)
Naphthalene	.49	(1)
Nitrobenzene	.49	(1)
Toluene	.031	(1)
1,1,1-Trichloroethane	.044	(1)
Trichloroethylene	.031	(1)
Xylenes	.015	(1)
Chromium (total)	(1)	0.094
Lead	(1)	.37

[1] Not applicable.

BDAT TREATMENT STANDARDS FOR K101
(Nonwastewaters]
[Low Arsenic Subcategory - less than 1% total arsenic]

Constituent	Maximum for any single grab sample	
	Total composition (mg/kg)	TCLP (mg/l)
Ortho-nitroaniline	14	(1)
Cadmium	(1)	0.066
Chromium (total)	(1)	5.2
Lead	(1)	.51
Nickel	(1)	.32

[1] Not applicable.

BDAT TREATMENT STANDARDS FOR K102
(Nonwastewaters]
[Low Arsenic Subcategory - less than 1% total arsenic]

Constituent	Maximum for any single grab sample	
	Total composition (mg/kg)	TCLP (mg/l)
Ortho-nitrophenol	13	(1)
Cadmium	(1)	0.066
Chromium (total)	(1)	5.2
Lead	(1)	.51
Nickel	(1)	.32

[1] Not applicable.

A new criterion that SARA added to CERCLA for use in determining cleanup priority concerns the contamination or potential contamination of the ambient air. Because stabilization/solidification often involves the treatment of organic constituents, this criterion is of primary concern in that the stabilization/solidification method selected must not release VOC's into the ambient air.

The second benefit from SARA—the requirement that EPA establish a coordinated research program aimed at developing alternate treatment technologies that will be applicable for onsite treatment—will prevent the potentially dangerous transport of wastes to other locations. The purpose of the Superfund Innovative Technology Evaluation (SITE) Program is to maximize the use of technologies other than land disposal for treating wastes found at unregulated hazardous waste sites. One way of accomplishing this is through the full-scale demonstration of favorable, innovative technologies that provide permanent cleanup solutions. The SITE program involves a collaborative arrangement with the private sector in which the EPA selects the technologies, locations, and wastes; the developers set up and demonstrate their process or equipment on the site; and EPA evaluates how the system performs. This approach lends credibility to effective innovative technologies and speeds up the process of commercialization (Skinner and Bassin 1988).

The State of California also sponsors an innovative technology program. Many of the California SITE projects involve the demonstration of a stabilization/solidification process (McGraw-Hill 1988). Several other States have established the mechanisms for SITE programs and expect to begin operating them soon. These States will either work jointly with the EPA to help select sites for cofunded research projects or as independently operated studies. It is hoped that, through these programs, new methods of stabilization/solidification will be developed to take their place alongside the current commercial processes as one of the most efficient, permanent, and cost-effective methods of treating appropriate hazardous wastes.

3. State of the Art Stabilization/Solidification Processes

This section provides a brief introduction to stabilization/solidification processes to promote a better understanding of these processes. Much of the stabilization/solidification that now occurs in the United States is based on the chemistry of lime or cement. Although not currently used on a widespread basis, some organic processes may have considerable use in the future or for specialized applications. The widespread use of organic processes for hazardous waste stabilization/solidification has been hampered by their high cost, their high energy consumption, or their specialized applications. They have been used in the nuclear industry, however, and may have application for certain hazardous wastes.

Other considerations that are important in an evaluation of a stabilization/solidification process include 1) pretreatment, i.e., whether the waste will benefit from an operation such as neutralization; 2) proprietary agent addition, i.e., whether the stabilized/solidified waste will benefit from the other agents; and 3) site considerations, i.e., what influence the site itself will have on treatment operations.

3.1 Process Overview of Stabilization/Solidification Technology

A brief overview of four stabilization/solidification technologies is presented in this section. These four technologies can be grouped as inorganic stabilization/solidification—cement-based and pozzolanic; and organic stabilization/solidification—thermoplastic and organic polymerization. Inorganic stabilization/solidification with cements and pozzolans has been used for hazardous waste more often than the other technologies. Organic stabilization/solidification with thermoplastic binders and organic polymerization has been applied to select hazardous wastes; however, this has taken place on a much smaller scale than stabilization/solidification with cements and pozzolans. Examples of the types of wastes that have been stabilized/solidified with these four binders are included at the end of each description.

3.1.1 Inorganic Stabilization/Solidification

3.1.1.1 Cement-Based Stabilization/Solidification

Cement-based stabilization/solidification is a process in which waste materials are mixed with portland cement. Water is added to the mixture, if it is not already present in the waste material, to ensure the proper hydration reactions necessary for bonding the cement. The wastes are incorporated into the cement matrix and, in some cases, undergo physical-chemical changes that further reduce their mobility in the waste-cement matrix. Typically, hydroxides of metals are formed, which are much less soluble than other ionic species of the metals. Small amounts of fly ash, sodium silicate, bentonite, or proprietary additives are often added to the cement to enhance processing. The final product may vary from a granular, soil-like material to a cohesive solid, depending on the amount of reagent added and the types and amounts of wastes stabilized/solidified.

Cement-based stabilization/solidification has been applied to plating wastes containing various metals such as cadmium, chromium, copper, lead, nickel, and zinc. Cement has also been used with complex wastes containing PCBs, oils, and oil sludges; wastes containing vinyl chloride and ethylene dichloride; resins; stabilized/solidified plastics; asbestos; sulfides; and other materials (Jones 1986; Tittlebaum and Seals 1985). Studies performed under the BDAT program on contaminated soil showed cement-based stabilization/solidification effective for arsenic, lead, zinc, copper, cadmium, and nickel (Weitzman 1988). Effectiveness of this process on organics is not known.

3.1.1.2 Pozzolanic Stabilization/Solidification

Pozzolanic stabilization/solidification involves siliceous and aluminosilicate materials, which do not display cementing action alone, but form cementitious substances when combined with lime or cement and water at ambient temperatures. The primary containment mechanism is the physical entrapment of the contaminant in the pozzolan matrix. Examples of common pozzolans are fly ash, pumice, lime kiln dusts, and blast furnace slag. Pozzolans contain significant amounts of silicates, which distinguish them from the lime-based materials. The final product can vary from a soft fine-grained material to a hard cohesive material similar in appearance to cement. Pozzolanic reactions are generally much slower than cement reactions. Waste materials that have been stabilized/solidified with pozzolans include oil sludges, plating sludges containing various metals (aluminum, nickel, copper, lead, chromium, and arsenic), waste acids, and creosote.

3.1.2 Organic Stabilization/Solidification

3.1.2.1 Thermoplastic Stabilization/Solidification

Thermoplastic stabilization/solidification is a microencapsulation process in which the waste materials do not react chemically with the encapsulating material. In this technology, a thermoplastic material, such as asphalt (bitumen) or polyethylene, is used to bind the waste constituents into a stabilized/solidified mass. The asphalt binder may be heated before it is mixed with a dry waste material, or the asphalt may be applied as a cold mix. In the latter case, compaction is used to remove additional water from the surrounding aggregate/waste particles. Bitumen may have commercial application for stabilizing/solidifying oil- and gasoline-contaminated soils. In this application, the hydrocarbon-contaminated soil is used to dilute the bitumen, which is then used as paving or patching material for roads. The resulting consistency will vary depending on the density of the hydrocarbon mixed into the bitumen and the amount of aggregate added to the mixture. Thermoplastic encapsulation can also be applied to electroplating sludges, painting and refinery sludges containing metals and organics, dry incinerator ash, fabric filter dust, and radioactive wastes (Tittlebaum et al. 1985).

3.1.2.2 Macroencapsulation

Macroencapsulation is a process that places the waste in an overpack drum. The waste is generally stabilized/solidified and (sometimes) microencapsulated prior to drumming. At one time it was believed that liquids could be stored in these drums because of their excellent integrity and shock resistance. The banning of liquids from landfills has effectively eliminated this option. Most macroencapsulated work stems from that originally conducted by Lubowitz and Wiles (1979) and fully developed by Environmental Protection Polymers (Unger et al., undated). A second macroencapsulated process was developed by Bondico, Inc. (Shaw 1987).

Typical wastes that Environmental Protection Polymers believe to be appropriate for macroencapsulation include low-level nuclear waste, electroplating sludge, and coal scrubber sludge (Unger et al.). Other appropriate candidates are highly concentrated or toxic materials such as incinerator ash, PCBs, and dioxins. The Environmental Protection Polymers process chemically stabilizes/solidifies a sludge with lime or cement. This material is then agglomerated by a polybutadiene binder (actually representing microencapsulation). Environmental Protection Polymers chose polybutadiene for agglomeration because of its high tolerance for a variety of pollutants and because it yields tough products. The polybutadiene agglomerates the waste by a thermosetting reaction in a polymerization process at temperatures of 120° to 200°C.

After the agglomeration step, the waste matrix is encapsulated. Encapsulation has been accomplished with polyethylene resin. Environmental Protection Polymers has actually molded this material onto some of its agglomerated products. Much effort has also been devoted to the development of a spin-welding process, which consists of four basic steps:

1) Polyethylene overpack drums are filled with agglomerated waste and then transferred to the loading area of the welding unit.

2) The operator activates the drum-positioning mechanism to position the drum for welding.

3) The lid is welded to the rotating drum.

4) The operator unclamps the drum and removes the encapsulated material.

Bondico, Inc., also developed a macroencapsulation system in which an electrical current fused the polyethylene lid and container together (Shaw 1987).

When considered on its technical merits alone, macroencapsulation appears to be an attractive stabilization/solidification technology. It has not been widely used, however, and is apparently unavailable for immediate use because Bondico is no longer in business. Some specialized applications may warrant the use of such a technology, and it is well documented.

3.1.2.3 Organic Polymerization Stabilization/Solidification

Organic polymerization stabilization/solidification relies on polymer formation to immobilize the constituents of concern. Urea formaldehyde is the most commonly used organic polymer for this purpose. Organic polymerization has been used primarily to stabilize/solidify radioactive wastes. This technology has been applied on a limited basis to hazardous wastes such as organic chlorides, phenols, paint sludges, cyanides, and arsenic. Polymerization can also be applied to flue gas desulfurization sludge, electroplating sludges, nickel/cadmium battery wastes, kepone-contaminated sludge, and chlorine product wastes that have been dewatered and dried (Kyles, Malinowski, and Staczyk 1987).

3.1.2.4 Organophilic Clay- Based Stabilization/Solidification Processes

This technology appears to be very promising in terms of binding organic wastes. Recent investigations (Gibbons and Soundararajan 1988) indicate that these organophilic binders truly bond with organic wastes, and when forced out, these waste molecules break down into smaller fragments. Further, this process appears to be cost-effective compared with incineration. Because this technology also uses cementitious materials, it can handle both organic and inorganic wastes.

3.2 Stabilization/Solidification Activity in the Various U.S. Environmental Protection Agency Regions

Table 3-1 lists sites where hazardous waste stabilization/solidification has been used, is in use, or is proposed to remediate hazardous wastes. This table presents pertinent information on each case study, including the site name, the company involved in treating the wastes, the types and concentrations of the contaminants encountered, the amount of material processed, the type of processing used, the disposal location, and whether the study is full-scale, pilot-scale, or bench-scale. Cement was used in the majority of the case studies presented, and pozzolanic materials were used in some.

Table 3-1. Solidification case studies.

Site/ contractor	Contaminant (concentration)	Treatment volume	Physical form	Chemical pretreatment Y/N	Binder	Percentage binder(s) added	Treatment (batch/ continuous in situ)	Disposal (onsite/ offsite)	Volume increase, %	Scale of operation
Independent, Nail, SC, Region IV	Zn, Cr, Cd, Ni	6,600 yd^3	Solid/soils	N	Portlant cement	20%	Batch plant	Onsite	>Small	Full scale (de-listing in progress
Midwest, U.S. Plating Company, Envirite	Cu, Cr, Ni	16,000 yd^3	Sludge	N	Portland cement	20%	In situ	Onsite	>0	Full scale
Unnamed, ENRECO	Pb/soil 2-100 ppm	7,000 yrd^3	Solid/soils	N	Portland cement and proprietary	Cement (15-20%) proprietary (5%)	In situ	Landfill	Mass >20% (volume >30-35%)	Full scale
Marathon Steel, Phoenix, AZ Silicat, Tech.	Pb, Cd	150,000 yd^3	Dry - landfill	N	Portland cement and silicates (Toxsorb)^{TM}	Varied 7-15% (cement)	Concrete batch plant	Landfill	NA	Full scale
Alaska Refinery HAZCON	Oil/oil sludges	2,300 yd^3	Sludges, variable	Y	Portland cement and proprietary	Varied 50+	Concrete batch plant	Onsite	>35%	Full scale
Unnamed, Kentucky, ENRECO	Vinyl chloride Ethylene dichloride	180,000 yd^3	Sludges, variable	Y	Portland cement and proprietary	Varied 25 +	In situ	Onsite (2 secure cells built on site)	>7-9%	Full scale
N.E. Refinery ENRECO	Oil sludges, Pb, Cr, As	100,000 yd^3	Sludges, variable	N	Kiln dust (high CaO content)	Varied, 15-30%	In situ	Onsite	>Varied, ~20% average	Full scale
Velsicol Chemical Memphis Env. Centre	Pesticides and organics (resins, etc.) up to 45% organic	20,000,000 gallons	Sludges, variable	N	Portland cement and kiln dust, proprietary	Varied (cement 5-15%)	In situ	Onsite	>Varied ~10% or less	Full scale
Amoco Wood River Chemfix	Oll/solids Cd, Cr, Pb	90,000,000 gallons	Sludges	Y	Chemfix proprietary	NA, proprietary	Continuous flow (pro-prietary process)	Onsite	Average 15%	Full scale (site delisted 1985)

(continued)

Table 3-1. (continued)

Site/ contractor	Contaminant (concentration)	Treatment volume	Physical form	Chemical pretreatment Y/N	Binder	Percentage binder(s) added	Treatment (batch/ continuous in situ)	Disposal (onsite/ offsite)	Volume increase, %	Scale of operation
Pepper Steel & Alloy, Miami, FL VFL Technology Corporation	Oil sat: soil Pb - 1000 ppm PCB's - 200 ppm As - 1-200 ppm	62,000 yd^3 (plus 5000 tons of surface debris)	Soils	Y	Pozzolanic and proprietary	~30%	Continuous feed (mixer proprietary) design	Onsite	~1%	Full scale
Vickery, Ohio Chemical Waste Management	Waste acid PCB's (<500 ppm), dioxins	~235,000 yd^3	Sludges (viscous)	Y	Lime and kiln dust	~15% CaO ~5% kiln dust	In situ	Onsite (TSCA cells)	>~9%+	Full scale
Wood Treating, Savannah, GA Geo-Con, Inc.	Creosote wastes	12,000 yd3	Sludges	Y	Kiln dust	20%	In situ	Onsite lined cells	>~14%	Full scale
Wyandotte, MI Treatment Plant Chem Met	Various/combined	20 million gal/yr	Various	N	Lime		Continuous	Offsite (secure landfill)		In-plant process
Chem Refinery, TX HAZCON	Combined metals, sulfur, oil sludges, etc.	90,000 gal. (445 yd^3)	Sludges (synthetic oil sludges)	N	Portland cement and proprietary	NA	Continuous flow	Onsite (secure landfill)	>Estimated 10%	Full scale
Chicago Waste Hauling, American Colloid	Metals: Cr, Pb, Ba, Hg, Ag	55 gallon/ batch (bench study)	Various		Proprietary	10-40%	Batch mix (pug mill)	NA	>Variable	Bench scale
API sep. sludge, Puerto Rico, HAZCON	API separator sludges	100 yd3	Sludges	N	Portland cement and proprietary	50% cement ~4% proprietary	Concrete batch plant	Offsite secure landfill	>~4-5%	Flull scale
Metalplating, WI, Geo-Con, Inc.	Al - 9500 ppm Ni - 750 ppm Cr - 220 ppm Cu - 2000 ppm	3000 yd3	Sludges	N	Lime	10-25%	In situ	Onsite landfill	>4-10%	Full scale
James River Site Virginia	Kepone contaminated sediments		Wet soil sludges	N	Cement-base, thermoplastic, polymer	Various	Various		NA	Bench scale only

(continued)

Table 3-1. (continued)

Site/ contractor	Contaminant (concentration)	Treatment volume	Physical form	Chemical pretreatment Y/N	Binder	Percentage binder(s) added	Treatment (batch/ continuous in situ)	Disposal (onsite/ offsite)	Volume increase, %	Scale of operation
Massachusetts, American Reclamation Corporation	Oil/gasoline contaminated soils	Variable	Wet soil	Y	Bitumen	Variable	Batch	Used as road patch/ paving materials	NA	Bench (pilot in process)
Saco Tannery Waste Pits, Maine/VFL Tech. Corporation	Cr (>50,000 ppm Pb (>1000 ppm) and organics	Varied	Sludge		Fly ash, quicklime	30% fly ash 10% quick lime	In-situ	Onsite	>15%	Pilot scale
Sand Springs Petrochemial, Complex, OK/Arco	Sulfuric acid, and organics		Sludge		Fly ash, quicklime	Varied	Batch	Onsite		
John's Sludge Pit, KS/Terracon Consultants, Inc.	Pb, Cr, acid		Sludge		Cement kiln dust and fly ash	Varied	Batch		>Variable	Bench scale
Gold Coast, FL	VOC's and metals	1500^a yd^3	Soil					Onsite		
Gurley Pit, AR	PCB's and organics	432,470^a yd^3	Soil					Onsite		
Liquid Disposal Landfill, MI	PCB's, VOC's, heavy metals		Soil					Onsite		
Northern Engraving, WI	VOC's, organic, and inorganics	4400^a yd^3	Sludge					Onsite		
Mid South, AR	PAH's, organics, and inorganics	45,750^a yd^3	Soil					Onsite		
Hialea, FL Geo-Con, Inc.	PCB's 0.800 ppm	300 yd^3 (7,000 yd^3 total)	Wet soil	N	HWT-20^{TM} (cement based)	15%	In situ	Onsite	>Small	Pilot scale

(continued)

Table 3-1. (continued)

Site/ contractor	Contaminant (concentration)	Treatment volume	Physical form	Chemical pretreatment Y/N	Binder	Percentage binder(s) added	Treatment (batch/ continuous in situ)	Disposal (onsite/ offsite)	Volume increase, %	Scale of operation
Douglassville, PA HAZCON	Zn, 30-50 ppb Pb, 24,000 ppm PCB's, 50-80 ppm Phenol, 100 µg/liter Oil and grease	250,000 yd^{3a}	Various soil/ sludges	N	Portland cement and proprietary	NA	Batch	NA	NA	Pilot scale
Portable Equipment, Clackamas, OR CHEMFIX	Pb, Cu, PCB's	40 yd^3	Soil	N	Cement, silicate	NA	Batch	NA	NA	Pilot scale
Imperial Oil Morganville, NJ Soliditech	PCB's	50 yd^3	Soil	N	Cement, proprietary	NA	Batch	NA	NA	Pilot scale

NA - Data not available
[a] Total volume onsite.

4. Physical Tests to Characterize Wastes Before and After Stabilization/Solidification

Physical testing is conducted to characterize and contrast waste before and after stabilization/solidification. It provides basic information on the treatability of the waste material and allows some estimate to be made of the cost of waste treatment and handling. Physical property characterization of unstabilized/unsolidified wastes focuses on treatability, excavation, transport, storage, and mixing considerations (Cullinane et al. 1986). Physical testing of stabilized/solidified wastes helps to demonstrate the relative success or failure of a stabilization/solidification process. The physical testing methods described in this section may be applicable to both untreated hazardous wastes and stabilized/solidified hazardous wastes; however, the tests were not developed for use on these wastes.

This section describes the more common physical tests used to evaluate waste stabilization/solidification processes. The physical tests listed in Table 4-1 include:

- _Index Property Tests_, which provide data that are used to relate general physical characteristics of a material (e.g., suspended solids) to process operational parameters (e.g., pumpability).

- _Density Tests_, which are used to determine weight to volume relationships of materials.

- _Permeability Tests_, which measure the relative ease with which fluids (water) will pass through a material.

- _Strength Tests_, which provide a means for judging the effectiveness of a stabilization/solidification process under mechanical stresses.

- _Durability Tests_, which determine how well a material withstands repeated wetting and drying or freezing and thawing cycles.

Individual values of waste properties derived from specific tests are used along with other available data to make informed engineering decisions. In several cases, the tests provide data useful only for comparing different stabilization/solidification methods.

It is important to note that many of these tests were originally developed for testing soils and cement-like materials for stability for construction projects. Ex-treme caution must be exercised when applying these tests to untreated and stabilized/solidified hazardous wastes, and in the subsequent data interpretation. Many of the tests involve frequent handling of the waste materials; therefore, due consideration must be given to personnel protection, sample handling and disposal requirements, and other factors associated with the presence of hazardous constituents in the samples.

4.1 Index Property Testing

Index property tests provide data that are used to relate general physical characteristics of a material to process operational parameters. These tests are most frequently performed on the unstabilized/unsolidified waste to determine the feasibility of various stabilization/solidification technologies. Stabilization/solidification process vendors have indicated that particle size distribution, pH, and moisture content are critical factors in choosing a stabilization/solidification process (Wiles 1987).

4.1.1 Particle Size Analysis (ASTM D422-63)

Particle Size Analysis is a quantitative determination of a material's particle size distribution and is broadly used in engineering classification of materials. The test provides data pertaining to material characteristics, such as:

- Relative proportions of gravel, sand, silts, and clay-sized particles (i.e. "texture")

- Uniformity, concavity, and average grain size

- Maximum and minimum size of particles

The grain size distribution of a material can also be qualitatively used to predict a fairly wide range of characteristics, including soil-water movement (i.e., permeability) and compressibility.

This analysis is usually performed prior to stabilization/solidification to suggest the feasibility of a particular process or the difficulties that could be encountered in processing.

4.1.1.1 Test Description

Grain-size analyses can be completed by sieve analysis or by hydrometer analysis. The sieve analysis is

Table 4-1. Physical testing methods.

Test procedure	Reference[a]	Purpose
Index property tests		
Particle size analysis	ASTM D422-63	To determine the particle size distribution of a material
Atterberg limits Liquid limit Plastic limit Plasticity index	ASTM D4318-84 ASTM D4318-84 ASTM D4318-84	To define the physical characteristics of a material as a function of its water content
Moisture content	ASTM D2216-80	To determine the percentage of free water in a material
Suspended solids	USEPA Method 208 C	To determine the amount of solids that do not settle from a column of liquids
Paint filter test	USEPA Method 9095-SW846	To determine the presence of free liquids in a representative sample of bulk or noncontainerized waste
Density testing		
Bulk density - drive cylinder method	ASTM D2937-83	To determine the in-place density of soils or soil-like material
Bulk density - sand-cone method	ASTM D1556-82	To determine the in-place density of soils or soil-like materials
Bulk density - nuclear methods	ASTM D2922-81	To determine the in-place density of soils or soil-like materials
Bulk density - stabilized waste		To determine the density of a monolithic stabilized waste
Compaction testing		
Moisture density relations of soil-cement mixtures	ASTM D558-82 ASTM 31557	To determine the relation between moisture content and density of a material
Permeability testing		
Falling head permeability	USEPA Method 9100-SW846	To measure the rate at which water will pass through a soil-like material
Constant head	USEPA Method 9100-SW846	To measure the rate at which water will pass through a soil-like material

(continued)

Table 4-1. (continued)

Test procedure	Reference[a]	Purpose
Strength testing		
Unconfined compressive strength of cohesive soils	ASTM D 2166-85	To evaluate how cohesive soil-like materials behave under mechanical stress
Unconfined compressive strength of cylindrical concrete specimens	ASTM D1633-84	To evaluate how cement-like materials behave under mechanical stress
Compressive strength of hydraulic cement mortars	ASTM C109-86	To measure the compressive strength of hydraulic cement mortars
Flexural strength	ASTM D1635-87	To evaluate a material's ability to withstand loads over a large area
Cone index	ASTM D3441-79	To evaluate a material's stability and bearing capacity
Durability testing		
Freeze-thaw durability	ASTM D4842	To determine how materials behave or degrade after repeated freeze-thaw cycles
Wet-dry durability	ASTM D4843	To determine how materials behave or degrade after repeated wet-dry cycles

[a]   ASTM - American Society for Testing and Materials.
Blank indicates no ASTM method or other reference method.

typically used to define the distribution of materials ranging from fine sand to gravel sizes (0.1 mm and larger). The hydrometer analysis is typically used to define the distribution of materials ranging from fine sand and silt to clay (0.1 mm and smaller).

The results of both of these tests are required for a complete evaluation of the distribution of particle sizes of soils containing both coarse- and fine-grained fractions. For materials containing less than 20 percent fines, the hydrometer analysis is often disregarded.

Evaluation of the particle size distribution of a material by sieve analysis is performed by first washing a dried and weighed sample through a No. 200 (0.075 mm) sieve to determine the percentage of fine-grained particles, and then redrying the sample and passing it through a series of circular screens with various mesh sizes. The sieving is usually done on a Ro-tap shaker, but sometimes by hand. The machine produces uniform circular and tapping motions that cause the particles to be sorted through the sieves. The apparatus can use as many as 13 screens at a time for maximum grain size differentiation. After sieving, the weight of the material on each screen is measured and is calculated as a percentage of the total. The results are plotted on a grading curve, as shown in Figure 4-1.

The use of this test is straightforward because the procedures are mechanical. Errors that do occur often relate to improper sampling of the material. These errors can be overcome by proper sampling techniques, such as quartering, or by using several samples and comparing their particle size distributions. Analysis may be difficult or unreliable when soil grains are contaminated with oils or other organic materials that have a tendency to agglomerate.

The hydrometer test is used to obtain an estimate of the distribution of particle sizes from the No. 200 sieve size (0.075 mm) to around 0.001 mm. This encompasses silt and clay-sized particles. According to Stoke's Law, which considers the terminal velocity of a single sphere falling in an infinity of liquid, the sizes calculated represent the diameter of spheres that would fall at the same rate as the soil particles (Note 14, ASTM D422-63).

In this test, a sample of approximately 50 grams of air-dried soil in 125 ml sodium hexametaphosphate ("Calgon") solution (40 g/liter) is dispersed to neutralize soil-particle charges. After it soaks in the dispersing solution, the sample is further dispersed by mixing it in a mechanical or air-jet mixing apparatus. The dispersed mix is then transferred to a sedimentation cylinder, and distilled water is added to a total volume of 1000 ml. The mixture is again agitated briefly, and a hydrometer is placed in the cylinder to monitor changes in specific gravity of the solution over time. The measured values of specific gravity of the solution and known (or assumed) specific weights of the solid and fluid phases are used to calculate the grain sizes still in suspension as larger sizes fall out of suspension. Although the specific weight of water is typically used for calculations, contaminants desorbed from the solids may affect this value to some degree.

Figure 4-1. Typical particle-size grading curve.

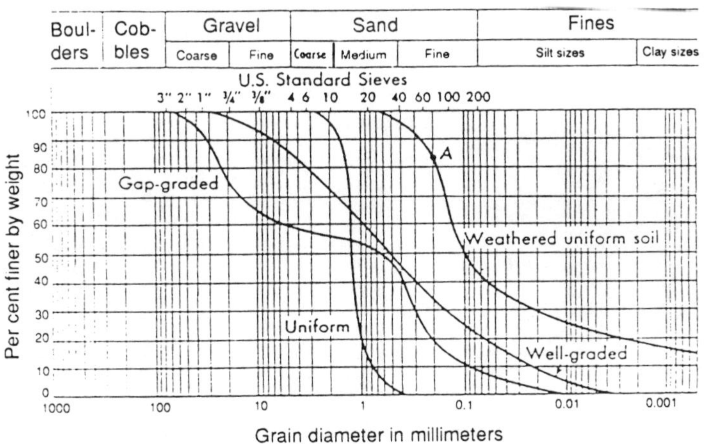

Source: Introductory Soil Mechanics and Foundations, by G. B. Sowers and G. F. Sowers. MacMillan Publishing Co., New York. 1970.

4.1.1.2 Interpretation and Application of Results

The shape of the gradation curve represents the particle size uniformity of the sample. A steep curve indicates a soil in which nearly all the grains are the same size. A flatter curve shows a wider variation in grain size and thus a well-graded soil. Figure 4-2 gives representations of well-graded (poorly sorted) and poorly graded (well-sorted) materials.

The absolute size and shape of the waste particles affect the feasibility of various stabilization/solidification processes and the ultimate strength of the stabilized/solidified product (Wiles 1987). Very fine or very coarse particles can increase the difficulty of stabilizing/solidifying wastes. For example, fine-grained wastes have been shown to produce poor stabilized/solidified materials because they lack the size needed to form a stabilized/solidified product with adequate durability (Cullinane et al. 1986). Vick et al. (1987), however, have shown that very hydrophobic contaminants such as dioxin tend to bind preferentially to small soil particles. Very large particles may also preclude the use of certain processing equipment or require particle size reduction prior to stabilization/solidification.

A well-graded soil that does not contain extremely large or extremely small particles is likely to show favorable physical characteristics after stabilization/solidification. The strong matrix formed by the interlocking of diversely sized particles often produces characteristics such as high strength, low permeability, and low leachability.

4.1.2 Atterberg Limits (ASTM D4318-84)

The Atterberg Limits are a simple and useful series of tests originally developed for classification and characterization of clays used in ceramics.

Atterberg Limits are the moisture contents that mark a material's liquid and plastic states. The liquid limit of a material is the moisture content at which it will flow as a viscous liquid. The plastic limit is the moisture content at the boundary between the plastic and brittle states. The plasticity index is the difference between the liquid limit and plastic limit. Atterberg Limit tests are applicable to fine-grained materials only, and their results are useful not only for classification, but also for correlation with a broad range of engineering properties and to indicate clay mineralogy.

Atterberg Limits are used to estimate such properties as compressibility, strength, and swelling characteristics, and to indicate how the material will behave when stresses are applied (Cullinane 1986). In addition, the plasticity index can be used to determine the proper amount of stabilization/solidification agent (e.g., lime) to be added to a waste. Atterberg Limit tests are applicable to any natural or artificial mixture of soil-like particles.

The tests are performed on both unstabilized/unsolidified and freshly stabilized/solidified soil-like wastes. The resulting data can be particularly useful during the compacting of friable materials after stabilization/solidification. To date, these tests have not been completed extensively on hazardous waste materials; however, they are expected to be used more frequently in the future.

Figure 4-2. Illustration of material gradation.

MATERIAL GRADATION

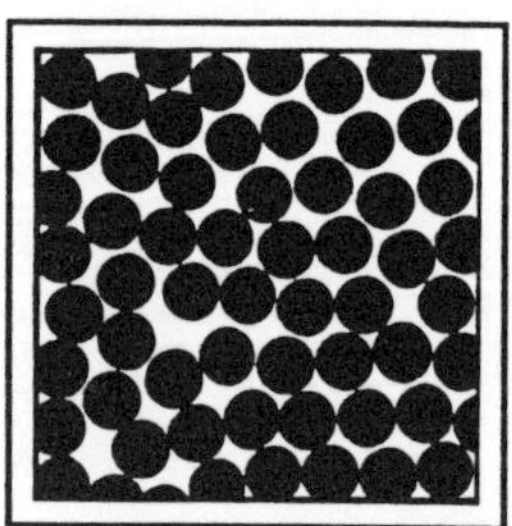

Poorly graded

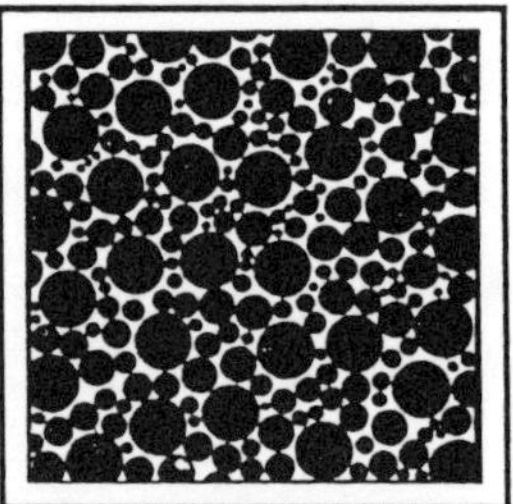

Well-graded

Source: Adapted from Caterpillar Performance Handbook, Caterpillar, Inc., Peoria, IL. October 1987.

4.1.2.1 Definitions

The liquid limit is the moisture content of a soil at the arbitrarily defined boundary between the liquid and plastic states. The liquid limit can be regarded as the water content at which soil no longer acts as a plastic, but begins to act as a liquid (i.e., flows).

The plastic limit is the moisture content of a soil at the boundary between the plastic and brittle states. The plastic limit is considered the water content at which a soil no longer acts brittle, but can be molded and deformed without falling apart.

The plasticity index is the difference between the liquid limit and the plastic limit. Reduction in the plasticity index is used to judge the effect of lime addition to clays.

4.1.2.2 Test Description

Liquid Limit Determination—The material is mixed with water, placed in a brass cup, and divided in two with a grooving tool. The liquid limit is the water content at which the divided sample flows together after the cup is dropped by an apparatus 25 times from a height of 10 centimeters. The water content of the material is determined by the method described in Subsection 4.1.3.

Plastic Limit Determination—A sample is alternately pressed and rolled into a 1/8-inch-diameter thread. The sample slowly dries out until the thread can no longer be pressed together or rerolled; that water content is the plastic limit.

Although procedurally defined, the definitions of the liquid and plastic limits are at least partially subjective. For good consistency, the same tester should run duplicate trials of identical samples.

4.1.2.3 Interpretation and Significance of the Test

The test is part of several engineering classification systems to characterize fine-grained fractions of a material. These test results characterize material-handling properties and the variation in the properties as a function of water content. Liquid limit and plastic limit are used individually or together with other soil properties to correlate with other engineering behaviors such as compressibility, permeability, compactibility, shrinking/swelling, and shear strength.

Liquid limits for various waste/stabilization/solidification agent mixtures typically range between 40 and 55 percent water content. Typical plastic limits for various waste/stabilization/solidification agent mixtures are between 20 and 50 percent water content (Morgan et al. 1984).

4.1.3 Moisture Content (ASTM D2216-80)

The moisture content test determines the amount of free water (or fluid) in a given amount of material. This test is often used to determine if pretreatment is necessary in the design of the stabilization/solidification process (Cullinane 1986). An example of waste pretreatment would be sludge drying, dewatering, or consolidation prior to stabilization/solidification. Moisture in a waste is not always detrimental. For example, the presence of water may be needed to provide a reaction mechanism (e.g., hydration) for correct stabilization of the waste.

In this test method, the term "water" refers to "free" or "pore" water, not waters of hydration. Also, water in discontinuous pores is not measured by this test.

It is also important to note that water is often not the only liquid-phase constituent in contaminated materials. The fluids may also include a broad range of liquid wastes present in solution or as nonaqueous phase liquids. This can have several effects on the performance and results of moisture content determinations. For example, if VOCs are present, samples should be aerated to allow volatilization of flammable VOCs before samples are oven-dried. The type and level of contamination may also influence the relationship between "free" and adsorbed water.

4.1.3.1 Test Description

A preweighed sample is dried in an oven for 24 hours at a constant temperature of 110°C or at a temperature below the dehydration temperature. The sample is allowed to cool to room temperature in a desiccator. The difference between the original sample weight and the weight after drying is used to determine the moisture content.

During the collection of samples for analysis, the moisture content (free water) must not change during handling. Samples must be placed in appropriate jars and sealed with plastic, foil, or wax. Maintenance of a constant temperature in the drying oven is also important for achieving accurate results. A 5°C range is allowable.

Moisture content results for soils high in organic matter require careful interpretation. Organic material, including VOCs, may be driven off during the drying process. Any weight loss due to the organic matter will be recorded as water loss.

This method does not give truly representative results for materials containing significant amounts of halloysite, montmorillonite, and gypsum; highly organic soils; or materials with pore waters containing dissolved solids. For these soils, vacuum drying at 60°C may be better. Another source of error in moisture content determination of stabilized materials is the accelerated hydration reactions that may occur at the elevated temperatures used during the test. These reductions reduce the amount of free water available for evaporation and cause erroneous weight measurements.*

4.1.3.2 Interpretation and Application of Results

The results of this test are usually expressed as fluid representing a percentage of total mass (Cullinane 1986). This basic information is required for the planning and execution of a stabilization/solidification project. Excess moisture may have to be removed by filtration. Low moisture content may indicate the need to add water to stabilize the waste correctly. For example, water is sometimes added to incinerator ash to stabilize/solidify the ash.**

In a study by Stegemann et al. (1988) of 69 wastes, water content in the untreated wastes ranged from close to 100 percent for wastewaters to less than 10 percent for soils. After waste stabilization/solidification, water content ranged from 64 percent to less than 1 percent. In this study, the water content was determined from weighing a ground sample before and after drying to a constant weight at 60°C.

4.1.4 Suspended Solids (USEPA Method 208C)

"Suspended solids" is a term originally used in the monitoring of the quality of wastewater (both the influent and effluent of a wastewater treatment plant) and, by inference, the performance of the plant. "Total solids" include all solids, suspended and dissolved, present in a sample. Suspended solids are those removed by filtration. The difference between total solids and suspended solids is known as the "dissolved solids." One must know the suspended solids content of supernatants in lagoons or other impoundments to plan dewatering operations.

4.1.4.1 Test Description

A well-mixed sample is filtered through a preweighed standard glass-fiber filter (Gelman Type A or equivalent). The filter is then dried to a constant weight at 103° to 105°C, cooled in a desiccator, and reweighed. Suspended solids, expressed as a percentage, is the increased weight of the filter for a known volume of sample.

4.1.4.2 Interpretation and Applicability

Suspended solids content is an important parameter for determining the materials handling requirements

for a waste material, i.e., to determine if the waste can be pumped. Suspended solids content also can be used to estimate the decrease in volume that can be achieved by dewatering. Table 4-2 presents consistency categories for various waste types based on approximate suspended solids content. Although these categories are approximate, they give an indication of how a waste can be handled and the operations that can occur in and on the material.

Table 4-2. Liquid waste consistency classification.

Consistency category	Characteristic properties
Liquid waste	Less than 1% suspended solids,* pumpable liquid, generally too dilute for sludge dewatering operations.
Pumpable waste	Less than 10% suspended solids,* pumpable liquid, generally suitable for sludge dewatering.
Flowable waste	Greater than 10% suspended solids,* not pumpable, will flow or release free liquid, will not support heavy equipment, will undergo extensive primary consolidation.
Nonflowable waste	Solid characteristics, will not flow or release free liquids, will support heavy equipment, may be 100% saturated, may undergo primary and secondary consolidation.

*Suspended solid ranges are approximate.
Source: Cullinane 1986.

4.1.5 Paint Filter Test (USEPA Method 9095-SW846)

The Paint Filter Test is used to determine the presence of "free liquids" in a representative sample of bulk (or noncontainerized) waste. The test is required by RCRA's 40 CFR 264.314 and 265.314 and is used to determine if a material releases free liquids.

The American Nuclear Society has a test similar to the Paint Filter Test, the Allowable Drainable Liquid Test (ANS 55.4). The EPA's Office of Solid Waste and Emergency Response is proposing that the Liquid Release Test be used in conjunction with the Paint Filter Test. In the December 24, 1986, <u>Federal Register</u>, the EPA proposed the use of the Liquid Release Test to test for release of liquids from nonbiodegradable absorbent mixtures when a waste is under compressive forces in a landfill. The proposed Liquid Release Test calls for the application of 50 psi pressure to the waste sample to determine if liquids will be released under compressive forces.

---

* Personal communication from Dr. Paul L. Bishop, University of Cincinnati, to M. Arozarena, PEI, September 30, 1988.
** Personal communication from M. John Cullinane, P.E., U.S. Army Corps of Engineers, Waterways Experimental Station, to T. Whipple, Earth Technology Corporation.

4.1.5.1 Test Description

In the Paint Filter Test, the material is placed in a paint filter, which rests in a funnel attached to a ring stand. If any portion of the material passes through and drops from the filter within the 5-minute test period, the material is deemed to contain free liquids.

4.1.5.2 Interpretation and Application of Results

As of May 8, 1985, the placement of bulk (or noncontainerized) liquid hazardous waste or hazardous waste containing free liquids in any landfill was prohibited [Land Ban HSWA Section 3004(c)(1)]. Absorbents (in contrast to stabilization/solidification agents) cannot be added to liquid wastes to achieve a temporary liquid-free status. In addition, no credit is given for the liner or leachate collection system that the landfill uses.

The Paint Filter Test may be performed after a waste is stabilized to determine whether the waste may be disposed of in a RCRA-authorized landfill. If it does not pass the Paint Filter Test, it must undergo further treatment before it can be disposed of in a landfill. (For further information on the Land Ban regulations, see Section 2.)

4.2 Density Testing

Bulk density is the ratio of the total weight (solids and water) to the total volume. Bulk density, along with specific gravity and moisture content measurements, can be used to calculate a material's porosity. More commonly, bulk density values are used to convert weight to volume for materials-handling calculations and are essential for characterizing the rates at which a soil can be excavated. In addition, bulk density data provide a comparison between stabilized and unstabilized waste. Calculated increases in volume of a material due to changes in bulk density after excavation may determine if the stabilized materials can be disposed of on site or must be shipped offsite.

Four methods of bulk density measurement are presented. The data from each are sufficiently accurate for calculating densities. Selection of a method is usually based on ease of use. Laboratory determination of specific gravity can supplement these measurements.

4.2.1 Bulk Density—Drive-Cylinder Method (ASTM D2937-83)

The Drive-Cylinder Method is designed to determine the in-place density and moisture content of soils or soil-like wastes. Because the test is not appropriate for nondeformable materials, it is limited to unstabilized, freshly stabilized, or soil-like stabilized waste that have been compacted by conventional earth-moving equipment.

A thin-walled cylinder is driven into the soil to obtain a sample. The weight per unit of volume of the sample in the cylinder is then determined.

4.2.2 Bulk Density—Sand-Cone Method (ASTM D1556-82)

For determination of bulk density via the Sand-Cone Method, a template is placed on a level surface of the unstabilized/unsolidified waste and the material is excavated through a hole in the template with small digging tools. A volume of roughly 1000 to 2500 cm³ is typically excavated through the 165-mm-diameter hole in the template. An apparatus filled with a sand of known density is weighed and placed on the template above the hole created by the excavation. Sand is released from the apparatus and allowed to fill the hole. The apparatus is then reweighed. From the sand's previously determined density and the weight of sand remaining in the apparatus, the volume of the sand needed to fill the hole is calculated. The ratio of the weight of material removed from the hole to the volume of sand needed to fill the hole is the material's bulk density. The moisture content is determined on a sample of the excavated material to evaluate dry density.

4.2.3 Bulk Density—Nuclear Method (ASTM D2922-81)

The density of soil and soil aggregate can also be measured in place by nuclear methods. The total or wet density of the material is determined by placing a gamma source and gamma detector either on, into, or adjacent to the material to be measured. Nuclear densiometry works on the principle of Compton scattering. The amplitude of back-scattered gamma radiation depends on the electron concentration in the material being measured, and this is, in turn, roughly proportional to the density of the material.

4.2.4 Bulk Density—Stabilized Waste

For determination of the bulk density of a monolithic stabilized waste, a sufficiently cured cube or cylinder of the solid is weighed and measured. The bulk density is then calculated by dividing the volume into the mass.

Bulk density measurements are used for materials-handling calculations such as volume removed, volume stabilized, and volume returned to the site. Bulk density measurements of stabilized materials are used to determine the amount of material that will need to be shipped offsite or to determine the degree of mounding that will occur as a result of stabilization/solidification. These applications are similar to cut-and-fill calculations for construction or road building.

For 69 stabilized/solidified products tested by Stegemann et al. (1988), bulk densities ranged from lighter than water (0.7 g/cm³ or 0.6 ton/yd³) to very dense (2.2 g/cm³ or 1.8 ton/yd³).

4.3 Compaction Testing—Moisture/ Density of Soil-Cement Mixtures

This test (ASTM D558-82) determines the relationship between the moisture content and density of soil-like materials. The test is normally completed on the stabilized waste before stabilization/solidification has occurred. The test determines the moisture content that allows maximum compaction to occur so as to achieve maximum density.

4.3.1 Test Description

The optimum moisture content is obtained by creating a series of samples for the particular waste-stabilization/solidification agent mix to be tested. Varying amounts of water are added to these samples, which are then placed in a standard mold with a volume of 1/30 ft³ and compacted in three equal lifts by use of a standard 5.5-lb rammer dropping from a height of 12 inches. After compaction, the samples are oven-dried for 12 hours, and their densities are then measured. The baseline moisture content (before the addition of water) is also noted.

Dry densities of the stabilized samples are plotted as a function of moisture content. This plot produces a curve as shown in Figure 4-3. Note: The moisture content of samples may reflect moisture added for testing purposes.

The moisture content corresponding to the peak of the curve is called the "optimum moisture content." The density corresponding to the peak of the curve is called the "maximum density."

4.3.2 Interpretation and Application of Results

The moisture content of a stabilized waste is applicable for soil-like stabilized wastes that must be re-compacted in place after they are stabilized. Experience in the construction fields shows that it is difficult to achieve proper compaction if materials are too wet or too dry. Water lubricates soil particles, which helps them slide into a denser position. The stabilized waste has an optimum moisture content at which a maximum density can be achieved upon compaction. It should be noted, however, that the optimum water content for compaction properties may not be the optimum water content for hydration reactions.*

---

* Personal communication from Dr. Paul L. Bishop, University of Cincinnati, on September 30, 1988.

Figure 4-3. Typical soil-cement moisture/density relationship.

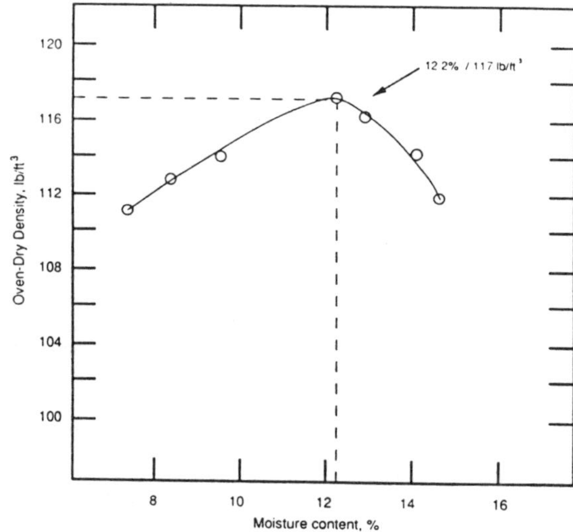

4.4 Permeability (Hydraulic Conductivity) Testing

Hydraulic conductivity, often referred to as permeability, is a measure of the resistance of a material to the passage of water. Permeability tests are performed to estimate the quantity and flow rates of water through a material under saturated conditions. Laboratory permeability testing consists of applying a hydraulic head of water to one end of a specimen and measuring the flow through the specimen (Carter 1983).

There are two basic types of permeability tests: constant-head and falling-head. The constant head test allows relatively large quantities of water to flow through the sample and be measured. This test is suitable for materials with a permeability greater than 10^{-6} cm/s. The falling-head test, which allows for more accurate measurement of small quantities of water, is more suitable for materials with a permeability of less than 10^{-6} cm/s (Carter 1983). The following are general descriptions of the testing techniques.

Falling-Head Permeability (EPA Method 9100-SW846) (Figure 4-4)

In this test, a specimen is sealed in an impermeable membrane and placed in a fluid-filled chamber, within which hydrostatic pressures can be applied to the specimen (a triaxial compression chamber with back pressure to ensure complete saturation is commonly used). Platens on the top and bottom of the specimen are connected through tubing to cylinders filled with air-free water. The water in the chambers is differentially pressurized to create a head difference across the specimen and to force flow through the specimen. The pressure differential must be specified. If the differential is very high, problems can result from expansion of internal pores and creation of a more continuous flow network, which results in erroneously high permeability readings.* The elapsed time for the water level to fall between calibrated marks on the glass cylinder is recorded. The test is usually repeated three or four times. The falling-head permeameter is usually equipped with a series of glass tubes of various diameters to allow a reasonable time interval to be used for the test. Selection of the correct size tube is based on judgment and accomplished by trial and error. Figure 4-4 is a schematic diagram of the apparatus required for this test.

Constant-Head Permeability (EPA Method 9100-SW846) (Figure 4-5)

In this test, the sample is placed in a water bath and is connected to a column of air-free water in a glass tube, with the head of the water maintained at a constant value by a

---

* Personal communication from Dr. Paul L. Bishop, University of Cincinnati, to M. Arozarena, PEI, September 30, 1988.

Figure 4-4. Basic layout of apparatus for falling-head permeability test.

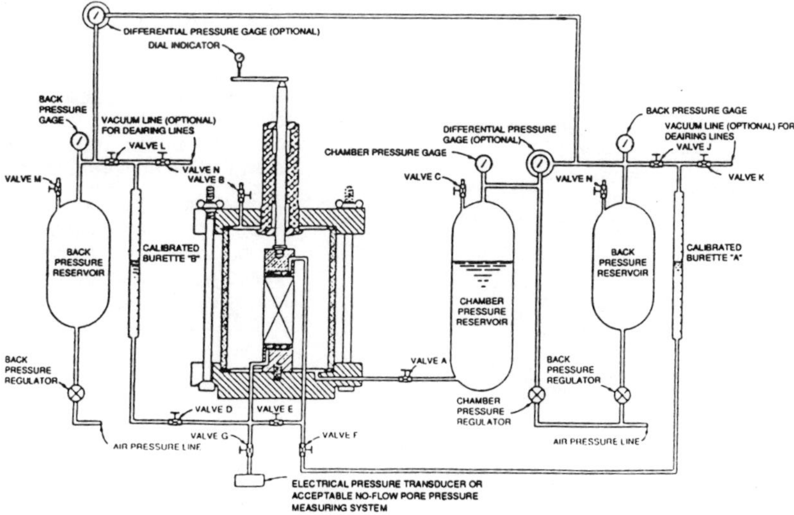

header tank and a balance tank. Water flows through the sample by gravity and is collected in a measuring cylinder. The time taken to collect a given quantity of water is recorded. The cylinder containing the soil sample usually has nipples at a number of points along the sides and is connected via rubber tubing to glass manometer tubes so that the piezometric head can be measured at various points throughout the specimen (Carter 1983). Figure 4-5 is a schematic diagram of the apparatus required for this test.

Figure 4-5. Basic layout of apparatus for constant-head

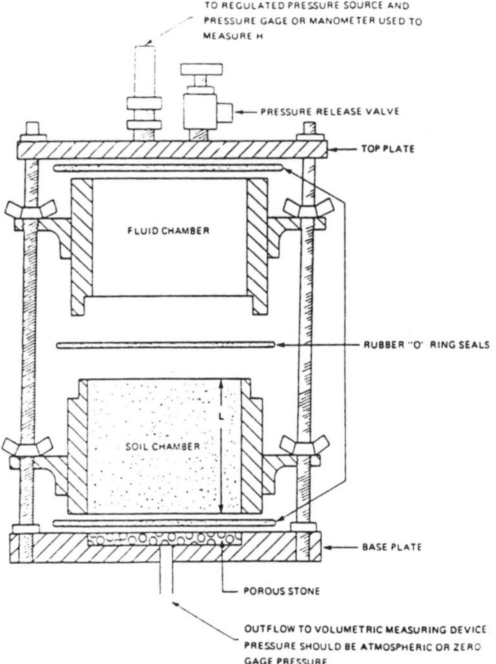

Both of these permeability tests are laboratory methods and are only considered accurate to within one order of magnitude. If a soil-like waste is stabilized and sampled in the field for testing in the laboratory, it is necessary to recompact the sample in accordance with USACE methods (USACE 1980) to obtain a representative sample.

The permeability of a stabilized waste is an important factor, as it indicates the ability of a material to permit the passage of water and to limit the loss of contamination from the stabilized waste to the environment. Permeability is examined in conjunction with leach test results to evaluate the potential of the stabilized waste to release contaminants into the environment.

The relevance of permeability measurements can be understood by comparing them with natural materials. Sand, a highly permeable material, has a hydraulic conductivity on the order of 10^{-2} cm/s. Clay, a material that is used to line lagoons and surface impoundments, can have hydraulic conductivity on the order of 10^{-6} cm/s or less and is considered relatively impermeable. Thus a stabilized waste with a permeability similar to clay is desirable because it will not permit the free passage of water though the stabilized waste. By slowing the contact of water with the waste, it reduces the possible transport of contaminants out of the waste. Typical hydraulic conductivities for stabilized wastes range from 10^{-4} to 10^{-8} cm/s. Hydraulic conductivities of less than 10^{-5} cm/s (upflow triaxial procedure) are recommended for stabilized waste destined for land burial (USEPA 1986e).

Highly permeable materials are not necessarily undesirable. If the stabilized waste does not readily leach contaminants into the water, a high permeability is not as important. High permeability values can also be addressed through engineering solutions such as construction of impermeable liners and covers. Finally, relative permeability can be just as important as actual permeability. The permeability of stabilized/solidified materials should be two orders of magnitude below that of the surrounding materials.

Permeability tests may also be completed on stabilized samples that have undergone durability testing. The durability tests mimic natural stresses that are applied to the stabilized samples (e.g., freeze/thaw weathering) (see Section 4.6). Permeability can potentially increase over time as a result of natural stresses.

4.5 Strength Testing

Strength-test values indicate how well a material will hold up under mechanical stresses created by overburden and earth-moving equipment. Strength-test data also are often used to provide a baseline comparison between unstabilized and stabilized wastes. Unstabilized waste materials generally do not exhibit good shear strength; however, if the waste is stabilized into a cement-like form, the strength characteristics can be expected to increase significantly. Correlation between strength and contaminant leachability has not been widely demonstrated.

Several other strength tests may be performed in addition to or in place of the tests described in the following subsections, depending on the intended use of the data. These include ASTM D1883-87, California Bearing Ratio of Laboratory-Compacted Soils, and ASTM C109-86, Compressive Strength of Hydraulic Cement Mortars.

4.5.1 Unconfined Compressive Strength of Cohesive Soils (ASTM D2166-85)

The Unconfined Compressive Strength measures the shear strength of a cohesive, soil-like material in unsaturated undrained conditions without lateral confinement on the sample. The test method provides an approximate value of the strength of cohesive soils (wastes) in terms of total stress. The soil (waste) may be undisturbed, remolded, or recompacted. The test is applicable to cohesive materials that do not expel water during the loading portion of the test. (Water is expelled from soil as a result of deformation or com-

paction.) Dry, crumbly, or fissured materials cannot be tested with this method.

4.5.1.1 Test Description

The test is completed by centering a specimen (prepared in accordance with ASTM D1632-87) on an apparatus containing upper and lower plates. The specimen is not supported laterally. The test is usually performed as a strain-controlled test, in which the specimen is subjected to a vertical strain rate of 0.5 to 2 percent per minute by applying axial load until the specimen fails. The test continues until the load values decrease with increasing strain, or until 15 percent strain is reached. The peak stress (at failure) is defined as the Unconfined Compressive Strength of the sample. For the assumptions made for this test (zero internal friction and zero minor principal stress), the shear strength equals the Unconfined Compressive Strength divided by two. The strain rate used for the test should be selected to achieve failure in less than 15 minutes. Load and deformation are recorded at sufficient time intervals to be able to plot a stress-strain curve (example in Figure 4-6). The moisture content of the sample is also determined (Subsection 4.1.3) because different moisture contents will change the shape of the curve.

4.5.1.2 Interpretation and Application of Results

Unconfined Compressive Strength results for unstabilized waste have limited application. They do, however, serve as a baseline for comparison with the Unconfined Compressive Strength results for stabi-

Figure 4-6. Stress-strain curve.

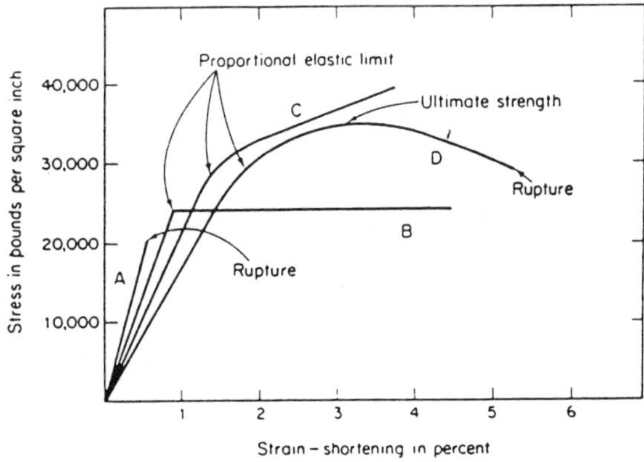

Source: <u>Structural Geology</u>, third edition, by Marland P. Billings. Prentice-Hall, Inc., Englewood Cliffs, NJ.

lized waste. If the unstabilized product is soil-like and the stabilized product is cement-like, there should be a marked increase in the Unconfined Compressive Strength (10 to 20 psi).

4.5.2 Unconfined Compressive Strength of Cylindrical Cement Specimens (ASTM D1633-84)

For stabilized cement-like wastes, the Unconfined Compressive Strength test can provide several pieces of useful information, including the following:

- The ability of the stabilized waste to withstand overburden loads.

- The optimum water/additive ratios and curing times for cement setting reactions.

- The improvement in strength characteristics from the unstabilized to the stabilized waste.

This test also is often conducted on samples subjected to durability tests (Subsection 4.6).

4.5.2.1 Test Description

The test is completed on a cylindrical sample of the materials (ASTM D1632-87). It can be completed with two different cylinder height-to-diameter ratios: 1.15 and 2.0 (Methods A and B). The cylindrical test specimen must be cured for a specified time in a room with 100 percent humidity. Typical curing times for cement are 1, 7, 14, and 28 days. The age of the tested sample should be noted.

The two height-to-diameter ratios cause several differences in testing and in results. Method A uses equipment more readily available in soil-testing laboratories; however, this test method may lead to more complex stress conditions during crushing. Therefore, Method A gives a relative measure of strength rather than a rigorous determination as found in Method B. Method A normally yields a higher compressive strength than an identical sample tested by Method B. Although no consistent preferences in test methods are noted in the literature, comparisons in Unconfined Compressive Strength should only be completed for samples tested by the same method.

The testing apparatus is commonly called a "Compression Testing Machine." The machine may take on several forms, but most important, it must be able to control the rate at which load (stress) is applied. The machine has an upper and lower plate, and the sample is placed upright (long axis vertically) on the lower plate.

The upper plate is lowered and brought into contact with the sample. Load is added continuously without shock by a screw. A sample fails when it loses its physical integrity by falling apart. The total load at failure of the test specimen is recorded to the nearest

10 pounds of force. The Unconfined Compressive Strength is the ratio of force applied at failure to the original cross-sectional area of the cylinder, usually expressed in pounds per square inch (psi). Unconfined Compressive Strength is often expressed in other units. Table 4-3 presents conversion factors for the more common units.

In its methods description, ASTM reports that the average strength difference of two duplicate samples is 8.1 percent. The samples should be tested by the same person to give the maximum precision when interpreting the failure of the sample.

4.5.2.2 Interpretation and Application of Results

The EPA considers a stabilized/solidified material with a strength of 50 psi to have a satisfactory Unconfined Compressive Strength (USEPA OWSER Directive, No. 9437.00-2A). This minimum guideline of 50 psi has been suggested to provide a stable foundation for materials placed upon it, including construction equipment and impermeable caps and cover material. A study by Stegemann et al. (1988) reported Unconfined Compressive Strength values for 69 stabilized/solidified wastes ranging from 10 to 2900 psi.

The minimum required Unconfined Compressive Strength for a stabilized/ solidified material should be evaluated on the basis of the design loads to which the material will be subjected. The anticipated overburden pressure and other loads, along with appropriate safety factors, can be used to calculate this.

Typical construction and compaction equipment can generate very high contact pressures of 1000 psi or more (e.g., sheepsfoot rollers), but surface contact pressures on the order of 50 to 100 psi are more common. This surface load is attenuated with depth so that bearing pressures are reduced to values on the order of 10 to 20 psi at a depth of 2 feet and 3 to 7 psi at a depth of 5 feet below grade. Overburden pressures will usually be on the order of 0.75 to 1.0 psi per foot of depth. If guidelines such as these are used, the stresses to which the stabilized/solidified waste will be subjected can be predicted, and design criteria can be selected accordingly.

Martin et al. (1987) suspect that one-dimensional compressibility may be a more useful indicator of mechanical stabilization/solidification in some situations than is Unconfined Compressive Strength. The one-dimensional compressibility test (e.g., ASTM D2435-80) allows a prediction of fluid expulsion during consolidation and evaluation of cover support. For stabilized/solidified waste forms that are relatively soft or ductile, this test can provide useful performance information; however, for stiff, cement-like, stabilized/solidified waste, the Unconfined Compressive Strength with a rational design criterion is probably

Table 4-3. Unconfined compressive strength conversion factors[a].

Units		psi	kg/cm^2	lb/ft^2	tons/ft^2	newtons/cm^2	pascals (newtons/m^2)
					Units		
1 psi	=	1	0.070	144	0.072	0.689	6895
1 kg/cm^2	=	14.2	1	2048	1.024	9.81	9.81×10^4
1 lb/ft^2	=	0.007	4.88×10^{-4}	1	0.001	0.005	47.9
1 ton/ft^2	=	13.89	0.976	2000	1	9.58	9.58×10^4
1 newton/cm^2	=	1.45	0.102	208.9	0.104	1	10^4
1 pascal (newton/m^2)	=	1.45×10^{-4}	1.02×10^{-5}	0.021	1.04×10^{-5}	10^{-4}	1

a   To convert the numerical value of a property expressed in one of the units in the left-hand column of the table to the numerical value of the same property expressed in one of the units in the top row of the table, multiply the former value by the factor in the block common to both units.

adequate for most situations.

Compressive Strength of Hydraulic Cement Mortars (ASTM C109-86) is often used in place of ASTM D1633-84. It is similar to ASTM D1633-84; however, cubic specimens, which require less material to form, are used instead of cylindrical specimens.

The shape of the stress-strain curve is indicative of the relative stiffness of the stabilized/solidified material. Thus, the shape of the curves can provide some insight as to whether problems with creep or consolidation of ductile behavior or cracking due to brittle or rigid behavior are possible.

4.5.3 Flexural Strength (ASTM D1635-87)

Flexural strength is a measurement of a material's ability to withstand loads applied in tension. During this testing, the loads are considered for the short axis of the sample. In contrast, the loads are applied to the long axis of the sample when the unconfined strength is measured.

4.5.3.1 Test Procedures

This test is completed on apparatus such as that illustrated in Figure 4-7. The specimen should be kept moist after curing and be tested as soon as possible after removal from the moist environment. The stabilized/solidified specimen (ASTM D1632-87) is placed on the apparatus, which supports the specimen at both ends. The head of the testing apparatus consists of a mass supported by two half-rods made of steel. The head is placed at the center of the specimen. The two end supports and the head produce four points of contact on the specimen. Load is applied to the head of the apparatus, which produces a stress at the center of the specimen. The force is applied perpendicularly to the long axis of the specimen to test the flexibility of the specimen. Load is added slowly to the specimen until failure. The load necessary to produce failure of the specimen is recorded to the nearest 10 pounds. If the machine uses a hydraulic press for loading stress, the rate of loading should not be greater than 100 psi per minute. After the specimen fails, the average width and depth of the specimen at the point of failure are measured to the nearest 0.01 inch.

4.5.3.2 Interpretation and Application of Results

The results are a measurement of the strength of a stabilized/solidified waste in terms of cracking in flexure. The results indicate resistance of the stabilized/solidified waste to cracking due to settlement of the underlying fill or due to surface loads. Flexural strength can be viewed as a performance criterion that is dependent on the disposal conditions. It is most appropriate when a material is subject to surface loads after placement or when differential settlement of subgrade materials is possible.

Use of this measurement would be appropriate if a waste were stabilized/solidified and compacted in a disposal cell in a series of lifts or layers. A problem could arise if the first lift were a stiff layer compacted over a soft bedding. In such a case, the compacted layer must have sufficient flexural strength to support the vehicle operations during the installation of the upper lifts. The compacted layer also must have enough strength to support the upper lifts without flexure. In either case, if the lift of compacted material does not have sufficient strength to support surface loads, the compacted material will crack and zones of high permeability and increased potential of leaching will be formed.

Figure 4-7. Schematic of apparatus for flexure test of soil-cement by third-point loading method.

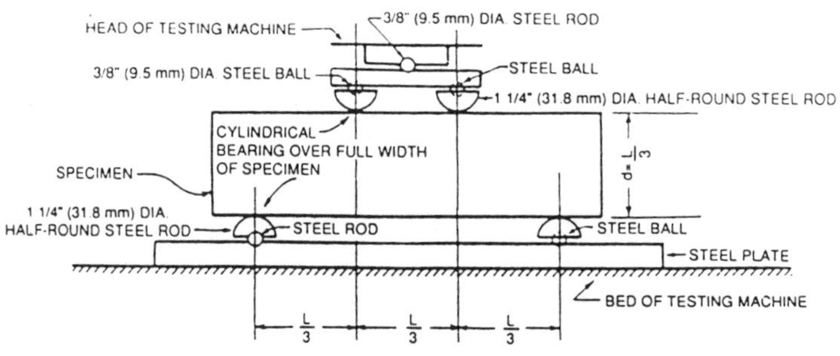

Source: The American Society for Testing and Materials, 1916 Race St., Philadelphia, PA 19103.

4.5.4 Cone Index (ASTM D3441-79)

A material's cone index is indicative of its stability and bearing capacity. The cone index test involves forcing a standard cone-shaped device into the stabilized/solidified waste being tested and measuring the penetration resistance offered by the material (Cullinane 1986).

Cone penetration tests are suited for testing stabilized/solidified sludges or other wastes for landfilling. The test helps to determine the types of vehicles needed to move and place material and the curing time required before other construction equipment can move over it [time before it can be used as subgrade (Tittlebaum and Seals 1985)].

4.5.4.1 Test Description

Various types of cone penetration tests have been used. One commonly used test involves placing a sample of soil-like stabilized/solidified waste in a cup at the bottom of a standard testing apparatus and scraping it level with the top of the cup with a palette knife. A standard metal cone on a shaft is held in place by a clamp. At the beginning of the test the tip of the cone is just touching the surface of the sample. When the clamp is released, the cone penetrates the sample. After 5 seconds, the clamp is closed and the penetration is halted. The distance the cone has penetrated the sample is measured to the nearest 0.1 mm with a dial gauge. The test is repeated to check precision.

4.5.4.2 Interpretation and Application of Results

The cone index was originally developed to determine the "trafficability" of compacted materials, which means it shows the ability of a compacted material to support construction equipment.* Test results are used to determine when a material is able to support the load of specific construction equipment. It is an important parameter in terms of logistical and cost considerations at a site. It is often imperative for equipment to be able to move over the stabilized/solidified waste as quickly as possible to continue work on other sections of the disposal area.

The cone penetration test is practical and inexpensive. Proponents of this test claim that its results can be used in lieu of the Unconfined Compressive Strength test in certain instances (e.g., when a rapid turnaround is required).* Exclusive use of the cone index test, however, would require laboratory correlation between it and the Unconfined Compressive Strength test. The test is also easy to complete in the field by use of a Pocket Penetrometer (Figure 4-8), which works on the same principle as the cone index apparatus just described.

4.6 Durability Testing

Durability testing evaluates the resistance of a stabilized/solidified waste mixture to degradation due to external environmental stresses. The tests are designed to mimic natural conditions by stressing the sample through 1) freezing and thawing, and 2) wetting and drying. The stabilized/solidified specimens undergo repeated cycling during the testing. Unconfined Compressive Strength, flexural strength, permeability, or other performance-based tests may be conducted on the stabilized/solidified samples after each cycle to determine how the physical properties of the stabilized/ solidified waste change as a result of simulated climatic stresses. The number of cycles a material can withstand without failing can be used to judge the mechanical integrity of the material.

Figure 4-8. Photograph of a pocket penetrometer.

Photograph courtesy Gilson Company, Inc., Worthington, Ohio.

---

* Personal communication from M. John Cullinane, U.S. Army Corps of Engineers, Waterways Experimental Station, August 3, 1988.

4.6.1 Freezing and Thawing Test of Solid Waste (ASTM D4842)

Seven molded samples (44 mm in diameter x 74 mm in length) are cured in moist containers for 28 days. One sample is selected for moisture content determination by drying to constant weight in accordance with ASTM D2216-80, revised to use a temperature of 60°C ± 3°. Three samples are subjected to testing and three are control samples. Each of the test specimens is weighed and subjected to 24 hours of freezing at -20°C ± 3°. The controls are kept in the moist containers for 24 hours.

After 24 hours, the samples are covered with distilled water and allowed to sit for another 23 hours. The samples are then removed from their beakers with tongs, and loosely attached particulates are removed by spraying distilled water from a wash bottle onto the surface of the specimen. The samples are then observed for physical deterioration and their weight loss is measured (solids content in beakers by evaporating water at 60°C ± 3° in drying oven). The freeze-thaw cycle is repeated for a total of 12 cycles or until the weight loss (corrected against the weight loss of control samples) of any of the specimens exceeds 30 percent.

4.6.2 Wetting and Drying Test of Solid Wastes (ASTM D4843)

Seven molded samples (44 mm in diameter x 74 mm in length) are cured in moist containers for 28 days. One sample is selected for moisture content determination by drying to constant weight in accordance with ASTM D2216-80, revised to use a temperature of 60°C ± 3°. Three samples are subjected to testing and three are control samples. Each of the test specimens is weighed and subjected to 24 hours of drying at 60°C ± 3°.

After 24 hours, the samples are allowed to sit for 1 hour and then covered with distilled water for another 23 hours. The samples are removed from their beakers with tongs, and loosely attached particulates are removed by spraying distilled water from a wash bottle onto the surface of the specimen. The samples are then observed for physical deterioration and their weight loss is measured (solids content in beakers by evaporating water at 60°C ± 3° in drying oven). The wetting and drying cycle is repeated for a total of 13 cycles or until the weight loss (corrected against the weight loss of control samples) of any of the specimens exceeds 30 percent.

4.6.3 Interpretation and Application of Results of Durability Tests

These tests relate to the long-term stability of the sample. If the results show low loss of materials and retention of physical integrity after testing, the stabilization/solidification process and agent-to-mix ratio are adequate. If the test results show a large loss of material and loss of physical integrity, a different waste-to-agent ratio or different stabilization/solidification agent should be used to provide the long-term stability needed.

No standards are currently established for determining whether stabilized material has passed durability testing; however, Vick et al. (1987) suggest that 15 percent weight loss is an acceptable amount. Because very few materials can withstand the full 12 cycles, the only true measure is a comparison of the results with another stabilization/solidification test (i.e., how many cycles can one mixture withstand versus a different mixture). Obviously, if a cement-like stabilized/solidified waste requires good strength characteristics for proper disposal and it loses its physical integrity during testing, one can conclude that a different stabilization/solidification agent or waste to agent ratio needs to be used.

Poor durability results often can be addressed by a change in design and should not be used as automatic grounds for exclusion. For example, materials that fail freeze-thaw durability testing can be placed below the frost line to mitigate their poor durability property.

5. Chemical Testing Procedures

This section is devoted to a discussion of leaching tests, as these are the tests most often used to evaluate the performance of stabilization/solidification as a treatment process for hazardous waste. Emphasis is on the appropriate selection of leaching tests and the interpretation of laboratory results. Also highlighted are the experimental conditions affecting the reproducibility of the laboratory data and the limitations in extrapolating test results to the field.

Chemical testing procedures applicable to untreated hazardous wastes may also have application to stabilized/solidified wastes. Table 5-1 lists general parameters, test methods, specific applications to untreated and stabilized/solidified wastes, and references.

5.1 Overview of Leaching Mechanisms and Leach Tests

In the field, leaching of hazardous constituents from stabilized/solidified wastes (waste forms) is a function of both the intrinsic properties of the waste form and the hydrologic and geochemical properties of the site. Although laboratory physical and chemical tests can be used to define the waste form's intrinsic properties, the controlled conditions of the laboratory environment are usually not equivalent to changing field conditions. At best, laboratory leaching data can simulate the behavior of waste forms under "ideal," static (conditions at one point in time), or "worst-case" field conditions. Presently, leach tests can be used to compare the effectiveness of various stabilization/solidification processes, but they have not been verified for determining the long-term leachability of the waste.

5.1.1 Leaching Mechanisms

Leaching of a porous medium in the field is generally modeled by solute transport equations that incorporate the following factors:

- Chemical composition of waste and leaching medium
- Physical and engineering properties (e.g., particle size, porosity, hydraulic conductivity) of the waste and surrounding materials
- Hydraulic gradient across the waste

The first factor includes the chemical reactions and kinetics between the leaching fluid and the waste that transform contaminants from an immobile form to a mobile form. The last two factors are used to define the transport of fluids and mobilized contaminants through the waste material.

The physical and engineering properties of the waste material and the hydraulic gradient determine how a leaching solution contacts the waste material. The hydraulic gradient, together with effective porosities and permeabilities, governs the velocity and quantity of the leaching solution migrating through the waste form. For example, if the waste is relatively impermeable (i.e., has low hydraulic conductivity) compared with the surrounding material, the leaching solution will tend to flow around the waste form. This can occur when an intact stabilized/solidified waste form is disposed of in a medium with a hydraulic conductivity 100 times greater (i.e., 10^{-6} to 10^{-4} cm/s). In such cases, most of the contact between the leaching solution and the waste form occurs at the geometrical surface of the waste form. The permeability of stabilized/solidified wastes can increase over time through physical and chemical weathering processes, however, and increase the amount of liquid flow through the waste. Therefore, in the long run, contact between the leaching solution and the waste will occur at the particle surface within the waste form.

The chemistry of the waste and the leaching solution defines the types and kinetics of the chemical reactions that mobilize or demobilize contaminants in the stabilized/solidified waste. Reactions that can mobilize contaminants adsorbed or precipitated within the waste form include dissolution and desorption. Under nonequilibrium conditions, these reactions compete with demobilizing reactions such as precipitation and adsorption. Nonequilibrium conditions generally develop when a stabilized/solidified waste is contacted by a leaching solution and can result in a net transfer, or leaching, of contaminants into the leaching solution.

The following chemical kinetic factors affect molecular diffusion of pollutants within the waste form:

Table 5-1. General chemical test methods that may also have application to stabilized/soldified wastes.

Parameter	Test method	Applicability to untreated and stabilized/solidified wastes	Reference[a]
pH	EPA Method SW-9045	Leachability of hazardous constituents (e.g., metals) may be governed by the pH of the solid	1
Oxidation/reduction potential (E_H)	ASTM D1498-76	Changes in E_H after treatment can change the leachability of many elements	2
Major oxides	ASTM C114	Mineralogy of the stabilized/solidified waste may aid in interpretation of leach test results	3
Total organic carbon (TOC)	Combustion Method	Used to approximate the nonpurgeable organic carbon in wastes and treated solids	4
Oil and grease	EPA Method 413.2	May be used to compare the leachable oil and grease from the treated and untreated wastes	5
Elemental analysis	EPA Method SW-846	Used to determine the fraction of metals leached to the total metals content of the untreated and stabilized/solidified wastes	1
Volatile organic compounds (VOCs)	EPA Method SW-846 (Methods 5030 and 8240)	Used to compare VOC concentrations in stabilized/solidified wastes and untreated wastes with the VOC concentrations in TCLP extracts to determine relative leachability of the treated and untreated wastes	1
Base, neutral, and acid compounds (BNA)	EPA Method SW-846 (Methods 3540, 3520 and 8270)	Used to compare BNA concentrations of leachates with respective concentrations in treated and untreated wastes to determine relative leachability of the treated and untreated wastes	1
Polychlorinated biphenyls (PCBs)	EPA Method SW-846 (Methods 3540, 3520, 680, and 8080)	Same as for VOC with respect to PCB leachability from treated and untreated wastes	6
Ion measurements	Std. Method No. 429	Used to determine leachate ionic species concentrations	7
Heat of hydration	ASTM C186-86	Measurement of temperature changes during current mixing will allow prediction of VOC emissions in the field	8
Alkalinity	Titrometry	Alkalinity changes in leachates may be used to determine changes in stabilized/solidified waste form	9,10

[a]   Reference:

1   U.S. Environmental Protection Agency. 1986. Test Methods for Evaluating Solid Waste. Volumes 1A-1C: Laboratory Manual, Physical/Chemical Methods; and Volume II: Field Manuals, Physical/Chemical Methods, SwW846, Third Edition, Office of Solid Waste. Document Control No. 955-001-00000-1.

2   ASTM D1498-76, Standard Practice for Oxidation-Reduction Potential of Water.

3   American Society for Testing and Materials. 1981. Standard Methods for Chemical Analysis of Hydraulic Cement, ASTM Committee C-1 on Cement, Philadelphia, Pennsylvania. July 1981.

4   Perkin Elmer 240C Users Manual.

5   U.S. Environmental Protection Agency. 1979. Methods for the Chemical Analysis of Water and Wastes. Office of Research and Development. EPA-600 4-79-020, March 1979.

6   Albord-Stevens, Bellar, Erchelberger and Bubble, Method 680, November 1985, Determination of Pesticides, PCBs in Water and Sediment by GC-MS, Office of Research and Development, U.S. Environmental Protection Agency.

7   Standard Methods for the Examination of Water and Wastewater, 16th Edition, APHA, AWWA, WPCF, 1985. Washington, D.C.

8   ASTM C186-86. Standard Test Method for Heat of Hydration of Hydraulic Cement.

9   Bishop, P. L. 1986. Prediction of Heavy Metal Leaching Rates from Stabilized/Solidified Hazardous Wastes. In Toxic and Hazardous Wastes Proceedings of the 18th Mid-Atlantic Industrial Waste Conference.

10 APHA-AWWA-WPCF. 1975. Standard Methods for the Examination of Water and Wastewater, 14th Edition. Method 403. pp. 278-281.

- Accumulation of waste species in the pore solution at the particle surface.

- Concentration of reactive species (e.g., H^+, complexing agent) in the pore solution at the particle surface.

- Bulk chemical diffusion of the waste or reactive species within the leachate pore solution or waste form.

- Polarity of the leaching solution and waste species.

- Oxidation/reduction conditions and competing reaction kinetics.

Because laboratory leaching tests usually involve standardized aqueous solutions (neutral, buffered, or dilute acidic solutions) rather than site-specific solutions, the results of the laboratory tests may not directly duplicate leaching in the field. As previously mentioned, laboratory leaching tests run with standard solutions can be used to compare the relative leachability of waste constituents under similar test conditions and with similar leaching solutions.

Depending on the physical and chemical properties of the waste and the leaching solution, the kinetics of contaminant transport (or leaching) in a porous medium are controlled by advective or dispersive/diffusive mechanisms. Advection refers to the hydraulic flow and subsequent solute transport of highly soluble contaminants in response to a hydraulic gradient. Dispersion refers to the transport of contaminants via mechanical mixing in the pore solution and molecular diffusion (the transfer of mass between adjacent layers of fluid in laminar flow). Because of the low permeabilities of most stabilized/solidified wastes, the rate of contaminant transport for adsorbed or chemically bonded constituents is generally considered to be controlled by molecular diffusion at the particular surfaces within the waste form, rather than advection or dispersion.

The buildup of chemical potential at the interface between particles and pore solution is the driving force for the diffusion-controlled transport of waste constituents within the aqueous solution and the waste form (Cote et al. 1987). This nonequilibrium condition is controlled primarily by the chemistry and velocity of the leaching solution.

Figure 5-1 shows the effect of the velocity of the leaching solution on the leaching rate of contaminants that are leached at the particle surface. Leaching solution velocity (v) is defined as the volume of leaching solution (V) contacted with waste per unit of surface area (SA) per unit of time (T):

$$v = V/(SA \times T)$$

Leaching rate (L) is the mass of the waste species (M)

leached per unit of surface area per unit of time:

$$L = M/(SA \times T)$$

The slope of the leaching curve in Figure 5-1 is the leachate concentration of the waste species in the leaching solution, or M/V. As shown in Figure 5-1, at high leaching solution velocities (rapid flow through the waste form), the leaching rate approaches the maximum, Lr. In addition, under rapid leaching solution velocities, leachate concentrations are very low (approaching zero) if leaching of the waste species is diffusion-controlled. High leaching rates and low leachate concentrations occur at the particle surface under rapid leaching velocities because nonequilibrium conditions at the particle surface are maintained. Rapid leaching rates occur in laboratory studies when the leaching solution is constantly being replenished with fresh solutions.

At low leaching solution velocities (i.e., static hydraulic conditions), the amount of a species leached approaches the saturation limit, Sl, or the maximum leachate concentration. Low leaching solution velocities and maximum leachate concentrations occur when the leaching solution is not replenished, and the same leaching solution is allowed to equilibrate with the waste.

These relationships between leachate concentrations and leaching solution velocities are important to the understanding and interpretation of leaching test results because the tests vary dramatically with leaching solution velocities, contact surface areas, leaching solution volumes, and time of leaching. The chemistry of leaching solutions also varies widely from neutral solutions to very acidic or strong chelating solutions.

5.1.2 Leaching/Extraction Tests

Numerous leaching tests have been developed to test solid wastes, including those developed specifically for stabilized/solidified nuclear and hazardous wastes. Because these tests have been developed by several different groups and specialists, the terminology applied to leaching tests has not been well established. Therefore, an attempt is made to clarify the terms used in this handbook with regard to leaching tests.

Extraction (or batch extraction) tests refer to a leaching test that generally involves agitation of ground or pulverized waste forms in a leaching solution. The leaching solution may be acidic or neutral. Also, it may vary throughout the extraction test. Extraction tests may involve one-time or multiple extractions. In either case, leaching is assumed to reach equilibrium by the end of one extraction period; therefore, extraction tests are generally used to determine the maximum, or saturated, leachate concentrations under a given set of test conditions.

Figure 5-1. Relationship between velocity of leaching solution and the leaching rate.

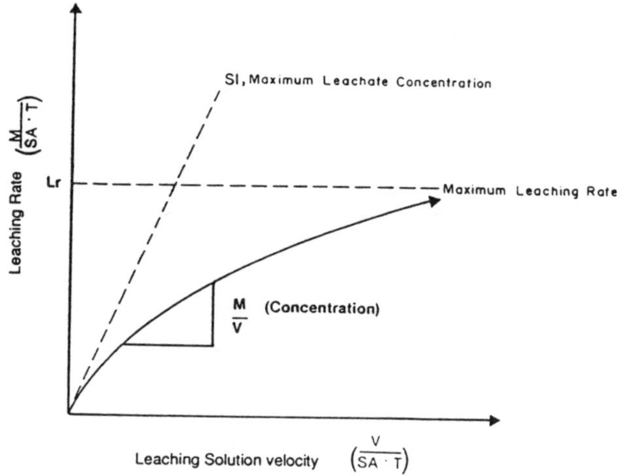

Source: Cote et al. 1987.

"Leach test," another type of leaching test, involves no agitation. The leaching of monolithic (instead of crushed) waste forms is evaluated in these tests. Leaching may occur under static or dynamic conditions, depending on the frequency of the leaching solution renewal. In static leach tests, the leaching solution is not replaced by a fresh solution; therefore, leaching takes place under static hydraulic conditions (low leaching velocities and maximum leachate concentrations for monolithic waste forms). In dynamic leach tests, the leaching solution is periodically replaced with new solution; therefore, this test simulates the leaching of a monolithic waste form under nonequilibrium conditions in which maximum saturation limits are not obtained and leaching rates are high. "Static" and "dynamic," therefore, refer to the velocity, not the chemistry of the leaching solution.

Results of dynamic leach tests are generally expressed in terms of a flux or mass transfer parameter (i.e., leaching rate), whereas data from extraction tests are expressed in terms of leachate concentration or cumulative fraction of total mass leached. Another key difference between these two leaching tests is that extraction tests are short-term tests lasting from hours to days, whereas leach tests generally take from weeks to years. Because of the crushed nature of the waste and the larger amount of surface area available for leaching, extraction tests (although short-term) are used to simulate "worst-case" leaching conditions. Leach tests on monolithic wastes (although longer in duration) are often used to simulate leaching under "well-managed," short-term scenarios in which the waste form is intact.

The column leach test is another type of laboratory leaching test. This test involves placing pulverized waste in a column, where it continuously contacts with a leaching solution at a specified rate. The leaching solution is generally pumped through the waste in an upflow column setup. Column tests are considered to be more representative of field leaching conditions than batch extraction tests because of the continuous flux of the leaching solution through the waste. This test is not often used, however, because of problems with the reproducibility of test results. These problems include channeling effects, nonuniform packing of the wastes, biological growth, and clogging of the column (Cote and Constable 1982). Nevertheless, column tests have been used along with extraction tests to study the effects of particle size on the leaching of heavy metals (Bishop 1986).

5.2   Leach Test Methods and Applications

This subsection presents a detailed review of the methods and general uses of several of the more common extraction and leaching tests. The following extraction tests are discussed:

- Toxicity Characteristic Leaching Procedure (TCLP)
- Extraction Procedure Toxicity Test (EP Tox)
- California Waste Extraction Test (Cal WET)

- Multiple Extraction Procedure (MEP)
- Monofilled Waste Extraction Procedure (MWEP)
- Equilibrium Leach Test (ELT)
- Acid Neutralization Capacity (ANC)
- Sequential Extraction Test (SET)
- Sequential Chemical Extraction (SCE)

The TCLP is used by EPA as the basis for the promulgation of best demonstrated available technologies (BDAT) treatment standards under the land disposal restrictions program. The EP Tox and Cal WET procedures are used by EPA and the State of California, respectively, for characterizing hazardous wastes. The remaining six tests provide useful information about maximum leachate concentrations under various conditions and the chemistry and waste constituents of the waste form.

In addition to the extraction tests, the following leach tests are also discussed:

- Materials Characterization Center Static Leach Test (MCC-1P)
- American Nuclear Society Leach Test (ANS-16.1)
- Dynamic Leach Test (DLT).

The MCC-1P and ANS-16.1 tests were developed for stabilized high- and low-radioactive wastes. The Dynamic Leach Test represents a modification of the ANS-16.1 leach test, which was developed for stabilized/solidified hazardous waste. These tests provide data for evaluating leaching rates (DLT and ANS-16.1) and maximum leachate concentrations (MCC-1P Static Leach Test) from intact waste forms leached with water.

Table 5-2 summarizes the main differences in test conditions among the nine extraction tests listed. The main experimental test variables in extraction tests are the leaching medium, the liquid-to-solid ratio, the particle sizes of the crushed waste sample, and the number and duration of the extractions. As shown in Table 5-2, leaching solutions in extraction tests vary from acids of different strengths and concentrations to distilled/deionized water. Liquid-to-solid ratios vary from as low as 3:1 (less acid added per gram of waste) to as high as 50:1 (more acid added per gram waste). Particle sizes range from less than 9.5 mm (larger particles and smaller contact surface area) to less than 0.15 mm (smaller particles and larger contact surface area). Extraction periods range from 2 hours to 48 hours, and the number of extractions range from 1 to 15. It is obvious that results from leach tests must be evaluated with an understanding of the differences in the experimental conditions.

5.2.1 Toxicity Characteristic Leaching Procedure (TCLP) (Federal Register 1986)

Waste samples are prepared by crushing the wastes to pass through a 9.5-mm screen, and liquids are separated from the solid phase by filtration through a 0.6- to 0.8-μm borosilicate glass-fiber filter under 50 psi pressure. Two choices of buffered acidic leaching solutions are offered under TCLP, depending on the alkalinity and the buffering capacity of the wastes. Both are acetate buffer solutions. Solution No. 1 has a pH of about 5; Solution No. 2 has a pH of about 3. The leaching solution is added to a Zero Headspace Extractor (ZHE) at a liquid:solid ratio of 20:1, and the sample is agitated with a National Bureau of Stan-

Table 5-2. Extraction tests.

Test method	Leaching medium	Liquid:solid ratio	Maximum particle size	Number of extractions	Time of extraction
TCLP	Acetic acid*	20:1	9.5 mm	1	18 hours
EP Tox	0.04 M acetic acid (pH = 5.0)	16:1	9.5 mm	1	24 hours
Cal. WET	0.2 M sodium citrate (pH = 5.0)	10:1	2.0 mm	1	48 hours
MEP	Same as EP Tox, then with synthetic acid rain (sulfuric acid: nitric acid in 60:40 wt.% mixture)	20:1	9.5 mm	9 (or more)	24 hours per extraction
MWEP	Distilled/deionized water or other for specific sit	10:1 per extraction	9.5 mm or monolith	4	18 hours per extraction
Equilibrium leach test	Distilled water	4:1	150 μm	1	7 days
Acid neutralization capacity	HNO$_3$ solutions of increasing strength	3:1	150 μm	1	48 hours per extraction
Sequential extraction tests	0.04 M acetic acid	50:1	9.5 mm	15	24 hours per extraction
Sequential chemical extraction	Five leaching solutions increasing in acidity	Varies from 16:1 to 40:1	150 μm	5	Varies from 2 to 24 hours

*Two acetate buffered solutions: 1) pH = 5, 2) pH = 3.0

dards (NBS) rotary tumbler at 30 rpm for 18 hours. The leaching solution is filtered and combined with the separated liquid waste fraction for analysis.

The EPA proposed this test 1) to replace EP Tox as the criterion for defining hazardous and nonhazardous wastes, and 2) to be used for some listed wastes as the standard criterion for hazardous waste treatment. With the ZHE apparatus, TCLP can be used to evaluate the leaching of volatile and semivolatile organic compounds. It has also been used to evaluate maximum "worst-case" leachate concentrations achievable in the field; however, several studies (Bishop 1986, Barich et al. 1987, U.S. EPA 1988c) show that TCLP leaching of cement-based waste forms may not necessarily yield maximum concentrations. Multiple extraction tests, such as MWEP or MEP, may be needed to assess maximum leachate concentrations under different pH conditions.

5.2.2 Extraction Procedure Toxicity Test (EP Tox) (U.S. EPA 1986)

Similar to TCLP in experimental design, EP Tox generally yields comparable results (Newcomer et al. 1986). Considerably different results may be observed if the more acidic TCLP extraction fluid No. 2 (pH = 2.88 ±0.05) is used. Table 5-3 summarizes the differences between the two test methods. The primary difference between EP Tox and TCLP is that the EP Tox leaching solution (only one solution of acetic acid at a pH of 5) is added periodically (up to a specified maximum acid addition) as needed to adjust the pH of the leaching solution during the course of the extraction. The TCLP solutions are buffered so that the leaching solution is added only once at the start of the extraction.

Table 5-3. Summary of procedural differences between EP TOX and TCLP.

Experimental parameter	TCLP	EP Tox
Filter size, μm	0.6-0.8	0.45
Filter pressure, psi	50	75
Leaching solution	Acetate buffered solution (pH ≈3 or 5)	Acetic acid (pH ≈5)
Period of extraction, h	18	24
Liquid:solid ratio	20:1	16:1

The EP Tox test also has been used to classify wastes as hazardous or nonhazardous. This test, however, is designed to determine semivolatile organic and heavy metal leachate concentrations; it does not include analysis of volatile organic compounds. Generally, EP Tox and TCLP yield similar leachate concentrations of metals. Studies by Newcomer, Blackburn, and Kimmell (1986) and Shively and Crawford (1987), however, indicate that TCLP extracts greater metal concentrations. The statistical mean TCLP leachate concentrations ranged from 1.0 to 3.0 times greater than those for EP Tox (Newcomer, Blackburn, and Kimmell 1986). Although EP Tox may also be used to evaluate maximum leachate concentrations, like TCLP, it should be used along with other extraction tests.

5.2.3 California Waste Extraction Test (Cal WET)*

As shown in Table 5-2, Cal WET differs from TCLP and EP Tox in the following parameters:

- Different leaching solution (sodium citrate buffered solution at pH of 5; or, for hexavalent chromium, distilled water)
- Smaller liquid:solid ratio (10:1)
- Smaller particle size (less than 2.0 mm)
- Longer extraction period (48 hours)

The State of California uses Cal WET to classify hazardous waste. Because of the different metal chelating ability of sodium citrate solution, Cal WET is a more stringent leach test than TCLP for some metals.

5.2.4 Multiple Extraction Procedure (MEP) (U.S. EPA 1986g)

Although MEP is not a regulatory leaching test, it has been used in some instances for delisting wastes. This test involves multiple (sequential) extractions of the crushed sample with a synthetic acid rain solution. The first extraction is performed with the acetic acid solution in accordance with EP Tox methods. The subsequent extractions are performed with the synthetic acid solution (concentrated sulfuric acid/nitric acid, 60/40 wt.%, diluted to a pH of 3). A total of nine extractions are usually performed; however, more extractions are possible if the last three extractions do not decrease the leachate concentrations.

Results obtained by MEP can be used to determine maximum leachate concentrations occurring under acidic conditions. This test can be used with EP Tox or with MWEP (multiple extraction test with water) to compare leachability of hazardous constituents under mild and acidic conditions.

---

* California Code, Title 22, Article 11, pp. 1800.75-1800.82.

5.2.5 Monofill Waste Extraction Procedure (MWEP) (U.S. EPA 1986f)

Formerly called the Solid Waste Leach Test (SWLT), MWEP involves multiple extractions of a monolith or crushed waste with distilled/deionized water. The sample is crushed to less than 9.5 mm, or it can be left intact if it passes the Structural Integrity Test (SW-846). Ultimately, however, monolithic samples of stabilized/solidified wastes can be crushed during the extraction test as a result of agitation by the rotary tumbler (Barich et al. 1987). The liquid:solid ratio is 10:1, and the sample is extracted with water four times (18 hours per extraction).

This test can be used to derive reasonable leachate compositions in monofilled disposal facilities, and this information can be used to assess waste-liner compatibility under mild leaching conditions. It also can be used with TCLP to determine delays in the release of hazardous constituents. In addition, results from MWEP can be compared with those from ELT to assess the maximum leachate concentrations achieved under mild leaching conditions.

5.2.6 Equilibrium Leach Test (ELT) (Environment Canada and Alberta Environmental Center 1986)

This leach test involves static leaching of hazardous constituents in distilled water. The particle size of the crushed sample (150 µm) is much smaller than that for TCLP and EP Tox to allow greater contact surface area and to reduce the time needed to achieve equilibrium conditions. Water is added once at a liquid:solid ratio of 4:1, and the sample is agitated for 7 days.

Like MWEP, ELT can be used to determine maximum leachate concentrations under mild leaching conditions. Although particle size and liquid:solid ratio are smaller for ELT, leachate concentrations from the two tests should be comparable if equilibrium conditions are achieved under both (Cote, Constable, and Moreira 1987). Sample heterogeneity and analytical limitations may cause differences.

5.2.7 Acid Neutralization Capacity (ANC) (Environment Canada and Alberta Environmental Center 1986)

Acid Neutralization Capacity (ANC) involves separate extractions of several predried, crushed, waste samples with leaching solutions of varying levels of acidity. The amount of sample is much smaller, the extraction is performed in test tubes and a rotary tumbler, and liquid-solid separation is accomplished by centrifuging instead of filtering. Particle size is less than 150 µm (-100 mesh), and the liquid-to-solid ratio of the extraction is 3:1. Ten samples (labeled 1 through 10) are extracted for 24 hours in one of 10 nitric acid solutions that increase incrementally in the number of equivalents of acid added per gram of dried solid.

The ANC test is used to determine the buffering capacity of the stabilized/solidified waste form. Figure 5-2 illustrates the change in pH as increasing amounts of acid are added for cement-based waste forms. Stegmann, Cote, and Hannak (1988) conclude that for a wide range of metal- and organic-bearing wastes stabilized with proprietary agents, the amount of acid required to bring the pH down to 9 (where many metals are soluble) varied between 2 and 10 milliequivalents per gram (meq/g) of waste. For cement-based wastes, the ANC is generally about 15 meq/g (Cote and Bridle 1987).

Figure 5-2. Acid neutralization capacity for several stabilized/solidified synthetic sludge samples.

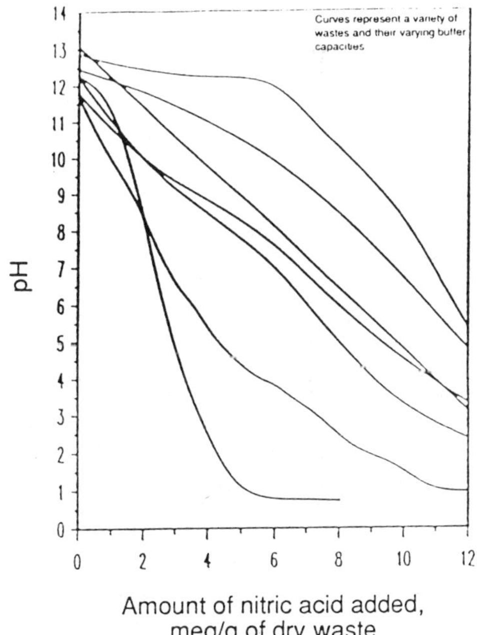

Amount of nitric acid added, meq/g of dry waste

The higher the buffering capacity of the waste, the greater the possibility of maintaining alkaline conditions and minimizing the amount of metals leached. The buffering capacity of the waste form, therefore, is very important in evaluating the amount and rate of metals leached from stabilized wastes in the field.

5.2.8 Sequential Extraction Test (SET) (Bishop 1986)

The Sequential Extraction Test is also designed to evaluate the buffering capacity of the waste form. Unlike ANC, SET involves 15 sequential extractions of one sample of crushed waste with particle sizes between 2.0 and 9.5 mm. Each extraction is performed on a shaker table for 24 hours with the same extraction solution (0.04 M acetic acid solution) and a liquid:solid ratio of 50:1. With each extraction, 2 meq/g of acid is added to the ground waste. The pH is measured and the leaching solution is filtered. At the end of the fifteenth extraction, the remaining solids are digested with three more extractions in which more concentrated acid solutions are used. These last three extractions are combined for analysis.

Bishop (1986) and Shively et al. (1986) used SET to evaluate the waste buffering capacity and alkalinity of cement-based waste forms. They also correlated analyses of pH, alkalinity, dissolved solids, and metals to determine metal speciation within the waste form. Shively et al. (1986) and Bishop (1988) observed that the pH of the leaching solution remained above 10 during the first three extractions, and only anionic arsenic (As) appeared in the leachate. The leachate pH rapidly dropped to 6 during the next three extractions and then leveled out at 4 during the last nine extractions (Figure 5-3a).

The buffering species is primarily calcium hydroxide, based on gran analysis (Bishop 1986). Also, analyses of the extracts from the last three digestions indicate that from 75 to 85 percent of chromium, lead, and silicon dioxide were not leached during the 15 extractions, whereas only 8 percent calcium remained after 15 extractions (Bishop 1986). Based on similar release curves, chromium and lead are believed to be bound to the silicate matrix (Figure 5-3b). These studies indicate that the buffering capacity for cement-based waste forms is about 18 meq/g.

5.2.9 Sequential Chemical Extraction (SCE) (Environmental Canada and Alberta Environmental Center 1986)

This test was developed specifically to evaluate the species of organic and inorganic waste constituents in a stabilized matrix. Like SET, the test involves sequential extraction of a sample. Unlike SET, however, the leaching solution increases in acidity from neutral to very acidic with each sequential extraction. The particle size of the sample is also very small (less than 45 µm), and the sample is agitated on a Burrell wrist-arm shaker.

Results from the SCE test are used to interpret the bonding nature of metals and organics in the stabilized waste matrix. Contaminants leached in Fractions A, B, and C are interpreted to be available for leaching, whereas those that are leached with concentrated acids (D and E Fractions) are considered to be unavailable for leaching. Studies by Bridle et al. (1987) and Stegmann, Cote, and Hannak (1988) indicate that metals are leached primarily in Fractions B and C, and thus are available for leaching in natural systems. They also conclude that this test cannot be used to determine speciation and bonding mechanisms for arsenic and mercury. Because of the lack of sensitivity indicated by the results from Fractions A through C, results from the SCE test should be used only to determine whether contaminants are potentially available or unavailable for leaching under the mild leaching conditions that may be encountered in the field (personal communication with P. L. Cote on July 14, 1988).

5.2.10 Materials Characterization Center Static Leach Test (MCC-1P) (MCC 1984)

This static leaching test was developed for high-level radioactive waste. It involves leaching of a monolithic waste form with water (ASTM Type I or II) at a ratio of volume of leaching solution to surface area solids (V/S) of between 10 and 200 cm. The period and temperature of extraction vary, depending on the schedule selected.

For organic and polymer stabilization/solidification processes, leachability should be evaluated by using MCC-1P or ANS- 16.1, as these waste forms cannot be ground. In general, however, MCC-1P test results can be combined with those from extraction tests (e.g., TCLP, MWEP) to determine a range of leachate concentrations in the short term (well-managed site with waste form intact) and the long run (waste matrix has been subjected to many years of environmental stress and is fractured).

5.2.11 American Nuclear Society Leach Test (ANS-16.1, 1986) (ANS 1986)

A "quasi-dynamic" leach test, ANS-16.1, is applied to stabilized/solidified low-level and hazardous wastes. A monolithic cylinder (length:diameter of 0.2 to 5.0) is leached with demineralized water applied at a V/S ratio of 10 cm under ambient temperatures. At the start of the experiment, the sample is rinsed to obtain zero contaminant concentration at the surface of the sample. Afterwards, the sample is immersed in water, which is replaced after 2 hours, 7 hours, 24 hours, 48 hours, 72 hours, 4 days, 5 days, 14 days, 28 days, 43 days, and 90 days.

The results of the leaching test are recorded in terms of cumulative fraction leached over the total mass in the waste form, F. Calculations are then used to derive an effective diffusion coefficient, De (cm²/s), and a leachability index (LX = -log De). The LX values range from 5 (De = 5-10, rapid diffusion) to 15 (De = 10-15, very slow diffusion).

Figure 5-3. Sequential extraction test results: (a) Leachate pH and cumulative alkalinity leached from solidified/stabilized subjected to multiple-batch extraction procedure. (b) Cumulative heavy metal, silicone, and alkalinity leached from solidified/stabilized samples subjected to multiple-batch extraction procedure.

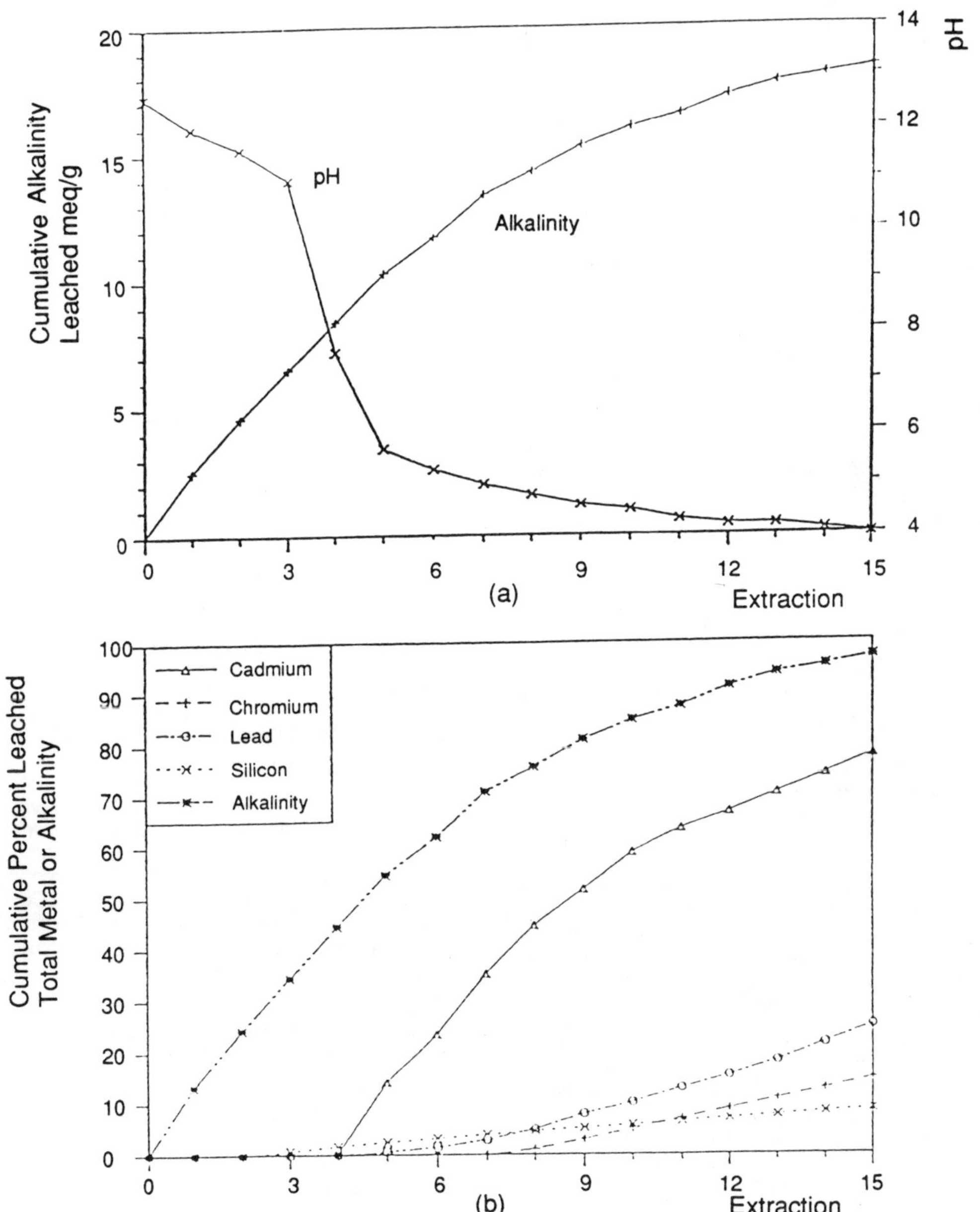

Interpretation of ANS-16.1 results assumes that leaching is diffusion-controlled and that nonequilibrium conditions are maintained during each leaching period. Because leaching rates for hazardous constituents vary widely and may not be diffusion-controlled, the leaching rates under an ANS-16.1 renewal schedule may actually be slower as leaching conditions become static (equilibrium conditions). Hence, the LX values obtained from ANS-16.1 results may be greater than those under the modified ANS-16.1 (Dynamic Leach) test, discussed in the next subsection.

5.2.12 Dynamic Leach Test (DLT) (Environmental Canada and Alberta Environmental Center 1986)

The Dynamic Leach Test is a modified version of the ANS-16.1. The only modification to the leach test is the renewal frequency of the leaching solution and the V/S ratio, which is based on known diffusion coefficients and results from batch extraction tests (e.g., ELT). The selected V/S ratio must ensure that the contaminant can be detected, and the renewal frequency for the leaching solution should ensure that nonequilibrium leaching conditions prevail. Otherwise, the extraction procedure, apparatus, and the subsequent calculations for deriving LX values are the same.

The leaching data must be checked for the following criteria:

- Cumulative fraction leached, F, increases linearly with T1/2, where T is time.

- Concentrations leached during one leaching period are less than the maximum, or saturated, concentrations.

- F is less than 0.2.

The linear relationship between F and the square root of time verifies that leaching rates are diffusion-controlled. If leaching is occurring through dissolution or convective mechanisms, the relationship would not be linear. If leachate concentrations are less than the saturated concentrations, leaching is occurring under nonequilibrium conditions and approaching maximum leaching rates. The final criteria, if met, allow leaching rates to be calculated by equations that are based on leaching from a semi-infinite medium.

Figure 5-4 illustrates potential relationships between cumulative fraction leached, F, and time of leaching √T. Curve A shows a nonlinear relationship between F and √T for a very soluble constituent. Curve B is a straight line with a Y-intercept (F > 0 at T = 0), which indicates that the contaminant was leached in the initial rinse (surface wash-off). Curve C is a straight

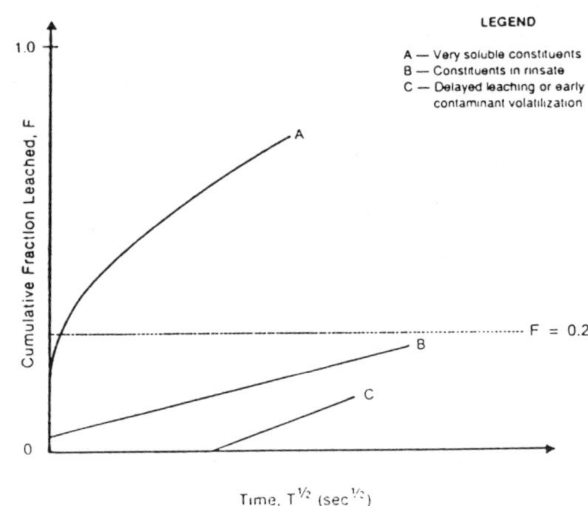

Figure 5-4. Schematic plot of cumulative fraction leached vs. the square root of time.

line with an X-intercept (F = 0 at T > 0), which indicates either delayed leaching or early volatilization of the contaminant from the surface of the waste form (Cote and Hamilton 1984).

Results from dynamic leach tests conducted on cement containing heavy metals with different initial concentrations and under different renewal schedules show that calculated De's varied within one order of magnitude (i.e., LX values varied within 1 unit) (Cote and Hamilton 1986). Studies on various waste types and various leaching solution stabilizers show that organic compounds tend to have LX values between 5 and 10 (higher leach rates), and metals tend to have higher LX values (slower leach rates) (Cote and Hamilton 1986; Stegmann, Cote, and Hannak 1988). Variability in test results for metals generally are attributable to analytical sensitivity, especially for relatively insoluble metals.

5.3   Experimental Factors Affecting Results and Interpretation

Results of both leaching and extraction tests vary, primarily because of sample nonhomogeneity. Curing time may also affect leach test results. The Corps of Engineers cures waste for 28 days, as opposed to the minimum curing time of 42 days recommended by

Bishop (1988). The relative standard deviations (%RSD) reported for TCLP, EP Tox and MCC-1P range from 10 to 100 percent (ANS 1986; Newcomer, Blackburn, and Kimmell 1986; Stegmann, Cote, and Hannak 1988). Numerous studies comparing the test results obtained by TCLP and EP Tox for their relative precision (summarized in Newcomer, Blackburn, and Kimmell 1986) show that, although neither is a precise test, TCLP is as precise or slightly more precise than EP Tox in inter- and intralaboratory tests with metal wastes. Because of the variability in leaching test results, more than one sample of stabilized/solidified hazardous waste should be leach-tested.

Other experimental parameters may also affect results of batch extraction tests (Cote and Constable, 1982; Newcomer, Blackburn, and Kimmell 1986). For TCLP, varying the acidity of the No. 2 leaching solution from 190 to 210 meq greatly influences test results for metals; other variables, such as liquid-to-solid ratio (19:1 and 21:1), extraction time (16 vs. 18 hours), and filter type, have relatively little impact on test results (Newcomer, Blackburn, and Kimmell 1986).

Cote and Constable (1982) conclude that rotary-tumbler agitators used in extraction tests yield better precision than shaker tables or wrist-arm agitators. In comparing results of several different extraction tests, Cote and Constable (1982) concluded that, of five test variables (leaching medium, liquid:solid ratio, number of extractions, particle size, and period of Figure 5-4 extraction), the first three have the greatest impact on extraction test results for metals. In general, cumulative mass fractions of metals leached are higher with more acidic leaching solutions, with larger liquid:solid ratios (with neutral solutions), or with multiple extractions (Cote and Constable 1982; Cote, Constable, and Moreira 1987). In theory, particle size and period of extraction have no effect on leachate concentrations if the extraction tests reach equilibrium conditions. Studies by Bishop (1986), however, show that although the opposite is observed initially, with multiple extractions, smaller particle sizes leach slightly more metals than do larger particles.

The effects of the liquid:solid ratio vary with the type of leaching solution and the solubility of the contaminant (Cote, Constable, and Moreira 1987). With neutral or slightly acidic leaching solutions and soluble waste species, an increase in the liquid:solid ratio increases the cumulative mass fraction leached, but the leachate concentrations remains constant (i.e., the ratio of mass leached to leaching solution volume remains the same). Leachate concentrations of insoluble species in the same medium decreased with increasing liquid:solid ratio. With more acidic leaching solutions, the leachate concentrations are controlled by the pH and chemistry of the leaching solution; therefore, leachate concentrations are not so easily predicted. In general, liquid:solid ratios must be large enough for analysis and small enough to allow detection of analytes above detection limits.

The impacts of these experimental parameters must be considered in an evaluation of the results of several leaching tests on a stabilized/solidified waste. Also, the pH and chemistry of the leaching solution must be monitored before and after the leaching test. For example, leaching tests on stabilized/ solidified metal wastes should include at least measuring the pH and alkalinity as well as total dissolved solids or conductivity of the leaching solution. Leaching tests conducted by Bishop (1988) include analysis for calcium, iron, aluminum, and other cement elements. These analyses aid in the interpretation of the leaching test data.

Other analytical techniques used to determine the speciation of waste constituents, particularly organic compounds, in the stabilized matrix are polarized light microscopy, energy-dispersive X-ray diffraction (EDX), scanning electron microscope (SEM), and electron probe microanalysis (EPMA) (Tittlebaum et al. 1987; Campbell et al. 1987). In addition, protocols for evaluating bonding between organic compounds and stabilizer in a stabilized/solidified waste have been developed by Dr. J. Soundararajan (PRC 1988). These protocols include four technologies: 1) Fourier transform infrared spectroscopy (FTIR); 2) thermal gravimetric analysis (TGA); 3) differential scanning colorimetry (DSC), and 4) DSC coupled with gas chromatography/mass spectroscopy (GC/MS).

5.4 Leaching Test Selection and Interpretation

As mentioned in the preceding discussions, leaching tests produce results that are not directly applicable to leaching behavior in the field. Nevertheless, the results of several leaching tests or of leaching tests combined with physical tests or microscopic techniques can be used as indicators of field performance and environmental impact.

When used for comparative purposes, results from several leaching tests can help to identify field conditions that result in higher concentrations of waste constituents. Therefore, these data may be used to site or design waste facilities that will minimize the leaching of hazardous constituents from the wastes. The data also may be used to predict the leaching of stabilized/solidified wastes at different stages in time. For example, leaching conditions of a well-managed operational facility where the monolithic stabilized/ solidified waste receives maximum precipitation infiltration may be simulated by the use of the DLT, as this test involves constant renewal of the leaching solution and a monolithic waste form. For a closed facility that has a cover which is maintained (i.e., a 30-year post-closure period) and minimizes precipitation infiltration, leaching conditions may be similar to those of the MCC-1P test (i.e., static hydraulic conditions). In the long run, the liners, cover, and waste form will degrade, and leaching conditions may be similar to those found in multiple extraction tests (such as MEP and MWEP) or equilibrium tests (such as ELT).

In the few cases where the actual field leaching solution is well known, use of this solution in the laboratory tests may yield more representative results. When the site leaching solution is used, however, the results may only be relevant to field leaching conditions in the short term because the site hydrogeochemistry will change over the long run.

6. Technology Screening

Technology screening is a tool used to organize relevant information for each hazardous waste site for the purpose of evaluating the applicability of stabilization/solidification. Such screening is required because of the imprecise nature of stabilization/solidification and because most wastes are far from homogeneous. This section outlines a series of steps for assembling information about stabilization/solidification. An understanding of the site and its waste, bench-test evaluations of the stabilized/solidified waste form through the tests described in Sections 4 and 5, and an understanding of the implementation and costs of stabilization/solidification will aid in making an informed judgment about product suitability. The choice of a final waste product will inevitably reflect a compromise based on chemical, physical, and cost considerations. This section does not discuss design considerations in the screening of stabilization/solidification technologies; rather, it discusses the feasibility of the technologies.

Selection of a stabilization/solidification treatment technology for hazardous waste site remediation depends on the nature of the waste, the physical containment properties required for the stabilized/solidified material, the applicability of the technology to the waste site, and cost (Wiles, undated). This technology may become increasingly sophisticated as a result of changing regulations, which require a controlled product with specific leachate characteristics (<u>Federal Register</u> 1988). Durability tests such as those described in Section 4 are also an important aspect of evaluating the stabilized/solidified product (Jones 1986).

Waste/binder incompatibility also presents a complication in the application of stabilization/solidification technology to hazardous wastes. Experience in the cement, asphalt, and nuclear waste industries demonstrates that small amounts of some compounds can significantly reduce the physical strength and containment characteristics of the waste/binder product (Jones 1986). Some of these interactions have been documented in stabilization/solidification case studies (Jones 1986; Tittlebaum and Seals 1985; Kyles, Malinowski, and Stanczyk 1987; JACA 1985; U.S. EPA 1988d). As of this writing, however, few data quantifying the effects of interfering compounds on a particular stabilization/solidification process are available (Jones 1986; Tittlebaum and Seals 1985). Protocols for processing hazardous waste materials by use of stabilization/solidification technologies have not been standardized.

6.1 Application of Technology Screening

Technology screening is a process in which both the limitations and applicability of a technology are systematically reviewed. With regard to stabilization/solidification, the technology screening process involves the following:

- Review of existing chemical and physical interferences between the waste and binder.

- Identification of previous studies of similar wastes.

- Identification of potential pretreatment options for the waste or the site that would extend the application of stabilization/solidification or improve the containment properties of the product.

- Assessment of site conditions that could affect the stabilization/ solidification of the waste or its ultimate disposal.

- Review of health and safety and process requirements.

- Deciding on the selection of a technology based on waste disposal requirements and overall costs.

Available technical information on interferences and applications of stabilization/solidification technologies to hazardous waste materials is often incomplete and sometimes conflicting. Table 6-1 presents a listing of interferences and applications of stabilization/solidification technologies for common hazardous waste materials. The information in this table should be viewed as a starting point, as stabilization/solidification has been applied to many wastes with varying contaminants and concentrations. Because of possible exceptions to the information presented, each waste should be evaluated individually.

Selection of a technology is often complicated by gaps in documented waste/binder interferences. Selection can be further complicated by site-specific conditions and disposal requirements. The technology screening methodology uses empirical data and

Table 6-1. Effect of waste components on stabilization/solidification.

Waste component	Cement-based	Pozzolan-based	Thermoplastic	Organic polymer
<u>Organics</u>				
Nonpolar as: oil and grease, aromatic hydrocarbons, halogenated hydrocarbons, PCBs	May impede setting. Decreases durability over a long time period. Volatiles may escape upon mixing. Demonstrated effectiveness under certain conditions.[a]	May impede setting. Decreases durability over a long time period. Volatiles may escape upon mixing. Demonstrated effectiveness under certain conditions.[b]	Organics may vaporize upon heating. Demonstrated effectiveness under certain conditions. [c]	May impede setting. Demonstrated effectiveness under certain conditions.[d]
Polar as: alcohols, phenols, organic acids, glycols	Phenol will significantly retard setting and will decrease durability in the short run. Decreases durability over a long time period.[e]	Phenol will significantly retard setting and will decrease durability in the short run. Alcohols may retard setting. Decreases durability over a long time period.	Organics may vaporize upon heating.	No significant effect on setting.
Acids as: hydrochloric acid, hydrofluoric acid	No significant effect on setting. Cement will neutralize acids. Types II and IV portland cement demonstrate better durability characteristics than Type I. Demonstrated effectiveness.[f,g]	No significant effect on setting. Compatible, will neutralize acids. Demonstrated effectiveness.[f,g]	Can be neutralized before incorporation.	Can be neutralized before incorporation. Ureaformaldehyde demonstrated to be effective.[f]
Oxidizers as: sodium hypochlorite, potassium permanganate, nitric acid, potassium dichromate	Compatible	Compatible	May cause matrix breakdown, fire.	May cause matrix breakdow, fire.

(continued)

Table 6-1. (continued)

Waste component	Cement-based	Pozzolan-based	Thermoplastic	Organic polymer
Salts as: sulfates, halides, nitrates, cyanides	Increase setting times. Decrease durability. Sulfates may retard setting and cause spalling unless special cement is used. Sulfates accelerate other reactions.	Halides are easily leached and retard setting. Halides may retard setting, most are easily leached. Sulfates can retard or accelerate reactions.	Sulfates and halides may dehydrate and rehydrate, causing splitting.	Compatible[h]
Heavy metals as: lead, chromium, cadmium, arsenic, mercury	Compatible. Can increase set time. Demonstrated effectiveness under certain conditions.[i]	Compatible. Demonstrated effectiveness on certain species (lead, cadmium, chromium). [d,j]	Compatible. Demonstrated effectiveness on certain species (copper, arsenic, chromium). [d]	Compatible. Demonstrated effectiveness with arsenic.[d]
Radioactive materials	Compatible	Compatible	Compatible	Compatible

a   Tittlebaum and Seals 1985; Van Keuren et al. 1987; JACA 1985; USEPA 1986; Jones 1986.

b   Musser and Smith 1984; USEPA 1984; Kyles, Malinowski, and Stanczyk 1987; Tittlebaum and Seals 1985.

c   Tittlebaum and Seals 1985; JACA 1985.

d   Tittlebaum and Seals 1985.

e   Kolvites and Bishop 1987.

f   JACA 1985.

g   Van Keuren et al. 1987.

h   Musser and Smith 1984.

i   Federal Register 1988; Kyles, Malinowski, and Stanczyk 1987.

j   USEPA 1984.

experience when such are available and addresses site- and waste-specific concerns as they apply to a certain stabilization/solidification technology. Thus, limitations and applications of a technology are previewed before large amounts of resources are committed to a site. The cost of hazardous waste treatment and disposal depends on several binder-, waste-,and site-specific conditions.

Site and waste modifications that aid or enhance stabilization/solidification methods should be considered. Specifically, binder additives can be used to reduce the negative effects of organic waste constituents (especially oils, greases, and solvents) on the stability of stabilized/solidified materials. Many of these additives are proprietary or experimental, and little direct information is usually available. In some cases, the wastes should be pretreated to increase their acceptability for stabilization/solidification. Examples of pretreatment technologies that have been used at hazardous waste sites and might be applicable to stabilization/solidification are neutralization, precipitation, adsorption, and chemical oxidation (JACA 1985; USEPA 1986; U.S. EPA 1988d). Site modifications (e.g., water table diversion or depression) may be able to be practiced at some stabilization/solidification sites. An important modification, which relates less to chemical properties than to physical properties, is mixing. Simply mixing a stratified lagoon, for example, can make processing more effective because it generates a uniform or constant material. This reduces the chance of errors related to process change. How frequently these site modification techniques are used is difficult to determine, but success stories have been reported (U.S. EPA 1988d).

As a result of site-specific conditions, different types of hazardous wastes, and increasing land disposal restrictions, a complete site characterization is needed for selection of an appropriate treatment regime for a particular site. In some cases, specific waste/binder incompatibilities will rule out a particular stabilization/solidification technology. Comparisons of various stabilization/solidification technologies with competing remediation measures such as soil washing are not straightforward. Selection of the proper remediation measure often is a compromise based on both engineering and scientific characteristics.

Figure 6-1 presents a technology screening flow chart that can be used to develop a treatment plan for a given waste site. As shown, this screening approach consists of five distinct phases, which may overlap or run concurrently:

1) *Waste Screening.* Obvious characteristics, such as the presence of large quantities of water or regulated materials, require the waste be pretreated to make it acceptable for stabilization/solidification.

2) *Waste Characterization.* The waste is analyzed for constituent compounds and the presence of inhibitors is established.

3) *Bench-Scale Testing.* Trial stabilized/solidified waste forms are created and tested by chemical and physical means to determine the proper agent type and quantity.

4) *Site Evaluation.* The applicability of the process to the particular site is reviewed. For example, the presence of a near-surface water table may require the ground water to be lowered for stabilization/solidification to proceed.

5) *Cost Evaluation.* A determination must be made as to whether stabilization/solidification can be implemented economically.

6.1.1 Waste Screening

The initial screening step entails identification of waste characteristics that may preclude the use of a technology. It may show that the complex hazardous waste contains a regulated constituent that cannot be landfilled (e.g., dioxin) or that the organic levels in the waste are in excess of vendor or literature specifications for successful stabilization/solidification. In some cases, engineering solutions may be available to overcome these initial problems.

The presence of oil or grease in the waste material slows (or sometimes stops) the setting and curing reactions of cementitious and pozzolanic stabilization/solidification. Vendors and empirical studies show conflicting results as to the amount of oil and grease that can be isolated within the solid inorganic product matrix without creating short- or long-term leachate or matrix problems. In low concentrations, oil and grease entrapment appears to take place and the matrix does not seem to be affected (Tittlebaum and Seals 1985; Rowe and API 1987). Some vendors indicate up to 25 percent oil and grease, by weight, can be accommodated by cement and silicate mixtures or pozzolanic mixtures without inhibiting the stabilization/solidification reactions (JACA 1985). Other studies (Kyles, Malinowski, and Stanczyk 1987) and vendors indicate success with lower or higher concentrations of oil and grease with their treatments. Depending on the oil and grease content of the material and its physical state, phase separators (oil/water separators, centrifuges, dissolved air flotation) or in situ biodegradation may be appropriate for removing excess amounts of oily material that are not easily treated by these inorganic stabilization/solidification technologies. The characteristic limitations and costs of each pretreatment technology will need to be identified before further consideration can be given to implementation. Bench-scale study will normally be required to confirm the response of the waste to pretreatment

Figure 6-1. Technology screening flowchart for stabilization/solidification.

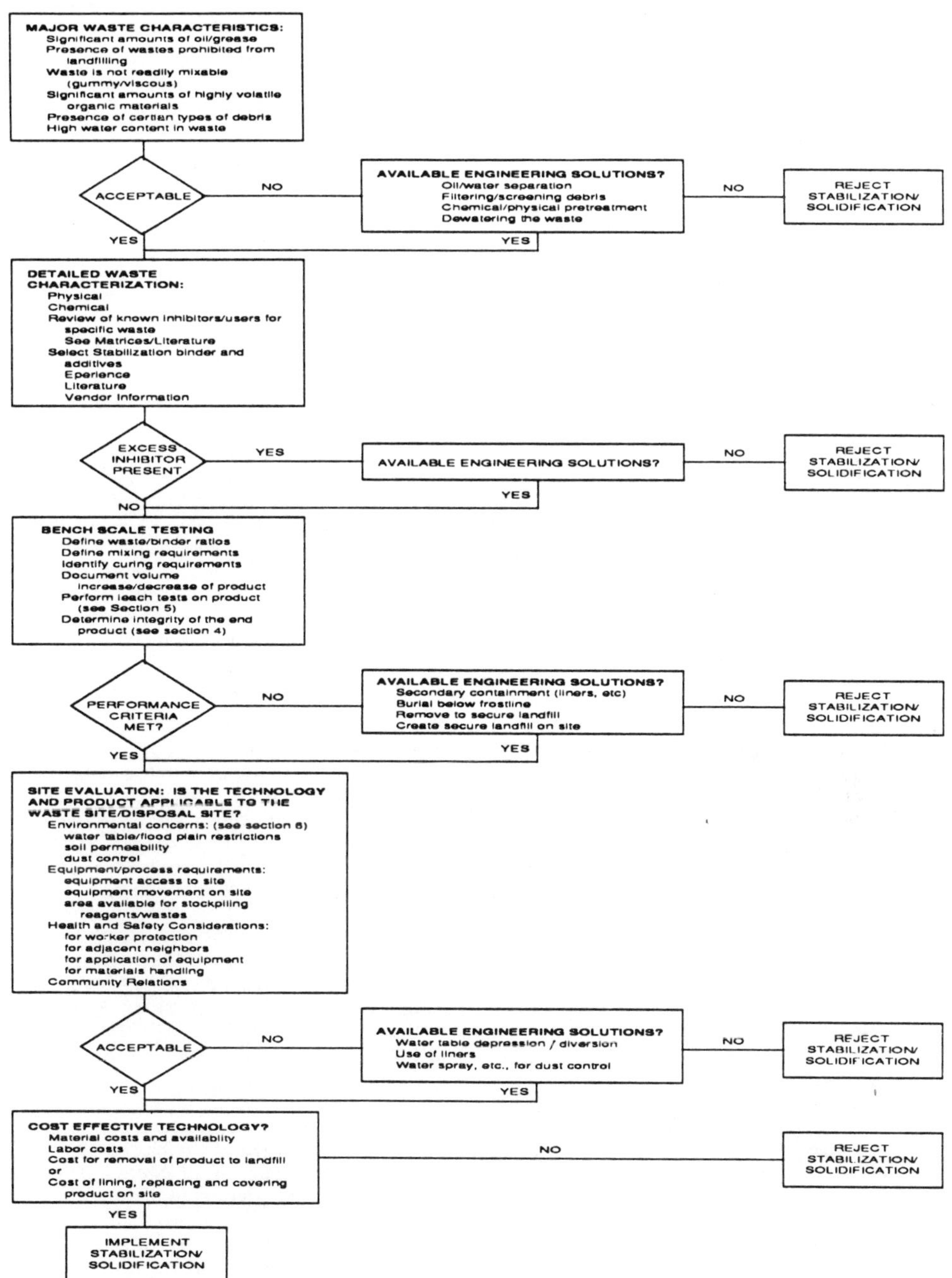

and to stabilization/solidification. Although interference from oil and grease is less of a problem when organic binders are used, the stabilization/solidification process will nevertheless vary with the type of organic binder and the level of oil and grease in the waste.

The capability of inorganic and organic binders to treat organic materials (e.g., pesticides, PCBs, and solvents) is a somewhat controversial issue in the hazardous waste literature. Organic chemicals can act as solvents for some organic binders (Tittlebaum and Seals 1985), and phenol or organic solvents can inhibit the setting and curing reactions of inorganic binders (JACA 1985). Different types of organics will affect the final product in various ways (Tittlebaum and Seals 1985; Kolvites and Bishop 1987), and the effect of the interfering material is usually disproportionate to the amount present in the waste (Jones 1986). Therefore, waste assessment must include individual chemical constituents and classes of organics rather than rely on total organic content of the waste material. Current literature (U.S. EPA 1986h) and vendor nformation often summarizes total organic concentrations, which are reported as being treatable with inorganic binders. One common range reported is 20 to 40 percent (U.S. EPA 1986h). This information cannot be directly applied to all classes of organic chemicals, however. It may be possible and cost-effective to treat or remove the interfering organic material or to modify the binder/additive mixture to achieve successful stabilization/solidification (Tittlebaum and Seals 1985; U.S. EPA 1986h). Depending on the type of organics present, several options are available for removal of a large fraction of the organic material or for controlling the effects of organic materials on the inorganic stabilization/solidification. These include:

- Soil washing
- Thermal removal
- Chemical oxidation
- Extraction removal
- Biodegradation
- Addition of adsorbent prior to mixing (limestone, diatomaceous clays, activated carbon, or fly ash)

For cement-based and pozzolanic processing, small concentrations of volatile organic carbons may be trapped within the inorganic matrix (Tittlebaum and Seals 1985); however, volatilization may occur during excavation, mixing, or curing because of the agitation and heat of hydration (Weitzman, Hamel, and Barth 1988). Air quality standards or health and safety precautions may require the control of volatile organics during processing. In addition, many organic binder technologies require preheating of the binder and waste as part of processing. For waste constituents that decompose or volatilize at the temperatures required for organic stabilization/solidification processes,

the use of these processes may be precluded. Process modifications (e.g., batch mixing with volatile off-gas scrubbing) may help to resolve this problem. When scrubbing is impractical, pretreatment to remove volatile organic carbon may be warranted.

Occasionally, the consistency of the waste may prevent mixing. For example, creosote wastes may be too viscous for easy excavation and mixing. Heating the material to reduce the viscosity and to improve mixing potential can result in noxious organic vapors. Excavation during the winter freeze is possible, but the wastes must be heated before they are mixed with the binder. Many organic binders are somewhat difficult to mix because of their viscous nature. Mixing problems can arise when materials having significantly different viscosities are combined.

When the wastes are very dilute, dewatering technologies may prove useful as a pretreatment method. Depending on the characteristics of the waste material, filtration (vacuum, belt filter press, chamber pressure filtration) or membrane separation may make stabilization/solidification treatment more feasible. Water may be required for the hydration of both cement- and pozzolan-based stabilization/solidification, however. Excess water is a liability for both organic and thermal binder technologies because it increases the cost of processing. In both cases, the water must be vaporized before or during processing, which requires more energy and thus greater cost. If a dewatering technology is used to remove excess water, the supernatant must be assessed for possible contamination and treatment prior to its disposal.

Removal of debris can be important for equipment protection and for process quality control. Documented cases of mixer or conveyor belt damage and problems with binder encapsulation of large articles of debris point up the need for this aspect of waste characterization (U.S. EPA 1988d). Debris can include common items such as tires and tree stumps that are buried with other hazardous waste. The physical separation and washing of these materials may be both feasible and economical. When debris is included with the waste material, processing difficulties may occur in the mixing and conveyer equipment. Debris can also interfere with quality control of the final product in all of the stabilization/solidification technologies.

6.1.2 Waste Characterization

Hazardous wastes are characterized through analysis of major physical and chemical properties. At this level of scrutiny, possible inhibitors are reviewed and binders, additives, and appropriate pretreatment options are chosen. Stabilization/solidification experience, vendor information, and literature reviews can serve as bases for the selection of the proper stabilization/solidification system, plausible waste/binder ratios, and other options. In some cases, waste pretreatment may be required to overcome specific waste problems.

Interferences from waste constituents can affect the physical properties of inorganic and organic binder/waste products. With regard to the inorganic binders, these constituents may interfere with the setting reactions, the water-to-binder ratio, the porosity, or the final strength of the product (Jones 1986). Interfering constituents may soften organic binders or require larger amounts of binder to overcome their negative effects. In addition, inhibitors exhibit various time-related effects. With regard to organic binders, some materials can slowly deteriorate the organic binder/waste product or prevent stabilization/solidification from occurring. Surface-active molecules work on the cement (or pozzolan) and water immediately. Set-controlling materials ionize and affect the chemical reaction between the cement (or pozzolan) and water after minutes or hours. Finely ground insoluble minerals affect the rheological behavior of fresh cement, but the chemical effects take several days or months to become manifested (Jones 1986). The following are examples of the physical manifestations that can occur when interfering constituents are present in the hazardous waste and stabilization/solidification is attempted with inorganic binders:

- Spalling and cracking
- Set retardation; hardening and waste containment are impeded.
- Flash set; mixing of binder and waste is incomplete as a result of a very quick set; equipment can be fouled.
- Chelated/complexed toxic constituents may accelerate leaching, even if the waste is successfully stabilized/solidified.
- Some waste constituents can react and cause swelling and disintegration of the stabilized/solidified mass long after setting reactions are complete (Jones 1986).

Oxidizing salts can cause slow deterioration of the organic binder matrix (Tittlebaum and Seals 1985). Large quantities of sulfur, calcium chloride, sodium arsenate, sodium borate, sodium phosphate, sodium iodate, sodium sulfide, as well as the soluble salts of magnesium, tin, zinc, copper, and lead all adversely affect inorganic binders (JACA 1985; U.S. EPA 1986h). Salts can often cause swelling and cracking within the inorganic matrix, which exposes more surface area to leaching. These effects can occur in the short term or the long term.

In addition to the specific chemical interferences listed in Table 6-1, other chemical factors have an impact on waste/binder compatibility. An alkaline pH is required for most inorganic binders to set and cure properly. In the high-pH environment generated with inorganic binders, amines or aminated compounds evolve as ammonia. In high concentrations, gaseous ammonia can cause immediate danger to life and health. Any environmental or waste condition that lowers the pH of the inorganic (e.g., lime, cement) crystalline matrix will eventually cause matrix deterioration. Flocculants such as ferric chloride can interfere with the setting of cements and pozzolans (JACA 1985). Phenols and nitrates cannot be immobilized with lime/fly ash, cement, and soluble silicates; fly ash and cement; or bentonite and cement (Stegmann, Cote, and Hannak 1988).

On hazardous waste sites, concentrations of the waste components usually vary with location and depth. Solid waste dumps and lagoons are often stratified or pocketed with varying waste materials. Records of the historic deposition of these wastes can reveal the types of materials that may be encountered at different levels or in different cells within a site.

Physical properties of the waste, such as volume, solids content, particle size distribution, water content, debris content, viscosity, and pH have an impact on waste processing. The volume, solids content, and particle size analysis aid in determining the waste-to-binder ratio. Moisture content is also important in determining whether water should be added or removed for processing.

6.1.3 Feasibility Testing

No set protocol has been established for determining the feasibility of stabilization/solidification at the bench-scale level. Testing addresses the leaching potential of the stabilized/solidified waste as well as its durability.

During bench-scale testing, different waste-to-binder ratios are used to stabilize/solidify waste samples. Selection of the optimum waste-to-binder ratio is usually based on leach-test and durability-test results. The desired properties of the stabilized/solidified product may vary with the geographic region and type of final disposal site. For example, EP Toxicity, unconfined compressive strength, wet-dry, and freeze-thaw tests may be among those required of a stabilized/solidified product at a given site. Incompatibilities in product quality may be resolved on a site-specific basis. For example, if the optimum EP Toxicity results are obtained with a stabilized/solidified waste that erodes readily during freeze-thaw testing, the solution may be to bury the waste below the frostline. Bench-scale tests can define process control requirements, including mixing requirements, curing time, and quality control parameters, which can then be further defined during pilot-scale testing. Defining these parameters is important and could save time and money during field operations. The processing specification is often recognized to be as important as the product specification.

Determining whether the volume of the stabilized/solidified product will increase or decrease is an important parameter, both for onsite and offsite disposal. At some remediation sites, the excavated original raw waste must be replaced with stabilized/solidified waste material. If the treatment results in a large increase in the volume of the stabilized/solidified product, some of it may have to be removed from the site or the waste volume must be reduced prior to stabilization/solidification treatment. When the waste product is transported to an approved landfill, an increase in its volume adds to the overall cost of remediation.

Whenever possible, waste materials from the actual waste site should be used for bench-scale testing. Differences in the physical and chemical characteristics of the actual waste (such as particle size distribution and complex chemical constituents) can affect waste/binder relationships and final product characteristics.

Bench-scale studies offer a significant advantage in the refinement of the overall system design, including the establishment of a quality control (QC) program. Bench-scale test mixtures can be used to verify or supplement the waste/additive design and provide both visual and numeric observations for establishing QC criteria.

During testing, the following information is important for an evaluation of inorganic and organic binder technologies:

- Leaching tests appropriate for the waste site.
- Durability tests as appropriate for the final disposition of the product (e.g., freeze-thaw testing has little relevance for a material that will be buried beneath the freeze line or that is in a part of the country where it does not freeze).
- Permeability tests of the final product, which should be two orders of magnitude below that of the surrounding materials.
- The waste-to-binder ratio needed to achieve leaching and durability characteristics appropriate for the waste site.
- Increase or decrease in the volume of the final product.
- Evolution of gases during processing, curing, or preprocess drying.
- The need for dewatering or adding water to the inorganic waste/binder mixture.
- Ability to mix the waste material and binder.

6.1.4 Site Evaluation

The fourth evaluation step involves a review of relevant environmental conditions that may affect waste containment on the site, during both excavation[*] and processing. Environmental parameters that should be considered include the depth to the water table; location of any sensitive environmental conditions within the study site; access to the site for equipment and material; processing space requirements; health and safety issues; and planning for potential community relations. Standard construction practices can be evaluated for such site modifications as water table depression/diversion or access road construction. Final disposal options are reviewed and confirmed during this phase. These options may include the construction of a liner on the site or offsite transportation and disposal. The availability of adequate soil for capping and lining the waste site also can be determined during this phase. Where applicable, the site evaluation should also include information on the placement and construction of monitoring wells.

Overall site-specific concerns with regard to a remedial action project are geared toward evaluating waste containment potential. Containment depends in part on the types of wastes, their characteristics, and their physical state. The proximity of a waste to the water table, onsite drainage tiles, or receiving streams offer significant migration potential before, during, and after onsite disposal of the waste material. Important site parameters in this regard include:

- Area of the site
- Permeability of the area soils, both for a review of leaching capabilities and also for possible liner/cap material
- Existing ground-water contamination
- Velocity and direction of both ground water and ambient air
- Site drainage
- Proximity to populated areas
- Access routes to the site
- Available work area/stockpiling area on the site
- Final disposal options and their site-specific implications
- Postremediation use of the site
- Sensitive environmental areas within the work site, such as flood plains or marshes.

---

[*] This discussion applies primarily to processes in which materials are excavated, processed on site, and returned to the excavated area. In situ processes do not require excavation and replacement of stabilized/solidified materials. This advantage could lead to the selection of an in situ process. Offsetting this advantage may be limitations due to poor mixing or depth. Section 7.1.8 describes these in situ processes. Other factors, e.g., the presence of a near-surface water table, relate to all stabilization/solidification processes.

- Waste product volume increase and its implications for the capacity of the site to contain final product if onsite disposal is required.
- Dusting and material stockpiling
- Ability to mix the materials adequately on the site
- Availability of the binder/additives in the amounts required for the entire site.

Access to the waste site for appropriate equipment for excavating and processing the waste materials should be evaluated on a site-by-site basis. Numerous standard construction practices provide solutions to the problem of difficult access. Sometimes, special roads must be built or existing roads modified to accommodate the needs of the site. Sometimes, contractors have modified their equipment to gain access. Such modifications add to the overall cost of the project, and customized equipment may be difficult to repair because of the unavailability of standard replacement parts.

In some cases, the waste site cannot provide sufficient area for the expected processing, binder stockpiling, and temporary or final waste disposal. Some kinds of processing require stockpiling of untreated excavated wastes, the processed wastes, and the binder, and these materials may have to be covered to reduce exposure to the elements. Binders increase the volume of the waste product, and this added volume could present difficulties where the waste/binder product is buried in the original waste site excavation. Solutions to problems posed by limited area must be developed on a site-specific basis. Delivery of preweighed amounts of the binder directly to the process site is one possible solution. The binder can then be added directly to the in-place mixing area rather than being stockpiled in bulk containers.

The presence of a water table in the waste area creates three problems: 1) a water table poses the possibility of existing ground-water contamination; 2) excess water (especially flowing water) can cause excavation difficulties; and 3) a water table creates the potential need for dewatering a saturated waste material prior to its processing. All three of these problems have significant cost implications and must be resolved before the final technology selection is made.

Some waste materials are difficult to excavate and handle. Waste pretreatment can consist of adding sorbents or inorganic binder to dewater the waste and to increase its handling ability before it is processed with different inorganic or organic binders. This adds both a cost factor and a volume factor to the overall operation. For selected wastes, excavation in the winter or summer will alter the consistency of the material sufficiently to allow its excavation.

Site environmental effects also can be critical. Seasonal timing to remove wastes while they have the most suitable viscosity level was discussed in Section 6.1.1. Freezing retards the overall reaction rate and may cause long-term damage of the matrix. Summer temperatures in excess of 100°F accelerate the reaction rate and may ruin the final product.

6.1.5 Cost Evaluation

In the fifth phase, an assessment is made of the costs involved with the entire technology as it will be applied to the waste and final disposal site. This includes the cost of binders, additives, pretreatment, excavation and moving equipment, labor, and final disposal of the waste. Disposal costs may include preparation for onsite burial (liners, caps, and possibly monitoring wells) or transportation and dumping fees for offsite disposal.

6.2 Summary

Choosing a waste disposal method is a complex decision. The costs, capabilities, and limitations of different technologies and their application to both the waste site and waste material must be weighed. Technology screening provides a systematic methodology for reviewing the most important elements of the waste material, the site, and the technology.

Few adequately documented studies exist that report the long-term performance of stabilized/solidified hazardous wastes. Performance in this context is determined by the physical and chemical stability of the waste/binder product and related leaching characteristics. Also, almost no published information is available on the nature, strength, and permanence of the bonds formed in the stabilization/solidification process. Some data have been compiled from the experience of the construction and nuclear waste treatment industries. These admixtures are managed in the construction trade by means of strict limitations on the amounts of additives allowed for a particular application. In addition, batch testing of a product is an integral part of the total construction project. During the screening of stabilization/solidification technologies for their adequacy for isolating toxic constituents and improving the handling characteristics of hazardous waste materials, a review of known inhibitors is recommended. The list of inhibitors is incomplete, however, and even the effects of a single component additive are still poorly understood. Interfering reactions with mixed or complex wastes cannot be predicted (Jones 1986; JACA 1985).

As long as this information is lacking, the practice of designing stabilization/solidification schemes for hazardous wastes will continue to be primarily empirical in nature. Such design practices are also likely to be based on the short-term performance of the stabilized/solidified product (Jones 1986).

7. Field Activities

This section covers the application of the stabilization/ solidification process at a site. It includes a discussion of stabilization/solidification functions and processes, lime stabilization techniques, and site conditions that can affect the stabilization/solidification process.

Many of the tests described in Sections 4 and 5 often relate to stabilized/solidified forms developed in the laboratory. Whereas these stabilized/solidified test forms may have excellent waste characteristics, duplicating these forms under field conditions can be difficult because of the equipment used for field stabilization/solidification and the conditions present at the site. Also, the volume of the materials to be stabilized/ solidified at a site often reaches thousands of cubic· yards. Equipment is usually geared to the characteristics of the site and its materials. For example, at many sites the material to be stabilized/solidified is in abandoned impoundments, and its removal may require large-scale earth-moving equipment such as tracked backhoes or draglines. At other sites, the soil itself needs to be stabilized/solidified, and equipment such as loaders and bulldozers is used to move this soil.

Regardless of the process chosen, adequate quality assurance/quality control (QA/QC) procedures should be implemented to be certain the material produced is near the desired composition. This seemingly simple task is difficult to accomplish and requires answers to questions such as the following: Is the material to be judged as it is produced? How much time is required to test the quality of the material? For large-scale operations in which the material could conceivably be reprocessed, the turnaround time for the QA/QC procedures must be quick so as to minimize the amount of material that must be recovered and to avoid the possibility of burying stabilized/solidified material that cannot be easily retrieved and reprocessed.

The operations described in this section frequently use earth-moving equipment. The manufacturers' literature contains descriptions of this equipment, such as the reach of backhoes, the capacity of loaders, and the associated cycle times of the equipment. The Caterpillar Performance Handbooks (Caterpillar 1987a, 1987b) are excellent sources of information for these operations. These easily understood handbooks provide information needed for estimating the costs of the basic operations required at a site.

With the exception of a few specialized processes, most hazardous waste stabilization/solidification operations fall into two categories: 1) lime/silicate-based stabilization/solidification, which results in a product similar to mortar; and 2) stabilization/solidification of a soil-like material. These two processes are quite different. One produces a material that hardens like mortar, and the process and its QA/QC procedures are comparable to those used in mortar production. The waste forms, however, differ greatly from mortar in physical characteristics and they are usually far less durable. The other process produces a material that must be described in terms of soil mechanics and soil handling.

7.1 Stabilization/Solidification Functions and Processes

Regardless of the stabilization/solidification process chosen, typically seven functions must be satisfied for its successful implementation:

1) Waste removal
2) Untreated waste transportation
3) Untreated waste storage
4) Chemical reagent storage
5) Waste/reagent mixing
6) Stabilized/solidified waste transportation
7) Stabilized/solidified waste replacement

In addition to the preceding, the waste may have to be stockpiled, moved, or further processed in pretreatment operations (e.g., dewatering or neutralization). Some processes may not require all of the steps listed. An example would be in situ stabilization/solidification, in which the reagent is brought to the waste and it is then stabilized/solidified in place.

7.1.1 Waste Removal

Waste removal generally involves the use of traditional earth-moving equipment such as tracked backhoes, draglines, bulldozers, and front-end loaders. Because this equipment has been in use for many years in the construction industry, its application to hazardous waste handling is merely one of adaptation. Nevertheless, the characteristics of the waste and the site must be considered.

Waste characteristics of importance are its corrosivity, its physical state (liquid or solid), and its potential hazard. Corrosivity is important with respect to the materials of construction for the removal equipment. Most construction equipment is fabricated of steel and is subject to acid attack. This problem can be eliminated by use of a neutralization pretreatment step. Neutralization should be performed carefully. The neutralization of acidic liquids generates hydrogen, an explosive gas. Slowing down the neutralizing process can control this generation. Also, neutralization should not take place in confined spaces. A final caution—chemicals should be added sparingly to avoid the incurrence of excess costs.

The physical state of the waste also may influence the sequencing of operations necessary for the site. For example, any liquid present above the waste (supernatant) can be managed by removing it and treating it as a separate waste.

The hazard posed by the waste is of concern to the equipment operators and other workers. Complete enclosure of the operator space in construction equipment and the provision of breathing air may be essential at sites where dangerous materials such as hydrogen sulfide or cyanide may be present.

Tracked backhoes and draglines are used for the remote removal of materials. Figure 7-1 shows tracked backhoes and draglines in operation at a hazardous waste site. Typically, a backhoe is used in areas where front-end loaders and bulldozers cannot be used because of the soil's instability. Backhoe operations are possible in hazardous waste sludges that ordinarily could not support equipment. During its operation, the tracked backhoe rests on mats constructed of wooden timbers. As the operation progresses, additional mats are used or the old mats are moved. The backhoe is limited by the distance and depth it can reach and by the cycle time of the operating sequence.* Distance and depth are available from the equipment manufacturer's manuals (Caterpillar 1987a, 1987b). Because these vary by machine size, the size of the site and the required reach of the backhoe determine the piece of equipment to be used.

The cycle time of the operating sequence depends on the operator, the material to be removed, the removal equipment, and the transportation equipment (e.g., dump trucks). An efficiently operated site considers all of these characteristics and their interrelation. For example, the rate at which waste is removed needs to be assessed so the proper number of dump trucks can be used at a site. The rate of material removal and transportation should be as evenly matched as possible (Caterpillar 1987a).

Draglines are applicable where reaches longer than 40 feet are necessary. They are delivered to a site in pieces, usually by truck, and assembled on site. This delivery and assembly must be considered as it may require several days. The assembly should take place before the arrival of workers whose activities are not related to its use. As shown in Figure 7-1, a dragline is a very large piece of equipment. Its size permits the removal of materials at a distance, but at the price of imprecision. Even a good dragline operator cannot assure removals more accurate than plus or minus a couple of feet in depth. Thus, excess material is usually removed during dragline operations. Although a dragline is often used to remove materials entrained with liquids, such removal is not efficient because the liquids escape from the bucket. Other considerations required in connection with the use of a dragline are similar to those for a tracked backhoe.

In contrast to backhoe operation, bulldozers and front-end loaders maneuver in and on the material that is being removed. The bulldozer pushes the material by scraping. In a stabilization/solidification operation, this movement of material may be to an area where stabilization/solidification takes place or to an area where it is loaded. Front-end loaders are capable of both removal and loading of solid material. This capability is advantageous when the material must be placed in a dump truck or a hopper for transportation or processing. Front-end loaders can also be operated in sludge material. The use of low-pressure tires (LPT) allows the equipment to be operated in soft (but not liquid) materials.

7.1.2 Untreated Waste Transportation

Depending on the nature of the waste and the site, waste can be transported by dump truck, pump and hose, or a fixed system such as a conveyor belt or screw auger system. Dump trucks are commonly used to transport solids, particularly when the liquid content is low and travel distance is more than one-fourth mile. As with the use of a backhoe, the carrying capacity of the trucks at a site must be matched with the capacity of the removal equipment. Spillage particularly must be considered because the waste is hazardous. Truck beds should be lined with plastic to prevent escape of waste when the waste contains liquid or when the trucks will travel offsite. The trucks can also be covered with tarpaulins; this is essential when the material consists of small particles that are subject to wind dispersion. As an alternative, the material can be sprinkled with water to reduce dusting. Finally, trucks are a commonly accepted means of conveying waste material when it must be taken out of a controlled area. When this occurs, the trucks

---

* Presence of overhead utilities greatly restricts the use of this equipment.

Figure 7-1. Photograph of tracked backhoes and draglines in operation at a hazardous waste site.

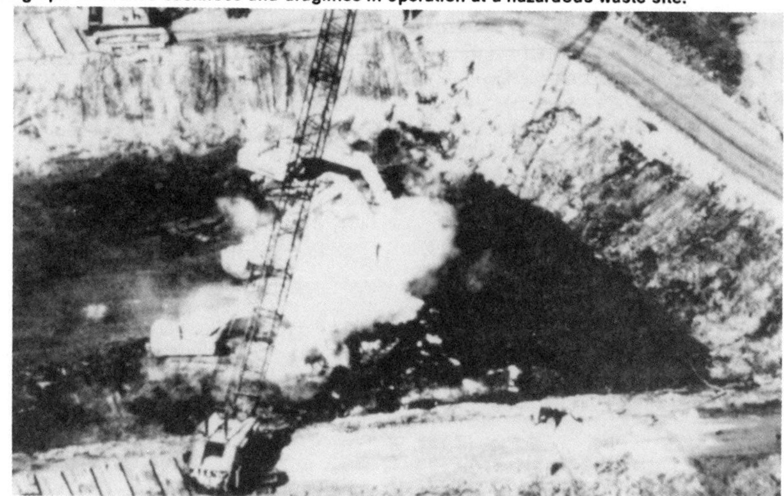

Source: Geo-Con Inc., Pittsburgh, PA.

must be thoroughly decontaminated upon exiting the site, and their beds must be decontaminated upon project completion. Decontamination usually produces large quantities of water, which in turn must be treated before its disposal.

Conveyor transportation is used at a site where large amounts of waste must be moved over a fixed distance for long periods of time. Their setup is complicated and costly, but the complexity and expense are offset by their ease of use once they are set up. Conveyors cannot move liquids, and spillage can be a problem if the material to be stabilized/solidified is not dewatered or its water decanted. Because a conveyor is a piece of moving equipment, time must be allotted for maintenance and repair. Also, the use of sidewalls may be necessary to prevent unwanted dispersal of hazardous materials.

Pumping of hazardous waste solids may be feasible if they are sludge-like in nature. In this instance the transportation function is also the removal function. The advantage of pumping is that the waste can be sent directly to the processing equipment.

Thixotropic sludges, which liquify upon movement, illustrate the need for thorough preparation (as discussed in Section 6, Technology Screening). These sludges can be transported without spillage by the use of gondola conveyors or protective displacement pumps. A special example of a pump, a sulfur pump, is equipped with a set of steam coils that melts the solid material on which it rests and enables the liquid to be pumped.

7.1.3 Untreated Waste Storage

Occasionally, untreated hazardous waste must be stored prior to stabilization/solidification. This storage should be evaluated for each individual site, but the following factors should be accounted for in any case:

- A 2 to 3 percent sloping of the storage area to provide for collection of liquids.

- Lining of the storage area with a high-density, polyethylene, flexible, membrane liner and sand to prevent contact between liquids and the soil and to allow for easy drainage of the liquids.

- Preparation of the underlying soils to remove rocks, cobbles, and vegetative matter that could puncture the liner.

- Installation of drainage pipes and a sump for final collection of liquids.

- Operation of a water-treatment system to remove hazardous constituents (from the collected liquids) and to reduce the volumes of liquids that must be stored.

- Covering of the wastes to prevent their dispersion.

- Berming of the sides of the area to resist slumping of the waste.

- Installation of a cement loading ramp for the unloading of wastes from trucks.

- Provision of a rubber-tired front-end loader to move wastes within the storage area.

Providing a temporary storage area is obviously a complex task. Inasmuch as the construction of this area can be both expensive and time-consuming, planning this aspect of the work is just as worthwhile as for other engineering projects. Untreated waste storage requirements can be minimized by matching excavation capacity with treatment capacity.

7.1.4 Chemical Reagent Storage

The satisfactory operation of a stabilization/solidification process requires some material stockpiling. Primary requirements are 1) sufficient storage of agent to prevent delays in operation; 2) ability to keep an agent dry, as virtually all stabilization/solidification agents are dry; and 3) ease of unloading upon delivery from the supplier and to the process. Bins and hoppers are the primary storage vessels for solid agents. The hoppers are sometimes transportable. On occasion, solid materials are delivered to the site in bags, which are palletted, or in bulk for storage in piles. Liquid storage is occasionally required, and chemical storage tanks are routinely used for this function.

7.1.5 Waste/Reagent Mixing

The heart of the stabilization/solidification process is the mixing of the hazardous waste and its stabilization/solidification agent. Unfortunately, this is also the area in which the process fails to meet expectations. Therefore, most QA/QC procedures are directed toward this operation.

The objective is to achieve mixing in a practical manner. Although ideally, all hazardous waste should be mixed and reacted with the stabilization/solidification agent, Tittlebaum et al. (1986) have shown that, even under ideal laboratory conditions, complete mixing is not achieved. Thus, a question of degree of mixing becomes a consideration before one piece of equipment is sent into the field.

Many of the stabilization/solidification activities that have occurred in the United States have used what is known as "area mixing." In area mixing, the stabilization/solidification agent is delivered to the area to be stabilized/solidified and mixed with the waste directly. Often, tracked backhoes, whose removal capabilities were described in Section 7.1.1, are used for the mixing. Figure 7-2 shows backhoes being used to mix sludge and lime. Typically, enough water is present in the waste to promote reaction with the stabilization/solidification agent and the waste. The waste is mixed until the operator judges the mixing to be complete. The benefits to this type of mixing are 1) the waste usually need not be removed from the site, 2) the operation requires only traditional earth-moving equip-

ment, and 3) it is inexpensive. Stabilization/solidification accomplished by backhoe mixing at Vickery, Ohio, has been well documented (Curry 1986).

Offsetting these benefits is the probability of incomplete and inadequate mixing of the waste. Virtually no QA/QC program has been enforced at many stabilization/solidification sites, and successful stabilization/solidification cannot be assured unless it can be demonstrated that proper mixing has occurred. In operations where area mixing has been used, the degree of mixing has been left to the operator's judgment.

Backhoe mixing is unlikely to result in complete mixing that assures that the waste is phyically or chemically trapped. Processing equipment, such as pug mills, ribbon blenders, etc., provides more efficient mixing.

A wide range of mixing equipment is available for use in industries that process solids. This equipment is described by Cullinane, Jones, and Malone (1986) in their handbook on stabilization/solidification of hazardous wastes. Typical mixing equipment suitable for use in stabilization/solidification operations includes pug mills, ribbon blenders, Mueller mixers, extruders, and screw conveyors (Green 1984). The waste and reagent can be metered into this equipment during mixing for quality control purposes.

Some waste stabilization/solidification contractors report the use of pug mills for mixing operations. Pug mills are manufactured by several companies and are occasionally modified by contractors to suit their special needs. Pug mills consist of double screws, which allow one material to be mixed with another on a flow-through basis. The materials are generally partially mixed before they are introduced into the pug mill. Figure 7-3 is a photograph of a pug mill operation. As shown in the photo, one solid is added at the top by a screw conveyor, mixed with a second solid, and then fed to the pug mill.

Many small generators drum inorganic wastes that are not suitable for incineration. These drummed liquid wastes can be stabilized/solidified by methods discussed by Cullinane, Jones, and Malone (1986). Although drum mixing has limited application, it may provide better mixing than other methods used to add stabilization/solidification reagents to hazardous waste.

7.1.6 Stabilized/Solidified Waste Transportation

Stabilized/solidified waste is transported by the same means (dump truck, conveyor) as unstabilized/unsolidified waste. Transporting stabilized/solidified waste is usually easier than transporting untreated waste because it should be water-free and less subject to wind dispersal. One disadvantage associated with transporting stabilized/solidified waste is that it may set up (as cement does) if held too long. This can be

Figure 7-2. Photographs of backhoe mixing of sludge and lime.

Photographs courtesy National Lime Association

Figure 7-3. Photograph of a soil-bentonite mixing plant using a pug mill.

Photograph courtesy National Lime Association

avoided by the proper addition of stabilization/solidification agent or by timely transportation to the stabilized/solidified waste disposal area. Figure 7-4 shows the loading of lime stabilized/solidified sludge into an unlined dump truck.

7.1.7 Stabilized/Solidified Waste Replacement

Replacement of the stabilized/solidified waste depends upon its form. For waste stabilized/solidified in situ, replacement is of no consequence. Cement-like materials should immediately be placed in their disposal area because they set up and harden.

The placement of soil-like materials should be performed in accordance with procedures used in the road-building industry, where the stabilized/ solidified waste is placed in lifts of 8 to 10 inches with earth-moving equipment such as graders or bulldozers. These procedures are described in detail in Section 7.2. Reportedly, the U.S. Army Corps of Engineers uses "roller-compacted concrete" in dam construction. This technique may also be applicable to hazardous waste stabilization/solidification.

After it is placed, the waste is compacted. Determination of the optimum moisture content of the material (see Subsection 4.1.3) is required to assure proper compaction. The optimum moisture content is that which gives the greatest density to a given material. Moisture must be added because too little moisture does not provide the lubrication required for the soil

Figure 7-4. Photograph of unlined dump truck being loaded with lime stabilized/solidified sludge.

Photograph courtesy National Lime Association.

particles to slide past one another during compaction. On the other hand, too much moisture causes the material to approach the density of water, which is presumably lower than that of the waste. Thus, for soil-like, stabilized/solidified, hazardous waste, the amount of water in the waste must be controlled as it is compacted.

A second factor that affects the compaction of the waste is the particle size distribution of the material. Poorly graded wastes that are predominantly of one particle size do not compact well. Well-graded materials compact better because the void volume of the larger particles is occupied by smaller particles, which gives a better fit. This factor is usually one that cannot be controlled unless the particle size distribution of the stabilization/solidification agent can be regulated. Figure 7-5 shows the equipment that should be used for compaction of various-sized materials (Caterpillar 1987a). Use of this information requires the determination of particle size distribution of the stabilized/solidified material in the laboratory. This, plus an assessment of whether the material has the characteristics of clay, silt, or sand, enables one to make a proper choice of compacting equipment.

7.1.8 In Situ Processes

Waste stabilization/solidification can also take place at the site itself. These in situ processes, which have been in use since 1980, were described by Cullinane et al. (1986). One major application for them has been in old lagoons where waste has been present for decades. Not all are designed for lagoon work, however, and their specialized application can be both cost-effective and technically sound. In all of these processes, the entire range of functions described in Section 7.1 are combined into one operation.

7.1.8.1 In situ Mixing: Geo-Con System

The Geo-Con Company of Pittsburgh, Pennsylvania, has developed a system of stabilizing/solidifying waste materials in situ. This process has been used in EPA's Superfund Innovative Technology Evaluation (SITE) demonstration project in Hialeah, Florida. The equipment also has other geotechnical applications, such as the stabilization/solidification of dam foundations and in the construction of slurry walls.

The process is based on a combination of an auger and caisson, which operates in the waste. The stabilization/solidification agent is fed into the auger and

Figure 7-5. Chart used to select equipment for compaction of various-sized media.

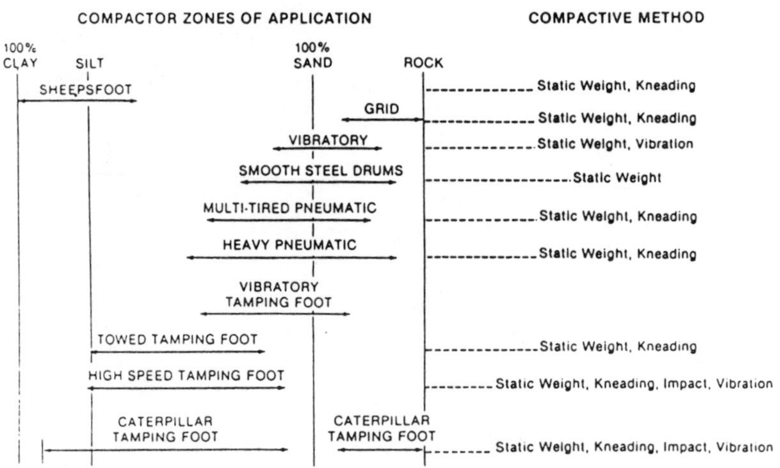

Source: Caterpillar, Inc. Caterpillar Performance Handbook, October 1987.

then into the waste through a hollow stem. Inside the caisson, the auger mixes the agent with the waste by a lifting and turning action. Geo-Con reports that it is possible to stabilize/solidify wastes to a depth of 100 feet. An important feature of the operation of this process is the overlapping bore patterns, which, if properly used, allow for complete coverage of the waste area.

The Geo-Con process is illustrated in Figure 7-6. The agent is fed from bulk storage tanks, usually in dry form. To date, Geo-Con has not specified the agents for use, but a variety of agents are available for this purpose. Prior to the mixing step, a liquid can be used to supply the water necessary for reaction between the stabilization/solidification agent and the waste and to improve mixing characteristics. A tracked backhoe can be used to carry the hydraulic drive that turns the auger. An exhaust fan draws air from the caisson through a dust collector and activated carbon for the removal of particulates and organics, respectively.

Typically, a week to 10 days is needed to mobilize this equipment. The process requires a crew of about eight people, and it is capable of processing 600 to 900 cubic yards of waste each day, depending on the

nature of the waste and the stabilization/solidification agent used.

The Geo-Con process has several advantages. First, the process is inherently less expensive than one that uses earth-moving equipment because the waste does not have to be excavated. Second, the water table does not have to be lowered. Third, fugitive emissions are controlled by the dust collector and the activated carbon tanks. Fourth, the quality control aspects of this process are good because the rate of addition of the stabilization/solidification agent can be regulated and the materials are well mixed. The one reported disadvantage of the process is its inability to penetrate masses containing boulders or other debris.

7.1.8.2 In Situ Mixing: ENRECO System

The ENRECO Company of Amarillo, Texas, uses an injection system for the stabilization/solidification of waste lagoons. The stabilization/solidification agent is fed to a lagoon through an injector on a backhoe. ENRECO manufactures its own equipment, and several variations of the basic design are available. Figure 7-7 is a photograph of ENRECO's equipment.

Figure 7-6. Crane-mounted mixing system advancing through unstabilized/unsolidified sludge layer.

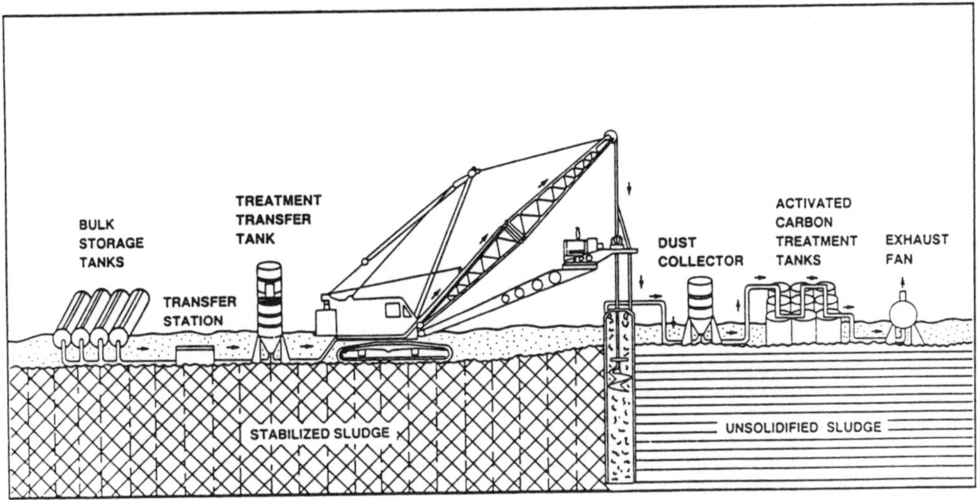

Source: Geo-Con Inc., Pittsburgh, PA.

Figure 7-7. Photograph of ENRECO backhoe equipped with stabilization/solidifying agent injector.

Photograph courtesy ENRECO Company.

The stabilizing/solidifying agent is delivered to the sites by truck, and it is often fed directly from a truck by a pneumatic system. Mixing is accomplished by the back and forth action of the injector and by the force of the pneumatic delivery.

The ENRECO system can stabilize/solidify to depths of 18 feet, although 10 feet is more common. Even greater depths have been reached by stabilizing/solidifying in 10-foot lifts, followed by removal. After the stabilization/solidification of a lift, the lagoons are solid enough to support the tracked backhoes that perform the removal. Typical processing rates are from 500 to 1500 cubic yards per day. A crew of three operators is required for each injection system. The stabilized/solidified material is usually left on site, but occasionally it is removed.

Typically, ENRECO uses cement, lime, or kiln dust as the reagent. These materials have been assembled from sources across the United States, so ENRECO is able to tailor the agent based on its source. Proprietary agents are also used, although infrequently. Volume is typically increased by 10 percent after stabilization/solidification. Reportedly, ENRECO has stabilized/solidified a variety of oil- and metal-bearing lagoons.

Quality control is maintained by sampling for unconfined compressive strength. A cone penetrometer or shear vane is used for the measurements. The EPA paint filter test is used to determine the presence of free liquids.

7.1.8.3 In Situ Mixing: Envirite System

Envirite of Plymouth Meeting, Pennsylvania, formerly American Resources Corporation, has 3 years' experience with in situ stabilization/solidification processing. Envirite uses two types of processing equipment, based on the depth of the waste. The first, called the HSS (High Solids Stabilization)System, consists of a rotary auger placed on the front end of a bulldozer. This system mixes the soil and stabilization/solidification agent in 8- to 10-inch lifts. The mixed waste is either compacted or transported elsewhere for disposal.

The second system, the PF-5 injector, is placed at the end of a tracked backhoe. This system can reach into a lagoon for remote access to the waste materials. The PF-5 unit uses five injection tubes, each with a separately controlled feed system, a dust control system, and a sludge/additive dispersion mixer. Impellers and augers at the outlet of the PF-5 have been designed to mix the waste and stabilization/solidification agent. The stabilization/solidification agent is blown into the PF-5, air-separated, transferred by screw conveyor, and mixed by the auger system. The PF-5 injector can stabilize/solidify material to a depth of approximately 10 feet; greater depths can be achieved with custom units.

Envirite uses cement, lime, or kiln-dust as a stabilization/solidification agent. These materials are custom-tested for each waste in Envirite's laboratory. Proprietary agents are not generally used by Envirite; however, the equipment is capable of accepting them. Equipment mobilization at a site requires several days. A crew of six is required to operate the equipment. Quality control is performed by sampling each 250 cubic yards of waste, which is measured for EP Toxicity (or TCLP) and unconfined compressive strength.

7.2 Stabilization/Solidification of Materials by Lime Stabilization Techniques

Lime stabilization is used to improve the characteristics of an area such as an airport runway or a highway roadbed (National Lime Association 1987; National Research Council 1987). Typically, lime and water are added to produce a material with increased strength and decreased plasticity index. Lime is generally added to clay materials, but other soils are also appropriate candidates. With respect to hazardous waste stabilization/solidification, the use of this technology is attractive for several reasons:

1) Lime is a base for many stabilization/solidification processes, either as a raw material itself or as a constituent of cement.

2) The equipment already exists for the process, and it can easily be adapted to hazardous waste use.

3) Much information has already been collected about the ancillary aspects of the process. For example, lime dusting and worker safety issues were addressed by the lime industry many years ago, and these issues can be easily implemented in the hazardous waste industry.

These stabilization/solidification techniques could be applied in several ways. First, and most closely related to the techniques of highway roadbed stabilization, would be the application of the stabilization/solidification agent to the uppermost layer of a soil, where it would be worked into the soil after being wetted. A second application of these techniques would be to place lifts of the waste and stabilization/solidification agent in a landfill. The stabilization/solidification agent would be reacted with the waste, possibly through the addition of water, and the mixture would be compacted so that additional lifts could be placed.

Another application of these techniques would be to use them with liquid or semiliquid wastes to which no water addition was required. Typical soil stabilization/solidification procedures include soil preparation, lime spreading, soil/lime mixing, compaction to the maximum practical density, and curing.

Soil preparation consists of those activities necessary to prepare the ground for acceptance of the lime stabilizer. Ground preparation would be required when a shallow layer of soil needed to be stabilized. In this instance, a grader-scarifier or disc harrow would be used for initial scarification. Soil pulverization also could be necessary. This is accomplished through the use of a disc harrow or rotary speed mixer.

Lime spreading requires the use of a variety of equipment. When shipment is by truck, self-unloading bulk tanker trucks are the most efficient for transporting and spreading the lime because no rehandling is involved. Unloading is effected pneumatically or by one or more integral screw conveyors. The spreading is accomplished with a mechanical spreader attached to the rear of the truck, or through metal downspouts or flexible rubber boots extending from each conveyor or airline.

Mixing is required to distribute the lime uniformly throughout the soil to the proper depth and width and to pulverize the soil to minus 2 inches. During this step, water may be added to raise the moisture content of the soil-lime mixture to the optimum. A rotary mixer is the best equipment for this purpose because of its uniform mixing, fine pulverization, and speed of operation. Blade mixing with a road grader is also satisfactory for some applications. Proper mixing is required for complete stabilization of all of the soil. The sharp contrast between the white lime and all soils permits visual determination of the uniformity of the lime and soil mixture.

The lime-soil mixture should be compacted to at least 95 percent of the maximum density. Various rollers and layer thicknesses have been used for compaction in lime stabilization. The most common practice is to compact in one lift by use of a sheepsfoot roller (Figure 7-8), followed by a multiple-wheel pneumatic roller (generally a 10-ton roller). A flat-wheel roller is then used for finishing. Single-lift compaction can also be accomplished with vibrating impact rollers or heavy pneumatic rollers; light pneumatic or steel rollers would then be used for finishing. Compaction should take place as soon as possible after mixing, but a delay of several days is not detrimental. During compaction, light sprinkling may be required to keep the soil at its optimum moisture content.

After it has been compacted, the soil-lime mixture must cure. Temperature, moisture, and time are required for adequate curing. The temperature should be greater than 50°F, and the moisture should be close to optimum.

7.3 Site Conditions

Certain site-specific conditions can determine if a stabilization/solidification project is feasible. These include waste type, land use, safety, and geographic considerations.

Figure 7-8. Photograph of a sheepsfoot roller being used for compaction.

Photograph courtesy National Lime Association

7.3.1 Waste Type

The types of wastes most likely to be stabilized/solidified are liquids, sludges, and solids. Liquids and sludges are found primarily in lagoons. Solids also may be present in these lagoons, or they may occur as soils that have been contaminated by liquids.

Pretreatment of lagoon materials may be necessary for stabilization/solidification operations. Typical pretreatment operations include neutralization and water removal. Water removal may be a costly and extensive operation, especially if the waste is present beneath the local water table. In this case, dewatering may occur throughout the entire project. The water produced by dewatering must be treated. Several companies offer water treatment operations for remote site operation. These operations can remove solids, dissolved organics, and biological matter.

Access to lagoon materials can be difficult. The distance to the waste will determine whether a backhoe or a dragline is used for removal. For operations that provide for remote removal of the waste, whether the dikes will support the weight of the equipment must be considered. Proper width of the roads/dikes also must be ascertained. Some equipment (e.g., front-end loaders and bulldozers) may actually operate in a lagoon. The feasibility of operating this equipment in the waste must be considered. One consideration would be whether the waste can support the vehicle. If not, the use of wide tracks or low-pressure tires might make this operation feasible. Another consideration is the steepness of the dikes or walls surrounding a lagoon. At the one site, front-end loaders could not be used to carry waste over a dike and into dump trucks for this reason. Instead, an extra operation was necessary, which involved the use of backhoes to transfer the waste from the lagoon bottom to its side.

Soil treatment operations can disturb ongoing operations, and stabilization/solidification operations should be planned to keep this disturbance to a minimum. Dusting of the soil may be a problem, especially if the contaminant is present in high concentrations or if a large receptor population is nearby. Both dust suppression and the stabilization/solidification operation require water. Planners should determine if a source of water is present at the site if it is free of additional contaminants. If the soil contains concrete or large rocks, jackhammers will be needed for their removal. In addition, some soils at a site may not be appropriate candidates for stabilization/solidification even though they are contaminated. For example, an aquitard at the Frontier Hard Chrome Site was contaminated with hexavalent chromium (Dames & Moore 1987). Removal of this aquitard for stabilization/ solidification could have inadvertently introduced contamination into the aquifer.

7.3.2 Land Use and Safety

Stabilization/solidification applications must include an appraisal of future site use. If the waste is very toxic, certain site uses (e.g., residential use) would be precluded after treatment. If the waste will not support structures, this option greatly limits the use of the site. One of the disadvantages of stabilization/solidification is that it increases the waste volume. A significant increase in an already large volume of waste could result in an unexpectedly large waste pile, in which case offsite disposal of a portion of the waste may be necessary.

Site operations require complete health and safety procedures for hazardous waste work. Workers in machinery may require supplied breathing air; others may require air-purifying respirators. Provision also must be made for adequate protective clothing (Level A, B, C, or D). Decontamination areas are essential for hazardous waste stabilization/solidification operations. The proximity of a site to nearby populations must be determined before the stabilization/solidification process is begun. Monitoring for dusts and organics may be necessary, and the presence of these materials in the atmosphere could prevent operations during unfavorable meteorological conditions.

7.3.3 Geography

This topic includes consideration of geology, surface waters, and meteorology. The principal geologic consideration is that relating to site aquifers. Waste that extends to aquifers may have to be dewatered as described earlier. Weather can have an effect on operations or the placement of the waste. For example, stabilization/solidification operations may not be feasible during freezing weather if the stabilized/solidified waste cannot cure. If the waste is subject to failure upon repeated freeze/thaw or wet/dry testing, an engineered solution can be required—i.e., simply placing the waste below the frost line. Finally, waste should never be placed in a floodplain.

7.3.4 General

Other considerations relate to the peculiarities of a site. For example, an area may be needed for stockpiling operations, and the absence of such an area could preclude stabilization/solidification as a remedial action. Power and utility sources for treatment operations are also important. If not available locally, electricity can be generated on site and water can be imported. Utilities can have an impact on field operations in another way as well; the presence of overhead power lines should be determined for safe operations at a site. Finally, any debris that is present in the waste may have to be removed. It is not unusual to encounter trees, large rocks, appliances, and other materials within a waste.

References

American Nuclear Society (ANS). 1986. ANSI/ANS-16.1-1986 American National Standard Measurement of the Leachability of Solidified Low-Level Radioactive Wastes by a Short-Term Test Procedure. Prepared by the American Nuclear Society Standards Committee Working Group ANS-16.1. Published by the American Nuclear Society, La Grange Park, Illinois. Approved April 14, 1986, by the American National Standard Institute, Inc.

Bishop, P. L. 1986. Prediction of Heavy Metal Leaching Rates From Stabilized/Solidified Hazardous Wastes. Toxic and Hazardous Wastes. In: Proceedings of the Eighteenth Mid-Atlantic Industrial Waste Conference, pp. 236-252.

Bishop, P. L. 1988. Leaching of Inorganic Hazardous Constituents From Stabilized/Solidified Hazardous Wastes. Hazardous Waste & Hazardous Materials, 5(2):129-144.

Bridle, T. R., et al. 1987. Evaluation of Heavy Metal Leachability From Solid Wastes. Water Science and Technology, Vol. 19, pp. 1029-1036.

California Code, Title 22, Article 11. Criteria for Identification of Hazardous and Extremely Hazardous Wastes, pp. 1800.75-1800.82.

Campbell, K., et al. 1987. Stabilization of Cadmium and Lead in Portland Cement Paste Using Synthetic Sea-water Leachant, Environmental Progress, 6(2).

Carter, M. 1983. Geotechnical Engineering Handbook. Chapman and Hall, New York.

Caterpillar, Inc. 1987a. Caterpillar Performance Handbook, Edition 18. Peoria, Illinois.

Caterpillar, Inc. 1987b. Caterpillar Performance Handbook: Hydraulic Excavators. Peoria, Illinois.

Cote, P. L., and T. W. Constable. 1982. Evaluation of Experimental Conditions in Batch Leaching Procedures. Resources and Conservation, Vol. 9, pp. 59-73.

Cote, P., and D. P. Hamilton. 1984. Leachability Comparison of Four Hazardous Waste Solidification Processes. In: Proceedings of the 38th Industrial Waste Conference, May 1983, Purdue University, West Lafayette, IN.

Cote, P. L., and T. R. Bridle. 1987a. Long-Term Leaching Scenarios for Cement-Based Waste Forms. Waste Management & Research, Vol. 5, pp. 55-66.

Cote, P. L., T. W. Constable, and A. Moreira. 1987b. An Evaluation of Cement-Based Waste Forms Using the Results of Approximately Two Years of Dynamic Leaching. Nuclear and Chemical Waste Management, Vol. 7, pp. 129-139.

Cullinane, M. J., Jr., L. W. Jones, and P. G. Malone. 1986. Handbook for Stabilization/Solidification of Hazardous Wastes. EPA/540/2-86/001. Hazardous Waste Engineering Research Laboratory, U.S. Environmental Protection Agency, Cincinnati, Ohio.

Curry, M. 1986. Fixation/Solidification of Hazardous Waste at Chemical Waste Management's Vicery, Ohio, Facility. In: Proceedings of Hazardous Materials Control Research Institute Conference, December 1986. pp. 297-302.

Dames & Moore. 1987. Feasibility Study of the Frontier Hard Chrome Site - Vancouver, Washington.

Earth Technology Corp. 1988. Closure Plan an Post-Closure Care Plan for the Municipal and Industrial Disposal Co.

The Earth Technology Corporation. 1988. Optimization of Soil Remediation Technologies Using Stabilization/Solidification.

Environment Canada and Alberta Environmental Center. 1986. Test Methods for Solidified Waste Characterization.

Federal Register, Vol. 51, No. 216, Friday, November 7, 1986, pp. 40643-40652.

Fitzpatrick, V. F., C. L. Timmerman, and J. L. Buelt. 1987. In Situ Vitrification - An Innovative Thermal Treatment Technology. Paper presented at Groundwater Treatment Conference, New York, September 16-18, 1987.

Gibbons, J. J., and R. Soundararajan. 1988. The Nature of Chemical Bonding Between Modified Clay Minerals and Organic Wastes Materials. American Laboratory, 20(2):38-46.

Gilliam, T. M., L. R. Dole, and E. W. McDaniel. 1986. Waste Immobilization in Cement-Based Grouts. Hazardous and Industrial Solid Waste Testing and Disposal. Sixth Volume. ASTM STP 933, D. Lorenzen, R. A. Conway, L. P. Jackson, A. Hamza, C. L. Perket, and W. J. Lacy, eds. American Society for Testing and Materials. Philadelphia. pp. 295-307.

Green, D. W., ed. 1984. "Processing Bulk Solids," in Perry's Chemical Engineers Handbook, 6th ed. McGraw-Hill, New York. pp. 21-3-10.

JACA Corporation. 1985. Critical Characteristics and Properties of Hazardous Waste Solidification/Stabilization. Unpublished Report. Prepared for U.S. Environmental Protection Agency, Hazardous Waste Engineering Research Laboratory, Cincinnati, Ohio.

Kolvites, B., and P. Bishop. 1987. Column Leach Testing of Phenol and Trichloroethylene Stabilized/ Solidified with Portland Cement. University of New Hampshire, Department of Civil Engineering, Presented at the ASTM 4th International Hazardous Waste Symposium, Atlanta, Georgia. p. 27.

Kyles, J. H., K. C. Malinowski, and T. F. Stanczyk. 1987. Solidification/ Stabilization of Hazardous Waste - A Comparison of Conventional and Novel Techniques, Toxic and Hazardous Wastes. In: Proceedings of the 19th Mid-Atlantic Industrial Waste Conference, June 21-23, 1987. Jeffrey C. Evans, ed. pp. 554-568.

Lubowitz, H. R., and C. C. Wiles. 1979. Encapsulation Technique for Control of Hazardous Waste Disposal. In Toxic and Hazardous Waste Disposal, Vol. 1. R. B. Pojasek, ed. Ann Arbor Science, Ann Arbor, Michigan.

Malone, P. G., L. W. Jones, and R. J. Larson. 1980. Guide to the Disposal of Chemically Stabilized and Solidified Waste. SW-872, Office of Water and Waste Management, U.S. Environmental Protection Agency, Washington, D.C.

Martin, J. P., et al. 1987. Stabilization of Petroleum Refining Wastes for an On-Site Landfill. In: Proceedings Oak Ridge National Laboratory DOE Model Conference, October 1987.

McGraw-Hill, Inc. 1988. Superfund Report, 11(4):15-16, February 17, 1988.

Morgan, D., et al. 1984. Oil Sludge Solidification Using Cement Kiln Dust. Journal of Environmental Engineering, 110(5), October.

Musser, D., ENRECO, and R. Smith, Raba-Kistner Consultants. 1984. Case Study: In Situ Solidification/Fixation of Oil Field Production Fluids - A Novel Technique. In: Proceedings of the 39th Industrial Waste Conference, Purdue University, West Lafayette, Indiana, May 1984.

National Lime Association. 1987. Lime Stabilization Construction Manual. Bulletin 326. Arlington, VA.

National Research Council. 1987. Lime Stabilization - Reactions, Properties, Design and Construction. Transportation Research Board, Washington, D.C.

Newcomer, L. R., W. B. Blackburn, and T. A. Kimmell. 1986. Performance of the Toxicity Characteristic Leaching Procedure. December 1986.

Rowe, G., and API Waste Treatment Task Force. 1987. Evaluation of Treatment Technologies for Listed Petroleum Refinery Wastes. Final Report, The American Petroleum Institute.

R. S. Means Co., Inc. 1987 Means Heavy Construction Cost Data 1987. Kingston, Massachusetts.

Shaw, M. D. 1987. Macroencapsulation: New Technology Provides Innovative Alternatives for Hazardous Materials Storage, Transportation, Treatment and Disposal. In: Proceedings of Hazardous Materials Control Research Institute Symposium. March 1987, Washington, D.C. pp. 96-100.

Shively, W., et al. 1986. Leaching Tests of Heavy Metals Stabilized With Portland Cement. Journal of Water Pollution Control Federation, 58(3):234-241.

Shively, W. E., and M. A. Crawford. Undated. EP Toxicity and TCLP Extractions of Industrial and Solidified Hazardous Waste.

Skinner, J. H., and N. J. Bassin. 1988. The Environmental Protection Agency's Hazardous Waste Research and Development Program. APCA J., 38(4):377-387, April 1988.

Stegmann, J., P. L. Cote, and P. Hannak. 1988. Preliminary Results of an International Government/Industry Cooperative Study of Waste Stabilization/Solidification. In: Hazardous Waste: Detection, Control, Treatment. R. Abbou, ed. Elsevier Science Publishers B.V., Amsterdam, The Netherlands. pp. 1539-1548.

Tittlebaum, M. E., et. al., 1985. "State of the Art on Stabilization of Hazardous Organic Liquid Wastes and Sludges." CRC Critical Reviews in Environmental Control, 15(2):191-193.

Tittlebaum, M. E., et al. 1986. Procedures for Characterizing Effects of Organics on Solidification/Stabilization of Hazardous Wastes, Hazardous and Industrial Solid Waste Testing and Disposal: Sixth Volume. ASTM STP 933. D. Lorenzen, R. A. Conway, L. P. Jackson, A. Hamza, C. L. Perket, and W. J. Lacy, eds. American Society for Testing and Material, Philadelphia. pp. 308-318.

Unger, S. L., et. al. Undated. Surface Encapsulation Process for Stabilizing Intractable Contaminants. Environmental Protection Polymers, Hawthorne, California.

U.S. Army Corps of Engineers. 1980. Engineering and Design Laboratory Soil Testing. Engineering Manual 1110-2-1906. May 1970, updated 1980.

U.S. Environmental Protection Agency. 1980. Hazardous Waste Management System: General Facility Standards. 45 FR 33066, May 19, 1980.

U.S. Environmental Protection Agency. 1982. Interim Status Standards for Owners and Operators of Hazardous Waste Treatment, Storage, and Disposal Facilities. 47 FR 8307, February 25, 1982.

U.S. Environmental Protection Agency. 1984. Case Studies - Remedial Response at Hazardous Waste Sites. EPA 540/2-84-002, February 1984. Trammel Crow Site and College Point Site.

U.S. Environmental Protection Agency. 1985. National Oil and Hazardous Substances Pollution Contingency Plan. 50 FR 47950, November 20, 1985.

U.S. Environmental Protection Agency. 1986a. Hazardous Waste Management System. 51 FR 46824, December 24, 1986.

U.S. Environmental Protection Agency. 1986b. Test Methods for Evaluating Solid Waste, Method 9095, SW-846, Third Edition. November 1986.

U.S. Environmental Protection Agency. 1986c. Prohibition on the Disposal of Bulk Liquid Hazardous Waste in Landfills - Statutory Interpretive Guidance. OSWER Policy Directive No. 9487.00-2A. EPA/530-SW-016, June 1986.

U.S. Environmental Protection Agency. 1986d. Hazardous Waste Management System; Land Disposal Restrictions. 51 FR 40572, November 7, 1986.

U.S. Environmental Protection Agency. 1986e. Handbook for Stabilization/ Solidification of Hazardous Waste. EPA/540/2-86/001.

U.S. Environmental Protection Agency. 1986f. A Procedure for Estimating Monofilled Solid Waste Leachate Composition. Technical Resource Document SW-924, 2nd Edition. Hazardous Waste Engineering Research Laboratory, Office of Research and Development, Cincinnati, Ohio, and Office of Solid Waste and Emergency Response, Washington, D.C.

U.S. Environmental Protection Agency. 1986g. SW-846 Test Methods for Evaluating Solid Waste, Vol. 1C: Laboratory Manual Physical/Chemical Methods, Third Edition. Office of Solid Waste and Emergency Response, Washington, D.C.

U.S. Environmental Protection Agency. 1986h. Mobile Treatment Technologies for Superfund Wastes. Office of Solid Waste and Emergency Response. EPA 540/2-86/003, September 1986.

U.S. Environmental Protection Agency. 1988a. Land Disposal Restriction for First Third Scheduled Wastes. 53 FR 11742, April 8, 1988.

U.S. Environmental Protection Agency. 1988b. Land Disposal Restrictions for First Third Scheduled Wastes. 53 FR 31138, August 17, 1988.

U.S. EPA. 1988c. Evaluation of Test Protocols for Stabilization/Solidification Technology Demonstrations. Revised Draft Report. U.S. Environmental Protection Agency, Office of Research and Development, Cincinnati, Ohio. Prepared by PRC Environmental Management, Inc. Contract No. 68-03--3484, April 25, 1988.

U.S. Environmental Protection Agency. 1988d. Stabilization/Solidification Case Studies. Unpublished document.

U.S. Environmental Protection Agency. 1989. Land Disposal Restrictions for Second Third Scheduled Wastes. Proposed Rules. 54 FR 1056, January 11, 1989.

Van Keuren, E., et al. 1987. Pilot Field Study of Hydrocarbon Waste Stabilization, Toxic and Hazardous Wastes. In: Proceedings of the 19th Mid--Atlantic Industrial Waste Conference, June 21-23, 1987. Jeffrey C. Evans, ed. pp. 330-341.

Vick, W., et al. 1987. Physical Stabilization Techniques for Mitigation of Environmental Pollution From Dioxin Contaminated Soils. Land Disposal, Remedial Action, Incineration and Treatment of Hazardous Waste. In: Proceedings of the 13th Annual Research Symposium, 1987. U.S. Environmental Protection Agency, Hazardous Waste Engineering Research Laboratory.

Weitzman, L., L. E. Hamel, and E. Barth. 1988. Evaluation of Solidification/ Stabilization as a Best Demonstrated Available Technology. Paper presented at 14th Annual Hazardous Waste Engineering Laboratory Conference, Cincinnati, Ohio, May 1988.

Wiles, C. 1987. A Review of Solidification/Stabilization Technology. Journal of Hazardous Materials, Vol. 14, 1987.

Wiles, C. C. Undated. Treatment of Hazardous Waste in the United States With Solidification/ Stabilization. U.S. Environmental Protection Agency. Hazardous Waste Engineering Research Laboratory, Cincinnati, Ohio.

40 CFR 268.10 "First Third," Federal Register, August 17, 1988.

Wiles, C., and H. Howard. Undated. USEPA Research in Solidification/Stabilization of Waste Material.

Part II

The information in Part II is from *Handbook for Stabilization/Solidification of Hazardous Wastes,* prepared by M.J. Cullinane, Jr., L.W. Jones, and P.G. Malone of the U.S. Army Waterways Experiment Station for the U.S. Environmental Protection Agency, June 1986.

ACKNOWLEDGEMENTS

This Handbook was developed by the Environmental Laboratory of the U. S. Army Engineer Waterways Experiment Station (WES) under the sponsorship of the U. S. Environmental Protection Agency (EPA). Authors were Mr. M. John Cullinane, Jr., Dr. Larry W. Jones, and Dr. Philip G. Malone. The Handbook was edited by Ms. Jamie W. Leach of the WES Publications and Graphic Arts Division. The project was conducted under the general supervision of Dr. John Harrison, Chief, Environmental Laboratory; Dr. Raymond L. Montgomery, Chief, Environmental Engineering Division; and Mr. Norman Francingues, Chief, Water Supply and Waste Treatment Group. Director of WES during the course of this work was Col. Allen F. Grum, USA. Technical Director was Dr. Robert W. Whalin.

Preparation of this Handbook was aided greatly by the constructive contributions of the following reviewers:

Carlton Wiles	EPA, HWERL
Roy Murphy	EPA, OWPE
Ann Tate	EPA, CERI
Richard Stanford	EPA, OERR
Andrew T. McCord	Snyder, N. Y. 14226
Tom Ponder	PEDCo. Environmental, Inc.
Radha Krishnan	PEDCo. Environmental, Inc.

Janet M. Houthoofd of the Land Pollution Control Division, Hazardous Waste Engineering Research Laboratory, was the EPA project officer.

A major part of this study included the evaluation of equipment and processes applied to the solidification/stabilization of hazardous materials. The information contained herein could not have been compiled without the valuable assistance of a number of representatives from industry. The following industries are acknowledged for providing information and assistance:

Albert H. Halff Associates, Inc.
Consulting Engineers and Scientists
8616 Northwest Plaza Drive
Dallas, TX 75225
(214) 739-0094

American Resources Corporation
P.O. Box 813
Valley Forge, PA 19482-0813
(215) 337-7373

Beardsley & Piper
Division of Pettibone Corp.
5501 W. Grand Avenue
Chicago, IL 60639
(312) 237-3700

BFI Waste Systems
P.O. Box 3151
Houston, TX 77001
(713) 870-7857

Charles Ross & Son Company
710 Old Willets Path
Hauppauge, NY   11787
(516) 234-0500

Chemfix Technologies, Inc.
1675 Airline Highway
P.O. Box 1572
Kenner, LA   70063
(504) 467-2800

The Gorman-Rupp Company
305 Bowman Street
P.O. Box 1217
Mansfield, OH   44903
(419) 755-1011

Hittman Nuclear & Development
  Corp.
9151 Rumsey Road
Columbia, MD   21045
(301) 730-7800

Mixing Equipment Co., Inc.
135 Mt. Read Blvd.
P.O. Box 1370
Rochester, NY   14603
(716) 436-5550

Rollins Environmental Services
P.O. Box 73877
Baton Rouge, LA   70807
(504) 778-1234

Soil Recovery, Inc.
101 Eisenhower Parkway
Roseland, NJ   07068
(201) 226-7330

Solidtek, Inc.
5371 Cook Road
P.O. Box 888
Morrow, GA   30260
(404) 361-6181

The Vaughan Pump Company, Inc.
364 Monte Elma Road
Montesano, WA   98563
(206) 249-402

The Vince Hagan Company
P.O. Box 5141
Dallas, TX   75222
(214) 339-7194

1. Introduction

1.1 Background and Definitions

The terms "stabilization" and "solidification" are used in this Handbook as defined in the EPA publication, "Guide to the Disposal of Chemically Stabilized and Solidified Waste" (Malone et al. 1980). Both stabilization and solidification refer to treatment processes that are designed to accomplish one or more of the following results: (1) improve the handling and physical characteristics of the waste, as in the sorption of free liquids; (2) decrease the surface area of the waste mass across which transfer or loss of contaminants can occur; and/or (3) limit the solubility of any hazardous constituents of the waste such as by pH adjustment or sorption phenomena.

Stabilization techniques are generally those whose beneficial action is primarily through limiting the solubility or mobility of the contaminants with or without change or improvement in the physical characteristics of the waste. Examples include the addition of lime or sulfide to a metal hydroxide waste to precipitate the metal ions or the addition of an absorbent to an organic waste. Stabilization usually involves adding materials which ensure that the hazardous constituents are maintained in their least mobile or toxic form.

Solidification implies that the beneficial results of treatment are obtained primarily, but not necessarily exclusively, through the production of a solid block of waste material which has high structural integrity--a product often referred to as a "monolith." The monolith can encompass the entire waste disposal site--called a "monofill"--or be as small as the contents of a steel drum. The contaminants do not necessarily interact chemically with reagents, but are mechanically locked within the solidified matrix--called "microencapsulation." Contaminant loss is limited largely by decreasing the surface area exposed to the environment and/or isolating the contaminants from environmental influences by microencapsulating the waste particles. Wastes can also be "macroencapsulated," that is, bonded to or surrounded by an impervious covering. These techniques are also considered to be stabilization/solidification processes.

The term "fixation" has fallen in and out of favor, but is widely used in the waste treatment field to mean any of the stabilization/solidification processes as described above; "fixed" wastes are those that have been treated in this manner.

Both solidification and chemical stabilization are usually included in commercial processes and result in the transformation of liquids or semi-solids into environmentally safer forms. For example, a metal-rich sludge would be considered stabilized if it were mixed with a dry absorber such as fly ash or dry soil. The benefits could be carried further if the sorbent and waste were then cemented into an impermeable, monolithic block. Or a waste would be considered chemically stabilized if the pH of the sludge were raised by the addition of lime ($Ca(OH)_2$) so that potential contaminants such as toxic metals were less soluble and thus less easily leached.

1.2 Purpose and Scope of this Handbook

This Handbook provides guidance for the evaluation, selection, and use of stabilization/solidification technology as a remedial action alternative at uncontrolled, hazardous wastes sites. The Handbook is designed to permit engineering personnel to proceed through concept development, determination of design requirements, and preliminary cost estimating for selected stabili-zation/solidification alternatives. A flow chart for evaluating considera-tions and procedures important to the stabilization/solidification option is shown in Figure 1-1.

The Handbook systematically reviews the technical basis for available stabilization/solidification systems, especially those suitable for onsite application at uncontrolled, hazardous waste sites. The general chemical systems involved in waste stabilization/solidification are discussed to pro-vide the background information necessary for the selection of the optimum treatment system for a specific waste. Also described are the testing and analysis techniques commonly used to characterize a waste to aid in the selection of pretreatment and stabilization/solidification processes. The compatibility of specific classes of wastes and additives, and the testing systems needed for the evaluation of the stabilized/solidified wastes once treated are also reviewed.

Specific materials and equipment that are used in waste stabilization/ solidification treatment and processing are discussed. Based on field surveys, four stabilization/solidification scenarios, including costs for materials, equipment, and operations associated with each, are developed and compared to provide a basis for planning-level cost evaluation of the many stabilization/solidification alternatives. Safety, environmental concerns, and cleanup and closure of waste processing and final disposal are considered.

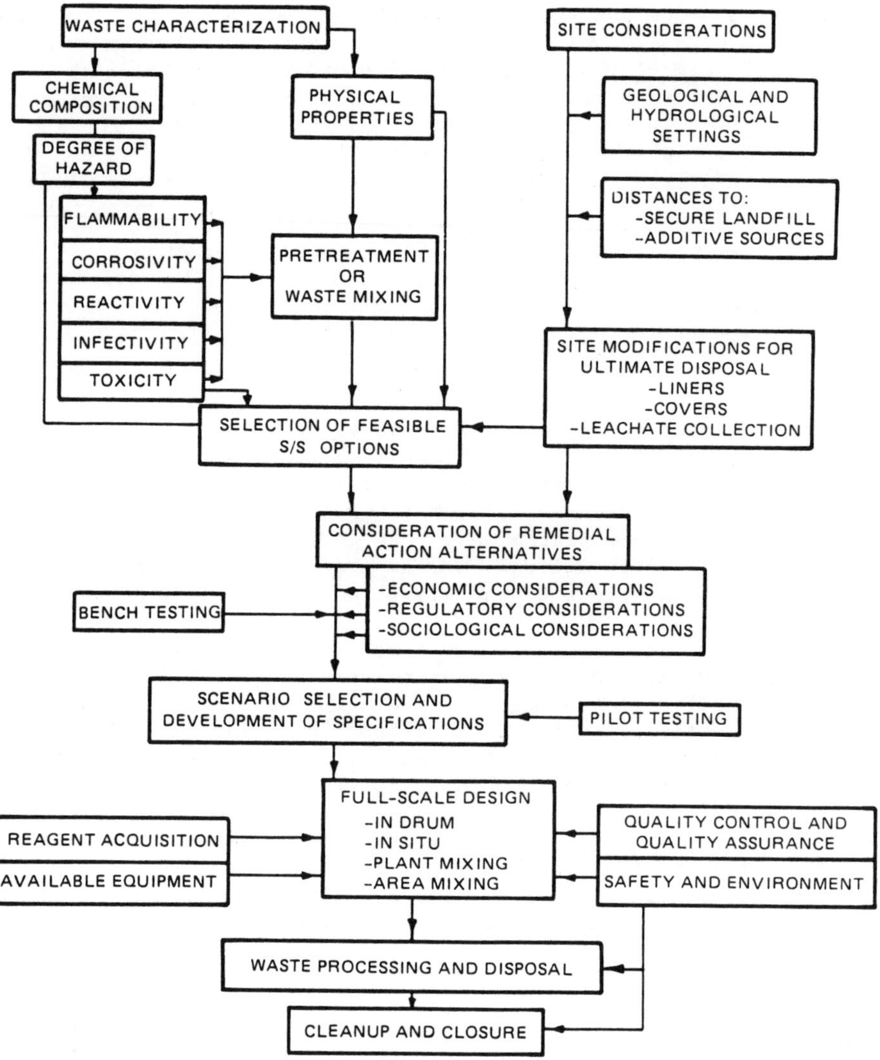

Figure 1-1. Flow chart for evaluating the stabilization/solidification (S/S) option.

1.3   Regulatory Basis for Use of

Stabilization/Solidification

The EPA hazardous site cleanup program, referred to as Superfund, was authorized and established in 1980 by the enactment of the Comprehensive Environmental Response, Compensation, and Liability Act (CERCLA), Public Law (PL) 96-510. This legislation allows the Federal government (and cooperating State governments) to respond directly to releases and threatened releases of hazardous substances and pollutants or contaminants that may endanger public health or welfare or the environment. Prior to the passage of PL 96-510, the Federal authority with respect to hazardous substances was mostly regulatory through the Resource Conservation and Recovery Act (RCRA) and the Clean Water Act and its predecessors. The general guidelines and provisions for implementing CERCLA are given in the National Oil and Hazardous Substances Contingency Plan (NCP) (Federal Register, 40 CFR 300, 1982).

Three classes of actions are available when direct government action is called for:

a. Immediate removals are allowed when a prompt response is needed to prevent harm to public health or welfare or to the environment. These are short-term actions usually limited to 6 months and a total expenditure of $1 million.

b. Planned removals are expedited, but not immediate, responses. These are intended to limit danger or exposure that would take place if longer term projects were implemented and responses were delayed.

c. Remedial actions are longer term activities undertaken to provide more complete remedies. Remedial actions are generally more expensive and can only be undertaken at sites appearing on the National Priorities List of the NCP.

Remedial actions may present technically complex problems that are expensive to resolve. The selection of technical measures takes place only after a full evaluation of all feasible alternatives based upon economic, engineering, environmental, public health, and institutional considerations. Offsite transportation and disposal of waste is generally an expensive option and is justified only when proven cost-effective, and then only in facilities that comply with current hazardous waste disposal regulations under Subtitle C of RCRA.

Waste stabilization is specifically included in the NCP as a method of remedying releases of hazardous materials and controlling release of waste to surface water. Solidification and encapsulation are mentioned as techniques available for onsite treatment of contaminated soils and sediments. Under the general requirement to evaluate all alternatives for remedial action, it

will be necessary to evaluate the cost effectiveness of stabilization/
solidification systems as applied to specific sites even if the technology is
not selected in the final analysis of remedial techniques. Costs and engineering considerations are critical to these evaluations.

The performance expected from stabilized/solidified waste must also be
as accurately assessed as possible. Cost estimates must take into consideration future expenditures needed to maintain the final waste disposal site
after response work is complete. The NCP emphasizes the selection of reliable, tested remedial technologies. Examples of successful applications are
an important part of any technical evaluation.

A further goal of this Handbook is to provide data necessary for the
technical decisions required by law and for preliminary cost estimates.
Other handbooks are available to supplement this document in developing plans
for specific site activities. Overall guidance on remedial action technologies, including a survey of stabilization/solidification, is provided in a
Technology Transfer Handbook by the EPA (U.S. EPA 1985a). The decision to
implement the stabilization/solidification option must be preceded by the detailed investigation of many variables. Both waste and site characteristics
must be evaluated to ensure that the stabilization/solidification alternative
is cost-effective and environmentally acceptable. The U.S. EPA (1985b,
1985c) has provided general guidance on the procedure to be followed in
selecting the most appropriate remedial actions.

REFERENCES

Federal Register. 1982. National Oil and Hazardous Substance Contingency
Plan. (40 CFR 300), Volume 47, No. 137, July 16, 1982.

Malone, P. G., L. W. Jones, and R. J. Larson. 1980. Guide to the Disposal
of Chemically Stabilized and Solidified Waste. SW-872, Office of Water and
Waste Management, U. S. Environmental Protection Agency, Washington, D.C.
126 pp.

U.S. EPA. 1985a. Remedial Action at Waste Disposal Sites (Revised).
EPA-625/6-85-006, U.S. Environmental Protection Agency, Cincinnati, Ohio.
497 pp.

U.S. EPA. 1985b. Guidance on Feasibility Studies under CERCLA.
EPA-540/G-85-003. Office of Emergency and Remedial Response, U.S. Environmental Protection Agency, Washington, D.C.

U.S. EPA. 1985c. Guidance on Remedial Investigations under CERCLA.
EPA-540/G-85-002. Office of Emergency and Remedial Response, U.S. Environmental Protection Agency, Washington, D.C.

2. Basis of Stabilization/Solidification Technology

Stabilization processes and solidification processes have different goals. Stabilization systems attempt to reduce the solubility or chemical reactivity of a waste by changing its chemical state or by physical entrapment (microencapsulation). Solidification systems attempt to convert the waste into an easily handled solid with reduced hazards from volatilization, leaching, or spillage. The two are discussed together because they have the common purpose of improving the containment of potential pollutants in treated wastes. Combined processes are often termed "waste fixation" or "encapsulation."

Solidification of waste materials is widely practiced in the disposal of radioactive waste. Many developments relating to solidification originated in low-level radioactive waste disposal. Regulations pertaining to disposal of radioactive waste require that the wastes be converted into a free-standing solid with a minor amount of free water. Most processes used for nuclear waste include a step in which granular, ion exchange waste and liquids are incorporated in a solid matrix using a cementing or binding agent (for example, Portland cement, organic polymers, or asphalt). The resulting block of waste, with relatively low permeability, reduces the surface area across which the transfer of pollutants can occur. No such requirement for producing a free-standing solid exists for hazardous waste disposal, and solidification usually involves only the addition of an absorbent (without a binding agent) to produce a finely particulate waste that has no free liquid.

Waste stabilization has also been practiced in radioactive waste disposal and has involved processes such as (1) selecting inert, nondegrading sorbents that take up and retain specific radionuclides, (2) adjusting pH and oxidation-reduction conditions in the waste to prevent waste solubilization in ground water, and (3) using zeolites rather than biodegradable organic polymers as ion exchange media.

In hazardous waste disposal, an effort is usually made to have the treated waste delisted, usually by passing the EPA Extraction Procedure (EP) leaching test. To accomplish this goal, a variety of strategies may be used to prevent contaminant leaching, including neutralization, oxidation/reduction, physical entrapment, chemical stabilization, and binding of the stabilized solid into a monolith. The development of an appropriate treatment strategy includes the following considerations:

   a. The waste should be treated to obtain the most inert and insoluble form chemically and economically feasible.

b. Media should be added to absorb any free liquid present.

c. When necessary, a binding agent should also be added.

The binding agent may be selected to stabilize the waste further, for example the addition of alkalinity in portland cement. In cases where the waste is extremely soluble or no suitable chemical binder can be found, the waste may be contained by encapsulation in some hydrophobic medium, such as asphalt or polyethylene. This may be done either by incorporating the waste directly in the partially molten material or by forming jackets of polymeric material around blocks of waste.

Several generic treatment systems have been developed for waste stabilization and solidification, but not all have been employed in remedial action on uncontrolled waste sites. The volumes of waste involved at uncontrolled waste sites generally require that only the least expensive systems that are effective be used. The large quantities and varieties of wastes that are usually present also require the use of adaptable systems that are effective over a wide range of conditions. The treatment systems that generally satisfy these needs are the pozzolan- or Portland-cement-based systems. Inexpensive absorber materials such as clay, native soil, fly ash, or kiln dust may also be added. Under specific circumstances, it may be necessary to select other systems that offer particular advantages such as improved waste containment or compatibility with specific wastes. This Handbook concentrates on the major stabilization/solidification systems that can be applied inexpensively to a wide variety of wastes. Systems that have limited application to mixed wastes (such as glassification or organic polymers) or systems that require specific waste materials (such as self-cementation in sulfate waste) are covered in other references such as Malone et al. (1980), Malone and Jones (1979), and Iadevaia and Kitchens (1980).

2.1 Types of Treatment Reagents and Processes

Most stabilization/solidification systems being marketed are proprietary processes involving the addition of absorbents and solidifying agents to a waste. Often the marketed process is changed to accommodate specific wastes. Since it is not possible to discuss completely all possible modifications to a process, discussions of most processes have to be related directly to generic process types. The exact degree of performance observed in a specific system may vary widely from its generic type, but the general characteristics of a process and its products can be discussed. Comprehensive general discussions of waste stabilization/solidification are given in Malone et al. (1980), Malone and Jones (1979), and Iadevaia and Kitchens (1980).

Waste stabilization/solidification systems that have potentially useful application in remedial action activities and are discussed in detail here include:

    a. Sorption

    b. Lime-fly ash pozzolan processes

    c. Pozzolan-portland cement systems

    d. Thermoplastic microencapsulation

    e. Macroencapsulation.

Other technologies such as fusing waste to a vitreous mass or using self-cementing material are too specialized or not sufficiently field applicable to be used at present (Malone et al. 1980).

Sorption involves adding a solid to soak up any liquid present, and it may produce a soil-like material. The major use of sorption is to eliminate all free liquid. Nonreactive, nonbiodegradable materials are most suitable for sorption. Typical examples are activated carbon, anhydrous sodium silicate, various forms of gypsum, celite, clays, expanded mica, and zeolites. Some sorbents are pretreated to increase their activity toward specific contaminants and many are sold as proprietary additives in commercial processes.

Lime/fly ash pozzolanic processes use a finely divided, noncrystalline silica in fly ash and the calcium in lime to produce low-strength cementation. The waste containment is produced by entrapping the waste in the pozzolan concrete matrix (microencapsulation).

Pozzolan-Portland systems use Portland cement and fly ash or other pozzolan materials to produce a stronger type of waste/concrete composite. The waste containment is produced by microencapsulation in the concrete matrix. Soluble silicates may be added to accelerate hardening and metal containment.

Thermoplastic microencapsulation involves blending fine particulate waste with melted asphalt or other matrix. Liquid and volatile phases associated with the wastes are driven off, and the wastes are isolated in a mass of cooled, hardened asphalt. The material can be buried with or without a container.

Macroencapsulation systems contain a waste by isolating large masses of waste using some type of jacketing material. The most carefully researched systems use a 208-ℓ drum or a polyethylene jacket fused over a monolithic block of solidified wastes.

2.1.1 Sorption

2.1.1.1 General

Most waste materials considered for stabilization/solidification are quids or sludges (semisolids). To prevent the loss of drainable liquid and improve the handling characteristics of the waste, a dry, solid absorbent is generally added to the waste. The sorbent may interact chemically with the waste or may simply be wetted by the liquid part of the waste (usually water) and retain the liquid as part of the capillary liquid. Figure 2-1 illustrates five common mechanisms by which sorbents can interact and immobilize small, polar molecules like water or charged ions on their surface or interstices, or react chemically to form new products.

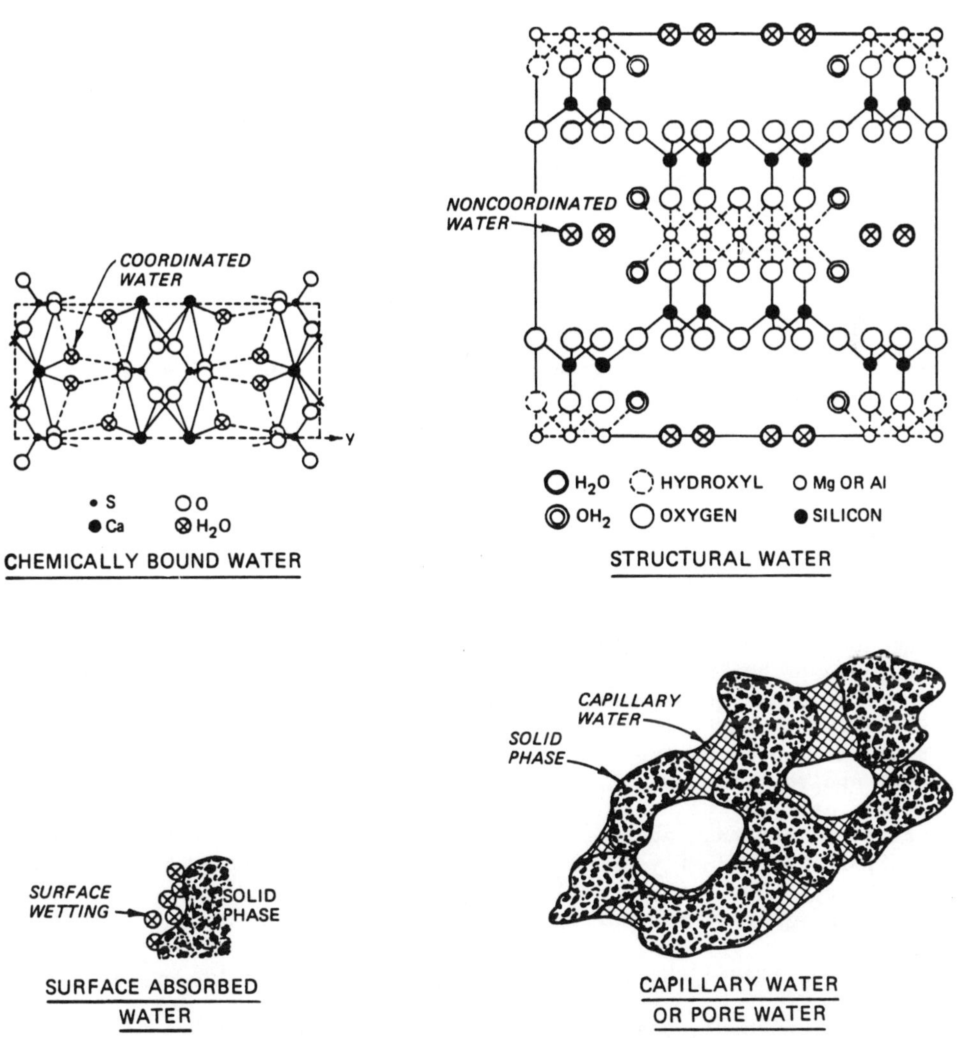

Figure 2-1. Mechanisms retaining water and ionic materials on and in solid phases.

The most common sorbents used with waste include soil and waste products such as bottom ash, fly ash, or kiln dust from cement and lime manufacture. In general, selection of sorbent materials involves trade-offs among chemical effects, costs, and amounts required to produce a solid product suitable for burial. Table 2-1 summarizes chemical binding properties of natural sorbents for selected waste leach liquids. Where the ability of a sorbent to bind particular contaminants is important to containment, sorbents with specific chemical affinities can be selected (Table 2-2). The pH of the waste strongly affects sorption/waste interactions, and pH control is an important part of any sorption process.

Artificial materials have also been advocated for use as sorbents in solidification; however, the relatively high cost of these materials has prevented their widespread use. Synthetic materials have generally found use where the binding of a specific contaminant in the waste is of paramount importance. Table 2-3 lists several synthetic sorbent materials that have been developed or tested for use with hazardous wastes.

Several major technical considerations are important in selecting a sorbent:

a. Quantity needed to satisfy the requirement for having no free liquid.

b. Compatibility or reactivity of the waste and the sorbent.

c. Level and character of contamination that might be introduced in the sorbent.

d. Chemical binding properties of sorbent for specific contaminants.

The quantity of absorbent necessary for sorbing all of the liquid in a waste to ensure that no free liquid is available varies widely depending on the nature of the liquid phase, the original solids content of the waste, the moisture level in the sorbent, and the availability of any chemical reactions that take up liquids during reaction. The high degree of variability seen in sorbents, and the changes in moisture content that can be brought about by storage and aging of sorbents, make it necessary to test sorbent batches on a bench scale rather than accepting specific ratios of sorbent-to-waste as constant. Typically when fly ash or kiln dust is being used to sorb an oil sludge (50% oil, 20% water), soil, fly ash, or kiln dust ratios of 1:1 (absorbent-to-sludge) up to 2.5:1 would be satisfactory. In field practice, extra sorbent is usually supplied. A program for testing sorbed waste for release of free liquid should be a standard part of sorption operations.

The ideal sorbent is an inert, nondegradable, nonreactive material. Though some sorbents are relatively inert, undesirable, or even hazardous, reactions can occur unless attention is paid to the potential for waste and sorbent to react. Table 2-4 lists a few of the possible reactions that should be considered when selecting sorbents.

TABLE 2-1.   TYPICAL PHYSICAL AND CHEMICAL PROPERTIES OF
COMMONLY USED NATURAL SORBENTS

Sorbent	Bulk density (kg/m^3)	Cation-exchange capacity (meq/100 gms)	Anion-exchange (meq/100 gms)	Slurry pH	Major mineral species present
Fly ash, acidic	1187	--	--	4-5	Amorphous silicates, hematite, quartz, mullite, free carbon.
Fly ash, basic	1187	--	--	9-10	Calcite, amorphous silicates, quartz, hematite, mullite, free carbon.
Kiln dust	641-890	--	--	9-11	Calcite, quartz, lime (CaO) anhydrite.
Limestone screenings	--	--	--	6-7	Calcite, dolomite.
Clay minerals (soils)	1519			--	Various (e.g., illite)
Kaolinite		5-15	6-20		Can be relatively pure kaolonite.
Vermiculite		100-500	4	--	Can be relatively pure.
Bentonite		100-120	--	--	Smectite, quartz, illite, gypsum, feldspar, kaolinite, calcite.
Zeolite	1543	100-300	--	--	Zeolite (e.g., heulondite, laumonite, stilbite, chabazite, etc.)

From: Sheih (1979), Haynes and Kramer (1982), Grim and Guven (1978).

TABLE 2-2. NATURAL SORBENTS AND THEIR CAPACITY FOR REMOVAL OF SPECIFIC CONTAMINANTS FROM LIQUID PHASES OF NEUTRAL, BASIC, AND ACIDIC WASTES

Contaminant	Neutral waste (calcium fluoride)		Basic waste (metal finishing sludge)		Acidic waste (petroleum sludge)	
Ca	Zeolite	(5054)*	Illite	(1280)	Zeolite	(1390)
	Kaolinite	(857)	Zeolite	(1240)	Illite	(721)
			Kaolinite	(733)	Kaolinite	(10.5)
Cu	Zeolite	(8.2)	Zeolite	(85)	Zeolite	(5.2)
	Kaolinite	(6.7)	Kaolinite	(24)	Acidic F.A.	(2.4)
	Acidic F.A.†	(2.1)	Acidic F.A.	(13)	Kaolinite	(0)
Mg	Basic F.A.	(155)	Zeolite	(1328)	Zeolite	(746)
			Illite	(1122)	Illite	(110)
			Basic F.A.	(176)	Basic F.A.	(1.7)
Zn					Zeolite	(10.8)
					Vermiculite	(4.5)
					Basic F.A.	(1.7)
Ni			Zeolite	(13.5)		
			Illite	(5.1)		
			Acidic F.A.	(3.8)		
F	Illite	(175)	Kaolinite	(2.6)	Illite	(9.3)
	Kaolinite	(132)	Illite	(2.2)	Acidic F.A.	(8.7)
	Acidic F.A.	(102)			Kaolinite	(3.5)
Total CN−					Illite	(12.1)
					Vermiculite	(7.6)
					Acidic F.A.	(2.7)
COD	Acidic F.A.	(690)	Illite	(1744)	Vermiculite	(6654)
	Illite	(108)	Acidic F.A.	(1080)	Illite	(4807)
			Vermiculite	(244)	Acidic F.A.	(3818)

* Bracket represents sorbent capacity in micrograms of contaminant removed per gram of sorbent used. After Sheih (1979) and Chan et al. (1978).
† F.A. = fly ash.

TABLE 2-3.   SYNTHETIC SORBENTS USED WITH HAZARDOUS WASTES

Sorbent	Waste treated effectively
Activated alumina	Sorbs fluoride in neutral wastes
Activated carbon	Sorbs dissolved organics
Hazorb*	Sorbs water and organics
Locksorb†	Oil emulsions
Imbiber beads‡	Inert spirits-type liquids (cyclohexane)

* Product of Diamond Shamrock Corp.

† Product of Radecca Corp., Austin, TX.

‡ Product of Dow Chemical Co., Midland, MI.

Sources: Product literature, Pilie et al. (1975), and Shieh (1979).

TABLE 2-4.   UNDESIRABLE SORBENT/WASTE REACTIONS

Sorbent	Waste type	Reaction
Acidic sorbent	Metal hydroxide	Solubilizes metal
Acidic sorbent	Cyanide	Releases hydrogen cyanide
Acidic sorbent	Sulfide	Releases hydrogen sulfide
Alkaline sorbent	Ammonium compounds	Releases ammonia gas
Alkaline sorbent (with carbonates such as calcite or dolomite)	Acid waste	Releases carbon dioxide, which can cause frothing
Carbonaceous sorbent (carbon, cellulose)	Oily waste	May create pyrophoric waste
Siliceous sorbent (soil, fly ash)	Hydrofluoric acid	May produce soluble fluorosilicates

2.1.1.2 Usefulness

Sorption has been widely used to eliminate free water and improve handling. Some sorbents have been used to limit the escape of volatile organic compounds. Sorbents may also be useful in waste containment when they modify the chemical environment and maintain the pH and redox potential to limit the solubility of the waste.

2.1.1.3 Limitations

Sorption eliminates the bulk flow of wastes from the site, but in many cases leaching of waste constituents from the sorbent can be a significant source of pollution. Sorbents are widely used in lined landfills to eliminate or control the pressure head on the liner, but the liner is the major protection for the surrounding environment.

2.1.1.4 Equipment Requirements

Sorption of wastes requires only that the waste be mixed with the sorbent. This can be done with nothing more than a mixing pit and a backhoe. More elaborate equipment such as a pug mill or ribbon blender can be used if better quality control is needed and if other materials handling equipment (pumps or conveyors) is available.

2.1.1.5 Applications

Most large, hazardous waste landfills are currently using sorption to satisfy requirements prohibiting burial of liquids. A discussion of successful application of sorption in waste disposal is presented in Morgan et al. (1982) and summarized in U.S.EPA (1984). Nineteen million liters (5 million gal) of oil sludge from a former refinery site was landfilled onsite after treatment with cement kiln dust. The process required 3.71×10^7 kg (40,939 tons) of kiln dust. The mixing was done primarily with standard construction equipment at a cost of approximately \$15 per cubic meter.

2.1.2 Lime/Fly Ash Pozzolan Treatment Process

2.1.2.1 General

Pozzolanic materials are those that set to a solid mass when mixed with hydrated lime. Natural pozzolanic materials (called pozzolana) consist of

either volcanic lava masses (tuff) or deposits of hydrated silicic acid of
mostly organic origin (e.g., diatomaceous earth); these are the "natural cements" used by the Romans to produce their famous, long-enduring aqueducts.
Artificial pozzolana are materials such as blast- furnace slag, ground brick,
and some fly ashes from powdered coal furnaces. A common feature of all
pozzolana is the presence of silicic acid (i.e. silicic mineral components
that can react with lime) and frequently appreciable levels of aluminum
oxide. Portland cement differs from pozzolana in that it is a defined mixture of powdered oxides of calcium, silica, aluminum, and iron which result
from the kiln burning (at 1400–1500° C) of raw material such as limestone and
clay (marl).

Solidification/stabilization of waste using lime and pozzolanic material
requires that the waste be mixed with a carefully selected, reactive fly ash
(or other pozzolanic material) to a pasty consistency. Hydrated lime (calcium hydroxide) is blended into the waste-fly ash mixture. Typically, 20 to
30% lime is needed to produce a strong pozzolan. The resulting moist material is packed or compressed into a mold to cure or is placed in the landfill
and rolled.

Standard testing systems (ANSI/ASTM C-311-77) and standard specifications (ASTM C618-80) exist for pozzolanic materials, especially for fly ash
(ASTM 1973). The specifications take into account both the chemical composition ($\%SiO_2$, $\%SO_3$, and moisture content) and physical properties (fineness,
pozzolanic activity with lime, specific gravity). By using fly ash that
meets the specification for a bituminous coal fly ash (Type F) or a sub-
bituminous coal fly ash (Type C), pozzolanic activity greater than a specified minimum can be guaranteed. Type C fly ashes have enough lime (more than
10% $Ca(OH)_2$) that they are not only pozzolanic but are also self-cementing.

2.1.2.2 Usefulness

Lime/fly ash treatment is relatively inexpensive, and with careful selection of materials an excellent solid product can be prepared. In general,
fly ash/lime solidified wastes are not considered as durable as pozzolan-
Portland cement composites (Malone et al. 1980). Leaching losses from
pozzolan-waste materials have been considered to be relatively high compared
with those for pozzolan-Portland cement waste products (Malone et al. 1983).
In diffusion-type leach testing of a variety of solidified waste produced
from a standard metal-rich waste, the lime-fly ash based material prepared
from a metal solution or a liquid sludge showed levels of containment that
were as good as any pozzolan-Portland cement treated waste. However, the
sample of lime/fly-ash-treated waste disintegrated in the leaching solution
(Cote and Hamilton 1983).

Table 2-5 estimates the quantity of additives required per unit volume
of waste for adequate treatment of six different waste types. This table is
furnished to provide an example of an application, not design information.

Note that when waste lime was used, the materials requirement increased 60%
to 70%. Bentonite addition reduced substantially the amount of fly ash
required.

TABLE 2-5. APPROXIMATE REAGENT REQUIREMENTS FOR SOLIDIFICATION OF VARIOUS
WASTE TYPES USING LIME AND FLY ASH*

Waste	Commercial lime (kg/ℓ)	Waste lime (kg/ℓ)	Lime, fly ash, and bentonite† (kg/ℓ)
Spent brine	3.2	5.4	2.2
Metal hydroxide sludge	2.9	5.6	1.1
Copper pickle liquor sludge	1.8	2.6	0.7
$FeCl_2$ pickle liquor sludge (>1.5% HCl)	2.5	4.0	1.9
Sulfuric acid plating waste (>15% H_2SO_4)	3.0	5.2	2.3
Oily metal sludge (oil and grease)	0.6	0.84	0.54

* After Stanczyk et al. (1982).
† Proportions not specified.

2.1.2.3 Limitations

Common problems with lime-pozzolan reactions involve interference with
the cementitious reaction that prevents bonding of materials. The bonds in
pozzolan reactions depend on the formation of calcium silicate and aluminate
hydrates. A number of materials (such as sodium borate, calcium sulfate,
potassium bichromate, and carbohydrates) can interfere with this reaction.
Oils and greases can also physically interfere with bonding by coating waste
particles. The cementing system is strongly alkaline and can react with certain waste to release undesired materials such as gas or in leachate.

2.1.2.4 Equipment Requirements

The use of the lime/fly ash pozzolan processes requires more complex
equipment than systems using sorbent materials only. In one treatment system

used for open sludge ponds, fly ash is mixed with a waste using a backhoe to form a moist mass that can be easily handled with a shovel. The waste/fly ash mixture is then loaded onto a weighing conveyor, and a metered amount of lime is added. The mixture is run through a pug mill and loaded for placement in a landfill. Other systems pump the sludge directly into a pug mill or ribbon blender, where the reagents are blended; they then pump the treated product directly to the final disposal area.

2.1.2.5 Applications

Lime/fly ash stabilization/solidification systems have been successfully used in managing hazardous wastes. However the containment performance generally is such that a hazardous waste would still be classed as hazardous after processing. Lime/fly ash-sorbent-based landfills have been established using liner and monitoring systems to ensure safe disposal.

2.1.3 Pozzolan-Portland Cement Systems

2.1.3.1 General

A wide variety of treatment processes incorporate Portland cement as a binding agent. Pozzolanic products (materials with fine-grained, noncrystalline, reactive silica) are frequently added to Portland cement to react with any free calcium hydroxide and thus improve the strength and chemical resistance of the concrete-like product. In waste solidification, the pozzolanic materials (such as fly ash) are often used as sorbents. Much of the pozzolan in waste processing may be inactivated by the waste. Any reaction that does occur between the Portland cement and free silica from the pozzolan adds to the product strength and durability.

Waste-solidifying formulations based on Portland and pozzolan-Portland systems vary widely, and a variety of materials have been added to change performance characteristics. These include soluble silicates (Falcone et al. 1983), hydrated silica gels, and clays such as bentonite, illite, or attapulgite. Approximate reagent requirements for some example applications are given in Table 2-6.

The types of Portland cement used for solidification can be selected so as to emphasize a particular cementing reaction (Bogue 1955). Five major types of Portland cement are commonly produced:

a. Type I is the typical cement used in the construction industry. It constitutes more than 90% of the cement manufactured in the United States.

TABLE 2-6.   APPROXIMATE PORTLAND CEMENT AND FLY ASH REQUIRE-
MENTS FOR SOLIDIFICATION OF VARIOUS WASTE TYPES*

Waste	Cement/fly ash (kg/ℓ waste)
Spent brine	3.8
Metal hydroxide sludge	2.4
Copper pickle liquor sludge	1.9
$FeCl_2$ pickle liquor sludge (>1.5% HCl)	3.5
Sulfuric acid plating waste (>15% (H_2SO_4)	3.8
Oily metal sludge	0.96

* After Stanczyk et al. (1982). The proportion of portland
cement to fly ash was not given.

b.   Type II is designed to be used in the presence of moderate sulfate
concentrations (150 to 500 mg/kg), or where moderate heat of hydration is required. Type II has a low-alumina-content (less than 6%
Al_2O_3) cement.

c.   Type III has a high early strength and is used where a rapid set is
required.

d.   Type IV develops a low heat of hydration and is usually prescribed
for large-mass concrete work. This type typically has a long set
time.

e.   Type V is a special low-alumina, sulfate-resistant cement used with
high sulfate concentrations (i.e more than 1500 mg/kg).

Type I Portland cement is widely used for waste solidification due to
its availability and low cost. Types II and V have been used to a limited
extent. They offer the advantage of having relatively low tricalcium aluminate content. Higher aluminum-content cement can undergo a rapid reaction
with sulfates (Na_2SO_4, K_2SO_4, $(NH_4)_2SO_4$, and $MgSO_4$) from a waste or surrounding ground water to form crystals of hydrated calcium alumino-sulfate.
The reaction products occupy a much larger volume than the original

calcium aluminate hydrate and the expansion cracks the curing waste/concrete mass.

Cement/fly ash processes typically are used in conjunction with sorbents or other additives which decrease the loss of specific hazardous materials from the rather porous, solid products. Such adaptations of the technology are also often necessary because some materials inhibit the binding action in Portland cement. Additives used in Portland cement have included:

a.   Soluble silicates, such as sodium silicate or potassium silicate. These agents will generally "flash set" Portland cement to produce a low-strength concrete. Research with soluble silicates indicates that these materials are beneficial in reducing the interference from metal ions in the waste solution (Columbo and Neilson 1978; Falcone et al. 1983).

b.   Selected clays to absorb liquid and bind specific anions or cations. Work with bentonite as an additive indicates that they reduce the amount of absorbent required in low-solids mixtures (Stanczyk et al. 1982).

c.   Emulsifiers and surfactants to allow the incorporation of immiscible organic liquids. Research in the nuclear waste field has indicated that waste turbine oil and grease can be mixed into cement blends if dispersing agents are used and if the proper mixing system is employed, but process details were not discussed (Phillips 1981).

d.   Proprietary absorbents that selectively bind specific wastes. These materials include carbon, silicates, zeolitic materials, and cellulosic sorbents; they hold toxic constituents and are encapsulated with the waste.

e.   Lime (CaO) to raise the pH and the reaction temperature and thereby improve setting characteristics.

2.1.3.2 Usefulness

Cement-based solidification and stabilization systems have proved to be some of the most versatile and adaptable methods. Waste/concrete composites can be formed that have exceptional strength and excellent durability, and that retain wastes very effectively (Malone et al. 1980). The addition of selected sorbents and/or emulsifiers often overcomes the problem of pollutant migration through the rather porous solid matrix and consequently lowers the leaching losses from the treated wastes.

2.1.3.3 Limitations

Pozzolan-Portland cement wastes have limitations that relate to the effects of the waste on the setting (retardation from calcium sulfate, borates, carbohydrates, etc.) and stability of the silicates and aluminates that form when portland cement hydrates. Additionally other materials such as oil and grease or large amounts of soft, fine wastes can prevent bonding of particles in the waste and lower strength. Acidic or acid-producing materials such as sulfides can react with carbonate and hydroxides and destroy concrete after setting has occurred.

The very high alkalinity of hydrating Portland cement can cause the evolution of ammonia gas if ammonium ion is present in abundance in the waste. Some metals have increased solubility at the very high pH's that occur in the cement hydration reaction (e.g. nickel, lead, and zinc).

2.1.3.4 Equipment Requirements

Commercial cement mixing and handling equipment can generally be used with wastes. Weighing conveyors, metering cement hoppers, and mixers similar to concrete batching plants have been adapted in some operations. Unless severe corrosion occurs, no adaptation of equipment is required. Where extremely dangerous materials are being treated, remote-control, in-drum mixing equipment such as that used with nuclear waste can be employed.

2.1.3.5 Applications

A number of commercial solidification vendors are currently operating using variations of pozzolan-Portland cement systems. Many use specific sorbents, additives, and proprietary formulations developed to answer the needs of specific clients.

2.1.4 Thermoplastic Microencapsulation

2.1.4.1 General

Thermoplastic microencapsulation has been successfully used in nuclear waste disposal and can be adapted to special industrial wastes. The technique for isolating the waste involves drying and dispersing it through a heated, plastic matrix. The mixture is then permitted to cool to form a rigid but deformable solid. In most cases it is necessary to use a container such as a fiber or metal drum to give the material a convenient shape for transport. The most common material used for waste incorporation is asphalt; but other materials such as polyethylene, polypropylene, wax, or elemental sulfur can be employed for specific wastes where complete containment is important and cost is not a limiting factor.

2.1.4.2 Usefulness

The major advantage that thermoplastic (asphalt) encapsulation offers is the ability to solidify very soluble, toxic materials. This is a unique advantage that cement and pozzolan systems cannot claim. If, for example, the wastes are spray-dried salt, there are few useful alternatives to microencapsulation. The asphalt encapsulation process can be used with moist salt and the mixer-extruder can be used to remove (and recover, if necessary) water or other solvents associated with the wastes. Drying the waste results in a substantial weight reduction over the original material and partly compensates for the additional weight of the asphalt matrix.

2.1.4.3 Limitations

Compatibility of the waste and the matrix becomes a major consideration in using thermoplastic microencapsulation. Most matrices employed with wastes are reduced materials (solid hydrocarbons or sulfur) that can react (combust) when mixed with an oxidizer at elevated temperatures. The reaction can be self-sustaining or even explosive if perchlorates or nitrates are involved.

Other compatibility problems relate to unusual softening or hardening of the waste/matrix mix. Some solvents and greases can cause asphalt materials to soften and never become rigid solids. Borate salts can cause hardening at high temperatures and can stall or clog mixing equipment. Xylene and toluene diffuse quite rapidly through asphalt.

Salts that partially dehydrate at the elevated temperatures used in mixing can be a problem. Sodium sulfate hydrate, for example, will lose some water during asphalt incorporation and if the waste/asphalt mix containing the partially dehydrated salt is soaked in water, the mass will swell and crack due to rehydration. This outcome can be avoided by eliminating easily dehydrated salts or by coating the outside of the asphalt/waste mass with pure asphalt (Doyle 1979). Chelating and complexing agents (cyanides and ammonium compounds) in waste have been shown to seriously compromise the containment of heavy metal wastes (Rosencrance and Kulkarni 1979). If care is taken to pretreat the waste to eliminate oxidizers and destroy complexing agents, the containment of the waste in asphalt is superior to pozzolan or pozzolan-Portland cement solidification.

Thermoplastic encapsulation requires complex, specialized mixing equipment and a trained operations staff to ensure safe, consistent operation. The requirement for drying the waste and melting the matrix material makes

the power consumption for waste solidification quite high compared with that
for pozzolan and pozzolan-Portland cement systems.

2.1.4.4   Equipment Requirements

Specialized equipment is required to ensure thorough mixing of the viscous material under controlled temperature conditions. The mixers or extruders used in waste solidification are similar to those used in the plastic
industry where coloring and filler materials are generally added to raw
plastics. When hazardous wastes are treated, the waste materials replace the
filler. Temperatures ranging from 130° to 230° C are used during mixing.

Screw-extruders that are routinely used in preparation of plastics for
molding are the major type of equipment used in waste microencapsulation.
These systems have staged heating and kneading of the waste and matrix material to ensure homogeneous blending of waste and matrix. Waste treatment
systems are adapted from standard extruders by adding fume control, safety
equipment interlocks, and systems for handling wastes without exposing the
operators to undue hazard.

2.1.4.5   Applications

Thermoplastic microencapsulation has been widely used in nuclear waste
disposal, and application to industrial waste disposal has been projected,
for instance, in disposal of arsenical wastes. Success with nuclear waste
disposal has been well documented (Doyle 1979).

2.1.5   Macroencapsulation or Jacketing Systems

2.1.5.1   General

Macroencapsulation systems contain potential pollutants by bonding an
inert coating or jacket around a mass of cemented waste or by sealing them in
polyethylene-lined drums or containers. This type of waste stabilization is
often effective when others are not because the jacket or coating of the outside of the waste block completely isolates the waste from its surroundings.
The waste may be stabilized, microencapsulated, and/or solidified before
macroencapsulation so that the external jacket becomes a barrier designed to
overcome the shortcomings of available treatment systems.

A macroencapsulation system that has been proposed for use with hazardous wastes involves drying the wastes and bonding the dried material into a
compressed block using polybutadiene. Polymerization of the binder requires
heating the waste sample to 120°C to 200°C under slight pressure. The block
is placed in a mold and surrounded with powdered polyethylene. The

polyethylene is then fused into a solid jacket using heat and pressure. In the proposed system, a 3.5-mm-thick jacket would be fused over a 450-kg block. The polyethylene would amount to approximately 4% of the mass by weight (Lubowitz and Wiles 1978).

2.1.5.2 Usefulness

Macroencapsulation can be used to contain very soluble toxic wastes. Leaching of the waste can be eliminated for the life of the jacketing material. This process has been used at remedial sites as drum over-packs to contain weak or leaking drums and containers.

2.1.5.3 Limitations

In some systems, the wastes have to be dried before they are fused into a block, thus increasing the risk of the release of volatile toxics. Furthermore, the waste must not react with the binder or jacket materials at the elevated temperatures required for fusing and forming a jacket. The jacketing material may have to be protected from chemical or photo degradation or physical stresses after disposal. Equipment such as special molds on processing machinery is highly technical and requires highly skilled labor unless loose-fitting over-packs are used.

2.1.5.4 Equipment Requirements

Macroencapsulation requires special molds and heating equipment for fusing the waste and forming the jacket. Molding equipment would have to be custom fabricated for waste handling.

2.1.5.5 Application

Macroencapsulation has been bench tested on a number of different wastes, but it has not been tested in a full-scale operation (Lubowitz and Wiles 1979). Results of bench testing are encouraging, but larger-scale operations have not been pursued.

2.2 Compatibility of Wastes and Treatment Processes

The chemical reactivity of the waste generally controls the selection of waste stabilization/solidification options and its optimization. Table 2-7 summarizes the major chemical considerations that direct the selection of a

TABLE 2-7. COMPATIBILITY OF SELECTED WASTE CATEGORIES WITH DIFFERENT STABILIZATION/
SOLIDIFICATION TECHNIQUES

Waste component	Treatment Type			
	Cement based	Pozzolan based	Thermoplastic microencapsulation	Surface encapsulation
Organics				
Organic solvents and oils	May impede setting, may escape as vapor	May impede setting, may escape as vapor	Organics may vaporize on heating	Must first be absorbed on solid matrix
Solid organics (e.g., plastics, resins, tars)	Good--often increases durability	Good--often increases durability	Possible use as binding agent in this system	Compatible--many encapsulation materials are plastic
Inorganics				
Acid wastes	Cement will neutralize acids	Compatible, will neutralize acids	Can be neutralized before incorporation	Can be neutralized before incorporation
Oxidizers	Compatible	Compatible	May cause matrix breakdown, fire	May cause deterioration of encapsulation materials
Sulfates	May retard setting and cause spalling unless special cement is used	Compatible	May dehydrate and rehydrate causing splitting	Compatible
Halides	Easily leached from cement, may retard setting	May retard set, most are easily leached	May dehydrate and rehydrate	Compatible
Heavy metals	Compatible	Compatible	Compatible	Compatible
Radioactive materials	Compatible	Compatible	Compatible	Compatible

After Malone et al. (1980).

particular waste stabilization/solidification system. Most solidification systems will work under adverse circumstances if adaptations are made in the waste or the processing train. Many compatibility problems can be overcome by specifying pretreatment steps to destroy or tie up some undesirable waste constituent.

2.3 Pretreatment Techniques for Waste Solidification

Pretreatment systems, which overlap with stabilization and sorption processes, can be used to achieve a number of results that condition the waste to ensure better and more economical containment after the remaining materials have been stabilized and solidified. These include:

a. Destruction of materials (such as acids or oxidizers) that can react with solidification reagents (lime or Portland cement).

b. Reduction of the volume of waste to be solidified (using processes such as settling or dewatering).

c. Chemical binding of specific waste constituents to solid phases added to scavenge toxic materials from solution and hold them in solids.

d. Techniques for improving the scale on which waste processing can be done--for example, bulking and homogenizing waste to allow a single solidification system to be used without modification on a large volume of waste.

Neutralization, oxidation or reduction, and chemical scavenging stabilize the waste in that they bring the chemical waste into an inert or less soluble form. Dewatering, consolidation, and waste-to-waste blending are also useful pretreatment methods which reduce the waste volume or numbers of different waste forms requiring treatment.

2.3.1 Neutralization

Most binder systems can operate well with wastes that are approximately neutral (pH 7.0), though alkaline wastes are also desirable in many circumstances where it is necessary to minimize solubility. Many toxic metals are amphoteric (show increased solubility at both high and low pH's) and by adjusting the pH it is possible to produce a minimum amount of metal in the supernatant liquid (Figure 2-2). Depending upon the metals present, the optimum pH is usually between 9.5 and 11, which offers the advantage of requiring less treatment of the discharged water produced by subsequent dewatering.

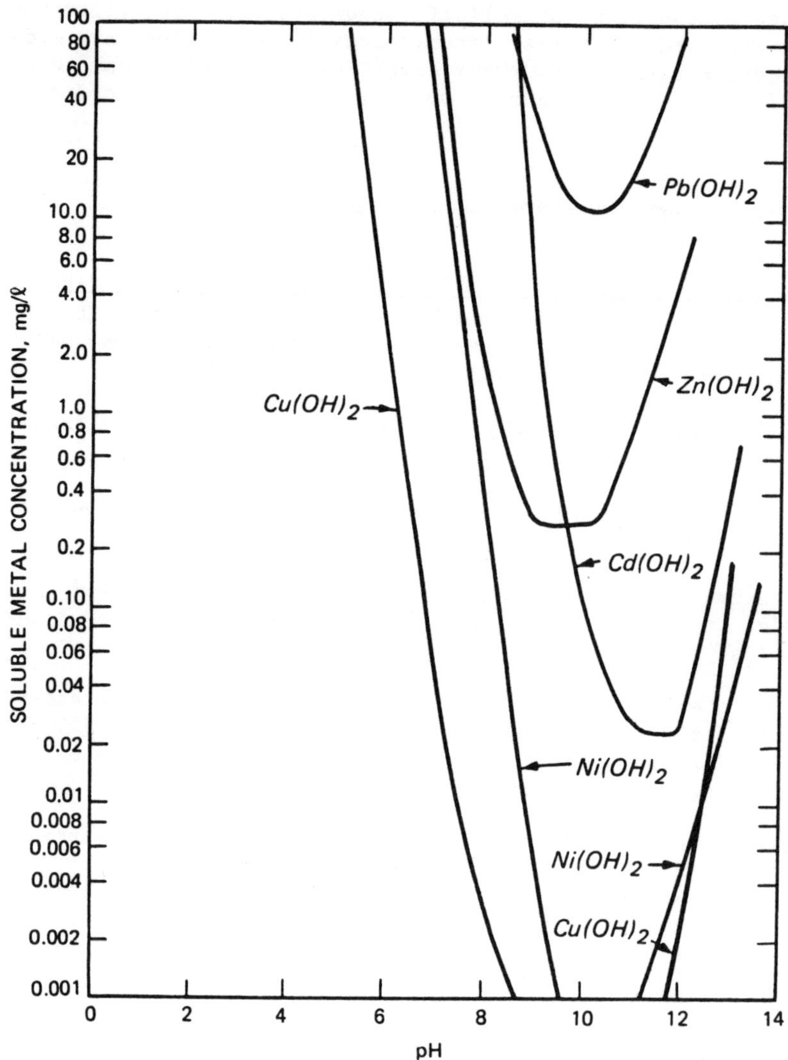

Figure 2-2. Theoretical solubilities of selected amphoteric metal hydroxides.

The selection of a neutralization agent is important in reducing the amount of leachable material in the waste. A common base used in neutralization is sodium hydroxide; however, resulting sodium salts typically have very high solubilities, and the supernatant liquid and sludge produced in neutralization will have higher levels of soluble materials than if other bases were used. Calcium hydroxide or calcium carbonate may be a better choice for neutralization because the resulting salts are generally less soluble than sodium salts. Calcium hydroxide and calcium carbonate also are available inexpensively in a relatively pure form.

Calcium carbonate offers the advantage that many carbonate metal salts are insoluble (for example, lead carbonate has a low solubility) and the carbonates are compatible with both Portland cement and pozzolan material. However, neutralization with carbonates can cause frothing due to evolution of carbon dioxide. Excess calcium hydroxide in Portland cement is thought to make the material more reactive to sulfate attack (Ramachandran 1976; Bogue 1955). In pozzolan materials, excess lime would react with free silica and should not pose a problem. DeRenzo (1978) and EPA (1982) discuss equipment needs and design for precipitation systems that use neutralization.

2.3.2 Oxidation/Reduction

In some cases, the most insoluble form of a toxic constituent is associated with a specific oxidation/reduction state. Iron, for example, is much less soluble at alkaline pH's in its oxidized state. Chromium in its oxidized state (Cr^{+6}) is more mobile than the reduced chromium (Cr^{+3}) in an alkaline solution.

The usual technique involved in oxidizing or reducing hazardous materials to a stable, insoluble state involves addition of hypochlorite, chlorate, persulfate chlorine or peroxide (oxidizers), or sulfides, ferrous salts, or sulfur dioxide gas (reducing agents). A discussion of oxidation-reduction systems along with equipment design is given in DeRenzo (1978), U.S. EPA (1982), and Nemerow (1971).

Oxidation of toxic organic constituents using UV-ozone or chemical oxidizers can lower the toxicity of the final product and the amount of fixation reagents required. And, of course, incineration can be considered an oxidative pretreatment because it usually generates a residue or scrubber sludge residual which often requires further treatment and disposal.

2.3.3 Chemical Scavenging

Chemical scavenging involves the use of some solid chemical agent to chemsorb or react with and bind up some specific waste constituent. This procedure is significantly different from adsorption, where the goal of the operation is to soak up free liquid and adsorb ions in solution. Chemical scavenging agents, many of which are proprietary, include chemically active adsorbents (for example, activated carbon), specific types of clays, ion exchange resins, natural and artificial zeolites, silica gels, and finely divided metal hydroxides (ferric hydroxide or aluminum hydroxide).

In all cases, an attempt should be made to ensure that the scavenging agent is compatible with the waste and the solidification reaction. Selected use of scavengers can greatly reduce the requirement to treat the discharge water after dewatering of the wastes. Scavenging can also assist in complicated treatment problems. For example, in the solidification of a paint stripping waste that contained phenol and a chrome (Cr^{+3}) paint pigment, attempts to oxidize the phenol with permanganate also oxidized the chromium and increased its leaching. Without treatment, the phenol leaching rate was unacceptable. A suitable scavenging material such as formaldehyde would be able to react with the phenol and reduce its leaching rate while leaving the chromium in its lower (less soluble) valance state.

Scavenger materials often improve solidification performance without adding appreciably to the volume of the waste. Scavenging materials, such as flocculating agents like polyelectrolytes or aluminum hydroxide or iron hydroxide, also assist in waste concentration or dewatering by improving the settling characteristics of fine-grained wastes in suspension.

2.3.4 Dewatering and Consolidation

Solidification systems can be made more economical by reducing the volume of waste to be treated by dewatering. Dewatering can also be used to lower the water content of the solidified waste which, in turn, lowers the leachability of the waste. A strong correlation is found between the leachability and the water content of solidified waste, which indicates diffusion of contaminants probably occurs through the pore liquid in solidified waste matrices (Cote and Hamilton 1983); thus a dryer, solidified product will have lower contaminant mobility.

Design of dewatering systems is discussed in DeRenzo (1978) and EPA (1982). A comparison of stabilization of dewatered and undewatered industrial sludge reported by Cote and Hamilton (1983) indicated the final volume after dewatering for a typical metal hydroxide waste was about 35% of the initial volume. Dewatering the metal waste increased containment (as measured by diffusion testing) and decreased costs due to lower fixation reagent requirements and less final product requiring disposal.

2.3.5 Waste-to-Waste Blending

Except in the case of extremely toxic wastes, it is generally not practical to set up stabilization/solidification systems to handle small volumes of waste, especially if the wastes vary significantly in their compatibility and containment performance in a selected process. At some point in the remedial action planning it is necessary to mix or bulk wastes in order to obtain sufficient volume for efficient pretreatment, stabilization, and/or solidification. If the nature of the waste permits bulk mixing before a treatment, then a simpler, large-scale pretreatment operation can be undertaken and a large mass of homogeneous material (feed stock) will be available for processing. Guidelines for mixing or bulking of wastes are given in Chemical Manufacturers Association (1982) and in Hatayma et al. (1981). The water separation and blending systems depend on identifying materials that have similar composition and pH and oxidation/reduction characteristics.

This same type of waste classification and blending is needed to develop feed stocks for pretreatment as well as to provide economy by processing large volumes.

When the reactions between different types of wastes (for example, acids and bases, or oxidizers and reducers) can be controlled and no unwanted side reactions occur (such as generation of H_2S or HCN gas), the waste blending becomes a treatment step where the wastes themselves are treatment reagents. Blended waste can then be further treated if additional pH adjustment or oxidation-reduction treatment is needed.

REFERENCES

American Society for Testing and Materials (ASTM). 1973. Annual Book of ASTM Standards, Part II. American Society for Testing and Materials, Philadelphia, Pennsylvania.

Bogue, R. H. 1955. The Chemistry of Portland Cement. Reinhold Publishing Corp., 2nd ed., 793 pp.

Chan, P. C., et al. 1978. Sorbents for Fluoride, Metal Finishing, and Petroleum Sludge Leachate Contaminant Control. EPA-600/2-78-024, U.S. Environmental Protection Agency, Cincinnati, Ohio. 94 pp.

Chemical Manufacturers Association. 1982. A Hazardous Waste Management Plan. Chemical Manufacturers Assoc., Washington, D.C., Loose-leaf.

Columbo, P., and R. M. Neilson. 1978. Properties of Wastes and Waste Containers. Progress Report No. 7. BNL-NUREG 50837, Brookhaven National Laboratory, Upton, New York.

Cote, P. L., and D. P. Hamilton. 1983. Leachability Comparison of Four Hazardous Waste Solidification Processes. Presented at the 38th Annual Purdue Industrial Waste Conference, LaFayette, Indiana, May 10, 11, 12, 1983. 17 pp.

DeRenzo, D. J. (ed). 1978. Unit Operations for Treatment of Hazardous Wastes. Noyes Data Corp., Park Ridge, New Jersey.

Doyle, R. D. 1979. Use of an Extruder/Evaporator to Stabilize and Solidify Hazardous Wastes. In: Pojasek, R. B (ed.), Toxic and Hazardous Waste Disposal, Vol. 1, Ann Arbor Science Publishers, Ann Arbor, Michigan. pp. 65-91.

Falcone, J. S., Jr., R. W. Spencer, and R. H. Reifsnyder. 1983. Chemical Interactions of Soluble Silicates in the Management of Hazardous Wastes. Draft Report. The PQ Corp., Lafayette Hill, Pennsylvania.

Grim, R. E., and N. Guven. 1978. Bentonites. Elsevier Scientific Publishing Co., New York. 256 pp.

Hatayma, H. K., et al. 1981. Hazardous Waste Compatibility Protocol. California Department of Health Services, Berkeley, Calif., Rept. on Grant R804692010, U. S. Environmental Protection Agency, Cincinnati, Ohio.

Haynes, B. W., and G. W. Kramer. 1982. Characterization of U. S. Cement Kiln Dust. Bureau of Mines Information Circ. 885, U. S. Dept. of Interior, Washington, D.C. 19 pp.

Iadevaia, Rosa, and J. F. Kitchens. 1980. Engineering and Development Support of General Decon Technology and the DARCOM Installation Restoration Program. Task 4. Draft Rept. Atlantic Research Corp. Alexandria, Virginia. 77 pp.

Lubowitz, H. R., and C. C. Wiles. 1978. Encapsulation Technique for Control of Hazardous Materials. In: Land Disposal of Hazardous Waste, Proceedings of 4th Annual Research Symposium, EPA-600/9-78-016, U. S. Environmental Protection Agency, Cincinnati, Ohio. pp. 342-356.

Lubowitz, H. R., and C. C. Wiles. 1979. Encapsulation Technique for Control of Hazardous Wastes. In: Pojasek, R. B. (ed.), Toxic and Hazardous Waste Disposal, Vol. 1, Ann Arbor Science Publishers, Inc., Ann Arbor, Michigan. pp. 198-232.

Malone, P. G., and L. W. Jones. 1979. Survey of Solidification/ Stabilization Technology for Hazardous Industrial Wastes. EPA-600/2-79-056, U. S. Environmental Protection Agency, Cincinnati, Ohio. 41 pp.

Malone, P. G., L. W. Jones, and J. P. Burkes. 1983. Application of Solidification/Stabilization Technology to Electroplating Wastes. In: Land Disposal of Hazardous Waste, Proceedings of the 9th Annual Research Symposium, U. S. Environmental Protection Agency, Cincinnati, Ohio. pp. 247-261.

Malone, P. G., L. W. Jones, and R. J. Larson. 1980. Guide to the Disposal of Chemically Stabilized and Solidified Waste. SW-872, Office of Water and Waste Management, U. S. Environmental Protection Agency, Washington, D.C. 126 pp.

Morgan, D. S., J. I. Novoa, and A. H. Halff. 1982. Solidification of Oil Sludge Surface Impoundments with Cement Kiln Dust (Draft Report). Albert Halff Associates, Inc., Dallas, Texas.

Nemerow, N. L. 1971. Liquid Waste of Industry: Theories, Practices, and Treatment. Addison-Wesley, Reading, Massachusetts. 584 pp.

Phillips, J. W. 1981. Applying Techniques for Solidification and Transportation of Radioactive Waste to Hazardous Waste. In: Proceedings of National Conference on Management of Uncontrolled Hazardous Waste Sites, Hazardous Materials Control Research Institute, Silver Spring, Maryland. pp. 206-211.

Pilie, R. J., et al. 1975. Methods to Treat, Control and Monitor Spilled Hazardous Materials. EPA-670/2-75-042, U. S. Environmental Protection Agency, Cincinnati, Ohio. 148 pp.

Ramachandran, V. S. 1976. Calcium Chloride in Concrete. Applied Science Publ. Ltd., London. 216 pp.

Rosencrance, A. B., and R. K. Kulkarni. 1979. Fixation of Tobyhanna Army Depot Electroplating Waste Samples by Asphalt Encapsulation Process. Technical Rept. 7902, U. S. Army Medical Research and Development Command, Ft. Detrick, Maryland. 23 pp.

Sheih, M. S. 1979. The Use of Natural Sorbents for the Treatment of Industrial Sludge Leachate. Ph.D. Dissertation, New Jersey Inst. of Tech., Newark, New Jersey. 144 pp.

Stanczyk, T. F., B. C. Senefelder, and J. H. Clarke. 1982. Solidification/ Stabilization Process Appropriate to Hazardous Chemicals and Waste Spills. In: 1982 Hazardous Materials Spills Conference, Government Institutes Inc., Rockville, Maryland. pp. 79-84.

U.S. EPA. 1985. Handbook for Remedial Action at Waste Disposal Sites (Revised). EPA-625/6-85-006, U.S. Environmental Protection Agency, Cincinnati, Ohio. 497 pp.

U.S. EPA. 1984. Case Studies 1-23: Remedial Response at Hazardous Waste Sites. EPA-540/2-84-002b, Office of Emergency and Remedial Response, U. S. Environmental Protection Agency, Washington, D.C. 637 pp.

3. Physical and Chemical Characterization of Untreated Wastes

3.1 Physical Characterization

The physical characteristics of a waste material are important in determining the handling requirements for a waste. The equipment and methods for
moving, storing, and mixing the waste will be determined by the range of
physical characteristics involved. In many cases initial testing will result
in a decision to introduce a preliminary dewatering or sorption step to provide a more easily handled solid with uniform physical characteristics. Physical characteristics that would be determined include:

   a. Percent moisture (water content)

   b. Suspended solids

   c. Bulk density

   d. Grain-size distribution

   e. Atterberg limits

   f. Cone index or California bearing ratio

   g. Unconfined compressive strength

Obviously some of these characteristics may not be useful because of the conditions of a particular material. If the waste is impounded, the testing
program should be designed to consider the condition of the waste after re-
suspension or partial dewatering or addition of an adsorbent. A detailed
discussion of a range of physical testing procedures applicable to solidification and stabilization of hazardous materials is presented in Bartos and
Palermo (1977).

3.1.1 Percent Moisture or Water Content

Water content is defined as the ratio of the weight of water to the
weight of solids and is expressed as a percentage. The percent moisture or
water content is used to develop requirements for pretreatment (settling,
flocculating, filtering, and absorbing) and for designing solidification

procedures for the treated materials. Procedures for determining water
content are given in Appendix I of U.S. Army (1972) and ASTM D2216-71 (ASTM
1973).

3.1.2 Suspended Solids

The amount of suspended solids is used to determine the materials handling requirements for the waste--that is, to determine if the waste can be
pumped or whether another conveying system should be used. The suspended
solids can also be used to predict volume decrease due to settling (primary
consolidation) or water removal. Table 3-1 gives a typical classification
system for the consistencies of slurried materials based on handling and
processing requirements (Wyss et al. 1980).

TABLE 3-1. HAZARDOUS WASTE CONSISTENCY CLASSIFICATION

Consistency category	Characteristics
Liquid waste	<1% suspended solids,* pumpable liquid, generally too dilute for sludge dewatering operation.
Pumpable waste	<10% suspended solids,* pumpable liquid, generally suitable for sludge dewatering.
Flowable waste	>10% suspended solids,* not pumpable, will flow or release free liquid, will not support heavy equipment, may support high flotation equipment, will undergo extensive primary consolidation.
Nonflowable waste	Solid characteristics, will not flow or release free liquids, will support heavy equipment, may be 100% saturated, may undergo primary and secondary consolidation.

* Suspended solids ranges are approximate.

From Wyss et al. (1980).

Suspended solids (or settleable matter) can be determined using
Method 224F(a) as given in APHA (1971). This method is equivalent to EPA
Standard Method for Settleable Matter (Storet No. 5008G) as given in U.S. EPA
(1979). Settleable matter is usually given in milliliters per liter volume
of waste suspension.

3.1.3 Bulk Density

The bulk density, or bulk unit weight, is the ratio of the total weight (solids and water) to the total volume. These basic data are needed to convert weight to volume in materials handling calculations. Procedures for determining bulk unit weight are given in Appendix II of U.S. Army (1972).

3.1.4 Grain-Size Distribution

The grain-size distribution of an industrial waste becomes important in designing remedial actions. Fine-grained wastes generally present more handling problems and are subject to wind dispersion. Fine-grained wastes also present problems in producing high-strength solidified waste. Large percentages of fines lower the ultimate strength developed in concrete/waste composites.

Grain-size analyses are performed using methods described in Appendix V of U.S. Army (1972) or ASTM D422-63 (ASTM 1973). Preparation of samples for grain-size analysis usually follows specifications given in ASTM D421-58 (ASTM 1973).

3.1.5 Atterberg Limits

The Atterberg limits test determines the water contents of the material at the boundaries between its plastic and liquid states. The plastic limit is the water content at which the waste will start to crumble when rolled into a 3-mm thread under the palm of the hand. The liquid limit is defined as the lowest water content at which the sludge will flow as a viscous liquid. The Atterberg limits are used in classifying fine-grained materials to estimate their properties such as compressibility, strength, and swelling characteristics; these provide an indication of how the material will react when stressed.

A full discussion of the test and the equipment involved is given in Appendices III and IIIA of U.S. Army (1972) and ASTM tests D424-59 and D423-66 (ASTM 1973).

3.1.6 Cone Index

These tests involve forcing a standard cone into a sample of soil or other granular material and determining the resistance offered by the medium being tested. These tests are typically used to examine the ability of a subgrade soil to support a load (trafficability), but they are equally valuable in examining the strength of in-place wastes. Details on the test

procedures and interpretation are given in Sowers and Sowers (1970) and
U.S. Army (1972).

3.1.7 Unconfined Compressive Strength

Unconfined compressive strength can only be measured on samples of cohesive or cemented waste. This type of test involves preparing a cylindrical
specimen and loading it axially to failure. The test load is applied at a
fixed rate of strain and compressive stresses are recorded as loading progresses. Unconfined compressive strength tests are used to determine bearing
capacity and shear strength of cohesive materials. Shear strength is an
important factor in determining the ultimate bearing capacity of the material, embankment stabilities, and pressures on retaining walls holding the
material in place.

The standard test procedure is given in Appendix XI of U.S. Army (1972)
and ASTM Standard Method D2166-66 (ASTM 1973). This type of testing requires
that an average value be determined from a series of multiple samples.

3.2 Chemical Characterization

The requirements for chemically characterizing wastes present at remedial action sites vary widely depending on preliminary information on the
types of waste involved. Any program of chemical analyses and testing should
be designed to discover the following:

a. The degree of hazard involved in handling and treating the wastes.
    These data are used to develop requirements for protective clothing
    and adaptations required for mixing and transporting equipment.

b. The presence of interfering materials that can complicate
    stabilization/solidification. These data are used to develop pre-
    treatment alternatives.

c. The compatibility of wastes that would permit the mixing and consol-
    idation of wastes for pretreatment and stabilization/
    solidification. This type of testing allows more economical opera-
    tion and continuous processing of bulked wastes.

Testing programs oriented toward defining the degree of hazard involved
in a waste material are outlined in U.S. EPA (1980). This type of testing
concentrates on quantification of potential toxicants and screening for priority pollutants. Any program designed to evaluate the containment developed
during stabilization/solidification must be based on consideration of the
bulk composition of the waste. Leach testing of the treated waste will generally concentrate on the most potentially dangerous or soluble compounds
discovered in the waste.

Chemical compounds that can present problems during stabilization/ solidification may be relatively common, nontoxic materials. Oil and grease may interfere with pozzolan-Portland cement based processes. High concentrations of sulfate can cause swelling and spalling of pozzolan-Portland cement solidified wastes. High sulfate concentrations can be reduced by lime addition. The testing and analysis program will vary with the solidification process or processes being considered for use. Table 2-7 lists some of the constituents that can affect the performance of different stabilized/ solidified waste materials and pretreatment options available to alleviate the problem.

Testing procedures for consolidating hazardous wastes have been developed to assist in segregating chemically compatible waste for storage and transportation. These same protocols can be adapted for screening hazardous waste for pretreatment and stabilization/solidification. A general system designed for consolidating drummed waste is given in Chemical Manufacturers Association (1982). A more general compatibility testing procedure and a waste compatibility matrix are available in Hatayma et al. (1981).

REFERENCES

American Public Health Association (APHA). 1971. Standard Methods for the
Examination of Water and Wastewater. Amer. Public Health Assoc., New York,
New York. 874 pp.

American Society of Testing and Materials (ASTM). 1973. Annual Book of ASTM
Standards, Part II, Philadelphia, Pennsylvania.

Bartos, M. J., Jr., and M. R. Palermo. 1977. Physical and Engineering
Properties of Hazardous Industrial Wastes and Sludges. EPA-600/ 2-77-139,
U.S. Environmental Protection Agency, Cincinnati, Ohio. 89 pp.

Chemical Manufacturers Association. 1982. A Hazardous Waste Management
Plan. Chemical Manufacturers Assoc., Washington, D.C., Loose-leaf.

Hatayma, H. K., et al. 1981. Hazardous Waste Compatibility Protocol.
California Department of Health Services, Berkeley, Calif., Rept. on Grant
R804692010, U.S. Environmental Protection Agency, Cincinnati, Ohio.

Sowers, C. B., and G. F. Sowers. 1970. Introductory Soil Mechanics and
Foundations. 3rd ed., The Macmillan Co., London.

U.S. Army, Office, Chief of Engineers. 1972. Laboratory Soils Testing,
Engineer Manual 1110-2-1906, U.S. Army Corps of Engineers, Washington, D.C.

U.S. EPA. 1979. Manual of Methods For Chemical Analysis of Water and
Wastes. EPA-600/4-79-020, U.S. Environmental Protection Agency, Cincinnati,
Ohio. 298 pp.

U.S. EPA. 1980. Test Methods for Evaluating Solid Waste. SW-846, U.S.
Environmental Protection Agency, Washington, D.C. Unpaginated.

Wyss, A. W., et al. 1980. Closure of Hazardous Waste Surface Impoundments.
SW-873, Office of Water and Waste Management, U.S. Environmental Protection
Agency, Cincinnati, Ohio. 92 pp.

4. Selection of Stabilization/Solidification Processes

4.1 Background

In undertaking any remedial action involving stabilization/
solidification at an uncontrolled waste site, a number of problem areas have
to be addressed. These include:

a. <u>Characteristics of the present waste disposal site.</u> The geologic
    and hydrologic setting of the site determines to a great degree the
    feasibility of leaving the treated waste material on the site. An
    action could involve closing a site in place or constructing a new
    facility to contain the solidified waste onsite. Stabilization and
    solidification always increases the volume and mass of material to
    be disposed; therefore, solidification and transportation offsite
    is generally a more expensive option than shipping untreated wastes
    to a hazardous waste landfill.

b. <u>Character and volume of the waste to be stabilized or solidified.</u>
    Wastes that are hazardous due to flammability, corrosivity, reactiv-
    ity, infectiousness, or other properties that would normally exclude
    secure land burial usually cannot be solidified and disposed of by
    landfilling without adequate pretreatment. Wastes which are hazar-
    dous due to toxicity as defined by the Extraction Procedure (EP)
    testing benefit by stabilization and solidification in that it can
    decrease the concentration of toxic material in the EP leachate.
    Wastes that present specific problems (such as escape of volatile
    organics) may not be effectively contained using any economical
    stabilization/solidification technique although new sorbents are
    being developed to overcome these difficulties. Mixed wastes that
    require several pretreatment steps to produce solidification can
    become too expensive to process when costs are compared with those
    for transportation and secure land burial in a RCRA-permitted site.
    Small volumes of waste are often not economical to solidify or
    stabilize. At some sites where the wastes can be most easily
    handled by transportation and burial in a secure landfill, the least
    contaminated residual materials, such as sludges and contaminated
    soils, can be stabilized/solidified and landfilled in place. In
    every case, a cost comparison is a prime concern in examining
    stabilization/solidification options.

c. <u>Degree of hazard involved in handling the waste</u>. The safety requirement for handling wastes in some circumstances is so great that
stabilization/solidification for onsite disposal must be passed over
to reduce long-term exposure to site personnel and inhabitants in a
local area. Again, in such cases, marginally contaminated, high
volume materials (soils or absorbents) may be the only material
solidified and left onsite, although the bulk of the waste may be
fixed to make its handling or its ultimate disposal safer and more
economical.

d. <u>Possible site modifications to provide for ultimate disposal</u>. Where
the waste site in an unmodified condition would be unacceptable due
to an undesirable geologic or hydrologic setting, engineering modification such as liners and drainage control may overcome site problems. Waste solidification can provide part of the required containment, and site modifications can complete the safe containment
program.

A definition of how stabilization/solidification is to be employed at a
specific remedial action site should result from these considerations. For
instance, the wastes may be solidified and ultimate disposal involve burial
onsite, or contaminated soils or absorbers may be solidified and buried onsite, while the waste themselves are transported. If solidification systems
alone do not provide a high enough degree of protection, it may be necessary
to modify the site to provide improved waste isolation.

Once decisions have been made on the role of solidification and the
types and quantities of material to be solidified, it is possible to develop
specifications for the stabilized/solidified waste. The nature of the waste
and the containment properties required of the stabilized/solidified material
determine the type of processing that can be used.

4.2 Specifications for Stabilized/Solidified Wastes

Specifications for stabilized/solidified wastes can include these
characteristics:

a. Leachability of waste components to contacting water.

b. Free liquid content of waste.

c. Physical stability of waste under burial conditions.

d. Reactivity of waste.

e. Ignitability or pyrophoricity.

f. Susceptibility to biodegradation.

g. Strength or bearing capacity of the waste.

h. Permeability of the waste.

i. Durability of the waste under conditions of surface exposure
    (freeze-thaw and wet-dry testing).

No standards for testing of stabilized/solidified waste have been developed. The specification and testing procedures outlined in this section are a minimum suggested testing program, and the specifications indicated are desirable but not mandated.

4.2.1 Leachability

A wide variety of extraction or leaching tests have been proposed for hazardous waste. None have been totally satisfactory for all types of stabilized/solidified wastes (Lowenbach 1978). Three major types of test procedures are usually involved in any evaluation procedure: Testing for regulatory purposes, testing for maximum hazard assessment, and testing for design of landfill facilities.

The regulatory testing procedures involve mixing the waste with some specified amount of extracting fluid (usually dilute acid or distilled water) and analyzing the resulting extractant for a required number of potential contaminants. Regulations may require that the waste be tested as a monolith or broken in a specific procedure such as the EPA Structural Integrity Procedure (Federal Register 1980, page 33128). The sample may or may not have to be sieved prior to testing. A set of criteria usually based on multiples of concentrations specified in the Primary or Secondary Drinking Water Standards are provided. Regulatory tests vary widely but the most accepted is the EPA Extraction Procedure or EP Toxicity Test Procedure (40 CFR 261.24, Appendix II, Federal Register 1980, page 33127). The maximum concentration of contaminants allowable in the EP leachate is 100 times the National Interim Primary Drinking Water Standard. Leachates containing greater than this level cause the waste to be defined as hazardous and be subject to all regulatory provisions; leachates with lower levels of all listed contaminants cause the waste to be classified as nonhazardous and thus not covered by these regulations.

Any test developed to assess the maximum hazard posed by a waste that is landfilled must be a generally flexible procedure that can handle a wide variety of wastes with a broad range of contaminant concentrations. This type of test assesses the maximum concentration of contaminants that can be developed in water contacting the wastes to be disposed. Procedures can be varied with the type of waste being tested. The waste is ground to a fine powder to ensure that a maximum surface area is presented to the contacting liquid. The ratio of waste to leaching medium is varied in such a way as to achieve a solution saturated with respect to compounds in the waste. Thus, the leach liquid may be separated from the waste and added to fresh wastes until the concentration of contaminants in the leachate no longer increases.

If the composition of the waste indicates that common ion effects are preventing some potential contaminants from appearing in equilibrium concentrations as they would if the waste contained only the pure contaminant compound, the waste can be leached with successive volumes of fresh leaching medium until a maximum concentration for the contaminant of interest is found. This type of test has no fixed level for rejection of the waste as hazardous, but concentrations of potential contaminants that go above the levels considered harmful to human health and the environment are noted. One example of this type of protocol is the Maximum Possible Concentration (MPC) Test outlined in Malone et al. (1980).

Leaching tests developed for engineering purposes attempt to develop leachate that duplicates that obtained from the landfilled wastes. This type of test is used to provide a basis for designing leachate treatment systems for proposed landfills and in evaluating the performance of treated (solidified) wastes developed for landfill disposal.

Several engineering tests have been proposed. The Solid Waste Leaching Procedure (SWLP) tumbles ground or monolithic waste samples in ten volumes of water per unit weight of sample (Garrett et al. 1981). A minimum of four successive extractions are performed to determine the changing character of the leachate.

Another proposed test for solidified industrial wastes is the Uniform Leaching Procedure (ULP) outlined in Malone et al. (1980) and discussed in detail in American Nuclear Society (1981), Cote and Isabel (1983), and Cote and Hamilton (1983). The ULP is a static leaching test that assumes that diffusion from the surface of a solidified waste is the major mechanism for contaminant transfer to surrounding water. A specific volume of waste is exposed to a fixed volume of water (or other leaching medium) that is changed on a regular schedule. If the surface area of the emplaced waste is known, estimates of the loss of contaminants from diffusion can be developed. Concentrations of contaminants in leachate can be used to postulate the environmental impact of the emplaced wastes.

The ULP and other static leaching tests for industrial wastes have been criticized because of the low reproducibility (only one order of magnitude in a leachability index) and the low levels of contaminants that must be quantified in the leachate (Cote and Isabel 1983). These problems can be overcome by concentrating contaminant from the leachate or by using tracer or surrogate compounds that can be added to the waste in appreciable quantities. Surrogate compounds can be selected that mimic the behavior of the toxic components in the waste and are easily determined at low concentrations; however, these newer methods of increasing the reliability of leaching tests have not been widely used or accepted.

Current guidelines for solidified low-level nuclear waste state that waste developed for land burial must have a leachability index greater than eight when measured using the standard static leaching test proposed by the American Nuclear Society (Nuclear Regulatory Commission 1983). Pozzolan-based and pozzolan-Portland-cement-based solidified industrial wastes prepared from dewatered industrial-type sludge all had leachability indices of ten and above for arsenic, cadmium, chromium, and lead (Cote and Hamilton 1983). Industrial waste can be prepared to meet the Nuclear Regulatory Commission criteria if the waste is properly pretreated to eliminate highly leachable constituents and solidified using carefully developed procedures.

Any procedures for evaluating the leachability of stabilized/solidified waste should include all three types of testing: regulatory, risk assessment, and engineering design tests. The data developed in each type of test are useful for specific purposes such as delisting the waste as nonhazardous and determining the degree of containment needed in the disposal site.

4.2.2 Free Liquid Content

Free liquids in solid wastes are defined as liquids which readily separate from the solid portion of a waste under ambient temperature and pressure. Current regulations prohibit disposal of solid waste containing free liquids in landfills without pretreatment (i.e. mixing with an absorbent material) or treatment by in-situ absorption in the landfill.

A number of tests for free liquid have been proposed or can be adapted from other testing operations. Many test protocols, such as the inclined plane test or a simple gravity drainage test, do not take into account the pressure of overburden on the waste at the bottom of a landfill. A review of the test procedures is given in SMC-Martin (1981). Most solidified wastes are designed to be landfilled to an appreciable depth (10 to 20 m) of material. Therefore, any test for free liquid should take into account the increased pressure due to the overburden. To simulate overburden, a sample of material can be subjected to pressure while it is in an apparatus that will permit any exuded liquid to be collected. SMC-Martin (1981) outlines large and small pressure cells developed to measure free liquid production in moist refuse produced by overburden pressure.

A very simple approach is to place a solidified waste sample of specific size and weight between weighed clean filter pads and load the block of waste to pressures comparable to those developed in landfilling (10-m depth = about 200 kPa, or 30 psi). The exuded liquid is collected on the filter pads and the weight difference of the pads before and after pressure is applied is used to quantify the amount of exudate.

Current EPA regulations indicate that no free water should be present in the waste. The Nuclear Regulatory Commission (1983) has specified that solidified low-level radioactive wastes must be free-standing monoliths and that no more than 0.5 percent of the waste volume can be free liquid.

4.2.3 Reactivity and Ignitability

Stabilized/solidified wastes that are to be disposed of in a landfill (onsite or offsite) should meet the criteria for landfilled hazardous waste in that due care must be exercised if the treated wastes are ignitable or reactive (40 CFR 265.312, Federal Register 1980). In most circumstances where stabilization and solidification are used, the waste can be rendered nonreactive or nonignitable in treatment. Tests for ignitability and pyrophoricity are given in Malone et al. (1980). Solidified/stabilized wastes developed for radioactive waste burial must not only be nonignitable, they must also be nonpyrophoric (i.e, will not support combustion if ignited) and must be nonreactive and nonexplosive (Nuclear Regulatory Commission 1983). Similar specifications for solidified industrial waste are desirable.

4.2.3.1 Reactivity.

Solidified wastes can contain reactive compounds that remain reactive after treatment. The wastes should be tested for compatibility with materials (absorbents, liners, other wastes) they would contact during landfilling. Procedures discussed in Hatayama et al. (1981) are useful for this purpose. Where possible, reactive materials should be destroyed or neutralized before stabilization.

If the potential for explosive reactions in waste exists, the Explosive Temperature Test (40 CFR 250.13) can be used to verify the hazard. Bureau of Explosives impact testing (49 CFR 173.53 (b), (c), (d) and (f)) can also be employed with solidified waste. Explosive and reactive wastes are not acceptable for landfilling.

4.2.3.2 Ignitability.

Solidified waste should not cause fires through friction, absorption of moisture, or spontaneous chemical changes. If ignited the material should not burn persistently (it should be self-extinguishing) or vigorously. Many biodegradable wastes produce methane under anaerobic conditions.

Many solidification systems which use cement and pozzolanic materials are inherently nonignitable and safe. Encapsulation systems using organic materials such as asphalt or polyethylene may require ignitability testing. Any liquid associated with the solid should be subjected to the test procedure given in ASTM Standard D-93-79 or D-3278-78 (ASTM 1973). Materials having flash-points less than 60° C are unacceptable. Any gases evolved from the waste should be nonignitable and nontoxic as specified in 49 CFR 173.300. The solid waste itself should not be capable of sustained burning if ignited. Tests such as ASTM F501 can be used to evaluate this property (Malone et al. 1980).

4.2.4 Physical Stability

Physical stability of the waste under conditions of burial is necessary
to ensure that the waste can support necessary construction equipment and
that, over the long run, it does not consolidate and cause the landfill cover
to collapse or fracture. Membrane covers can fail through shear if the
underlying waste consolidates or shrinks unevenly. Consolidation and shrink-
age are problems that occur most often in moist, organic-rich wastes.

The amount of settlement that can be tolerated depends on the type of
cover on the landfill and any future use of the filled area. If a soil cover
is used and no future construction occurs on the landfill, then extensive
settlement may not disrupt drainage or impair performance. If the final
cover includes a membrane cover, settlement should be limited to the lowest
achievable value. Table 4-1 lists the suggested test procedures for determining characteristics that relate to settlement of stabilized waste residuals. Some of these characteristics such as particle-size distribution and
compaction may not be measurable on strongly cemented wastes. Wyss et al.
(1980) discuss typical testing programs for consolidation.

TABLE 4-1. RECOMMENDED TESTING PROCEDURES FOR PHYSICAL
CHARACTERISTICS THAT RELATE TO WASTE SETTLEMENT

Test	Procedure
Particle-size distribution density	ASTM D422-63 or EM 1110-2-1906 Appendix II*
Compaction	ASTM D698-70†
Consolidation	ASTM D2435-70
Compressive strength Unconfined Triaxial shear Plate load	ASTM D2166-66 ASTM D2850-70 ASTM D1194-72
Permeability	ASTM 2434-68 or EM 1110-2-1906 Appendix III*

* U.S. Army (1972).
† ASTM (1973).

4.2.5   Biological Stability

Biological activity in stabilized/solidified wastes is usually not desirable. Many biological reactions, such as sulfide oxidation or decomposition of hydrocarbons, can produce acids that attack lime-based solidification processes and increase the potential for leaching from the wastes. Methane gas can also be produced in large quantities under anaerobic conditions. Tests such as ASTM G21 and ASTM G22 (ASTM 1973) can be used to directly determine the ability of the wastes to support biological activity. The Nuclear Regulatory Commission (1983) requires that nuclear waste solidified with cement-based processes support no biological growth. Bituminous materials are permitted if only one bacterial colony develops per sample, using a sample of the size specified in ASTM C39 or ASTM D621 (ASTM 1973).

Indirect measuring systems can also be used. In indirect systems samples of the waste are subjected to biological testing and then followed by strength testing so that any decrease in strength can be documented.

4.2.6   Strength or Bearing Capacity

The ability of the treated waste to support the cover material relates directly to the strength and bearing capacity of the waste. Most measurements made on waste have used standard procedures such as ASTM D2166-66 or ASTM C39, where a sample of brittle material is tested to failure. Where bituminous materials containing wastes are included in the test procedure, ASTM D621 Method A (Nuclear Regulatory Commission 1983) has been recommended.

The dutch cone test and plate load test have been suggested as supplementary systems of testing solidified wastes (Brown and Assoc. 1981). These tests yield less precision but are applicable in the field.

Unconfined compressive strengths measured on solidified wastes have ranged from 5.5 kPa (0.8 psi) to 3.1×10^4 kPa (4500 psi) (Bartos and Palermo 1977). The Nuclear Regulatory Commission (1983) guidelines call for a compressive strength of 103.5 N/sq cm (150 psi) for rigid materials. Bituminous materials must show less than 20 percent deformation at this pressure.

Where it is suspected that the increasing the water content of the waste causes the waste to lose strength, a program of testing unsaturated and saturated specimens can be undertaken. Where soluble cementing materials like $CaSO_4$ are being used, wet-dry cycling should be required to demonstrate that the solidified waste will not lose strength after placement.

4.2.7 Permeability

Solidified wastes normally require the use of a falling head permeability test conducted in a triaxial compression chamber with back pressure to ensure complete saturation (U.S. Army 1972). Permeabilities measured in solidified waste typically range from around 10^{-4} to 10^{-8} cm/sec. No standards related to permeability have been developed for solidified waste. Such low permeabilities indicate decreased mobility in the treated waste and a slower transfer of contaminants from the solid mass to leaching waters.

4.2.8 Durability

Most solidified wastes do not have high durability when subjected to standard freeze-thaw or wet-dry test procedures (Bartos and Palermo 1977). However, solidified wastes are generally buried and not subjected to varying conditions. An adequate cover usually can minimize temperature and moisture variations in the wastes. Durability testing becomes important where uncovering of the waste by erosion or human activity is likely or where long-term durability must be estimated.

Durability testing is usually done using soil-cement test protocols. These include ASTM D560-57 for freeze-thaw testing and D559-57 for wet-dry testing.

4.3 Example Specifications

To select or develop an optional solidification system, it is necessary to specify the performance required under the conditions of burial that are being considered. Table 4-2 is an example of a specification that might be developed for a solidified waste. Some features of the waste can only be specified as landfill design is evaluated. For example, the loading under which the free liquid test would be run would depend on the maximum depth or loading proposed in the landfill. The durability testing may be restricted to the expected number of cycles that might occur after waste placement and before cover placement.

TABLE 4-2.   EXAMPLE SPECIFICATIONS FOR SOLIDIFIED WASTE FOR LAND BURIAL

Characteristic	Recommended Value
Leachability	For major toxic components leachability is greater than 6 using ANS 16.1. Must pass EP test.
Free liquid content	No liquid exuded under maximum loading proposed in landfill design.
Physical stability	Will not allow unacceptable settlement under land-fill design conditions.
Reactivity of waste	Nonreactive.
Ignitability	Nonpyrophoric. Flash point below 60° C using ASTM D-93-79 or D3278-78.
Ability to support microbial growth	No microbial growth observed using ASTM G21 or G22.
Strength	Greater than 1000 kPa (150 psi) using ASTM 39 or ASTM D621.
Permeability	Less than 1×10^{-5} cm/sec when measured using upflow triaxial procedure.
Durability	As required by site design. Measured using ASTM D560-57 and ASTM D559-57.

REFERENCES

American Nuclear Society. 1981. Measurement of the Leachability of Solidified Low-Level Radioactive Wastes. Draft of Standard ANS-16.1. 47 pp.

American Society for Testing and Materials (ASTM). 1973. Annual Book of
ASTM Standards, Part II. Philadelphia, Pennsylvania.

Bartos, M. J., Jr., and M. R. Palermo. 1977. Physical and Engineering Properties of Hazardous Industrial Wastes and Sludges. EPA-600/2-77-139,
U.S. Environmental Protection Agency, Cincinnati, Ohio. 89 pp.

Cote, P. L., and D. Isabel. 1983. Application of a Static Leaching Test to
Solidified Hazardous Wastes. Presented at ASTM International Symposium on
Industrial and Hazardous Solid Wastes, Philadelphia, Pennsylvania, March 7-10,
1983.

Cote, P. L., and D. P. Hamilton. 1983. Leachability Comparison of Four
Hazardous Waste Solidification Processes. Presented at the 39th Annual
Purdue Industrial Waste Conference, Lafayette, Indiana, May 10-21, 1983.

Federal Register. 1980. Hazardous Waste and Consolidated Permit Regulations.
Vol 45, No. 98, Book 2, pp. 33063-33285, May 19, 1980.

Garrett, B. C., et al. 1981. Solid Waste Leaching Procedure Manual. Draft
Report Contract 68-03-2970, U.S. Environmental Protection Agency,
Cincinnati, Ohio. 53 pp.

Hatayma, H. K., et al. 1981. Hazardous Waste Compatibility Protocol.
California Department of Health Services, Berkeley, California, Report on
Grant R804692010, U.S Environmental Protection Agency, Cincinnati, Ohio.

Lowenbach, W. 1978. Compilation and Evaluation of Leaching Test Methods.
EPA-600/2-78-095, U.S. Environmental Protection Agency, Cincinnati, Ohio.
111 pp.

Malone, P. G., L. W. Jones, and R. J. Larson. 1980. Guide to the Disposal
of Chemically Stabilized and Solidified Wastes. SW- 872, Office of Water and
Waste Management, U.S. Environmental Protection Agency, Washington, D.C.
126 pp.

Nuclear Regulatory Commission. 1983. Branch Technical Position on Waste
Form. Document 204.1.5/TCJ/1/5/83, Nuclear Regulatory Commission,
Washington, D.C. 10 pp.

SMC-Martin. 1981. Test Protocol for Free Liquid Content of Hazardous Waste.
Phase I, Contract No. 68-01-3911, U.S. Environmental Protection Agency,
Washington, D.C. 128 pp.

U.S. Army, Office of Engineers. 1972. Laboratory Soils Testing. Engineer Manual 1110-2-1906, U.S. Army Corps of Engineers, Washington, D.C.

Wyss, A. W., et al. 1980. Closure of Hazardous Waste Surface Impoundments. SW-873, Office of Water and Waste Management, U.S. Environmental Protection Agency, Washington, D.C. 92 pp.

5. Bench- and Pilot-Scale Screening of Selected Treatment Processes

After preliminary selection of a stabilization/solidification system, a pilot-scale or bench-scale study can be developed to obtain detailed information on factors such as:

a. Safety problems in handling waste.

b. Waste uniformity and mixing and pumping properties.

c. Development of processing parameters and the level of processing control required.

d. Volume increases associated with processing.

Safety problems on larger scale stabilization/solidification operations may involve fuming, heat development, and volatilization of organic materials. Allowance may have to be made to adapt equipment for vapor control or cooling of reaction areas. Rapid addition of a reactive solidification agent (such as unhydrated lime) can cause rapid volatilization of organic compounds having low boiling points, with the possibility of a flash fire occurring. A fire believed to be caused this way occurred when lime was added to a sludge pit at Utica, Michigan, in 1983.

Heat transfer characteristics may be very different as a treatment or reaction system is scaled up and dimensions increase. With lower heat losses, temperatures rise, reaction rates are accelerated, and the solidification processes can become self-promoting. This is a common problem in operating with any large exothermic reaction such as hydration of Portland cement or the solidification of some organic polymers. Standard test procedures for heat of hydration of cements can be used in bench- and pilot-scale evaluation to predict heat generation and calculate temperature increases. A typical bench-scale procedure would be ASTM C 186, Test for Heat of Hydration of Hydraulic Cements (U.S. Army 1949).

A larger pilot-scale test involving 0.22 m^3 (8 cu ft) of cement or pozzolan is given in the Corps of Engineers Test of Temperature Rise in Concrete (CRD-C 38 in U.S. Army 1949). Adaptations of this test, such as fume collection and temperature monitoring, may be made to allow the effects of volatilization of organic compounds to be considered. The insulated block may have to be vented to simulate loss of low-boiling-point waste components.

When fumes from a solidifying waste are anticipated to be a problem, it is necessary to examine the headspace gases that develop in a closed container such as partially filled drums containing solidifying wastes. Standard organic vapor or gas monitoring equipment can be used to estimate the severity of the problem. Hatayama et al. (1981) outline the usual procedures that would be used to determine whether a potentially hazardous reaction will occur when solidification or stabilization reagents are added to a waste. Typical equipment includes organic vapor analyzers of the gas chromatograph or infra-red absorption types or detectors based on colorimetric systems. The objective of testing would be to determine the peak concentrations of irritating or toxic volatiles that might be produced with an addition of a given reagent. If the concentration of toxic volatiles obtained exceeds safety standards (after assuming a reasonable dilution for the site), then an enclosed or vent-controlled mixing and reaction system may be required.

Mixing and pumping problems can arise from variations in the pumpability of the waste onsite (c.f., Table 3-1). Mixing can become a problem if the solidifying waste changes viscosity rapidly during setting. If a specific mixing or pumping technique is to be used in the field, pilot testing can be used to evaluate the performance of mixers and pumps. Standard test CRD-C 55-78 outlines techniques to be used in evaluating concrete mixer performance (U.S. Army 1949).

Where the flowability or pumpability of a waste/solidifier mix is required, tests such as CRD-C 611-80 would be appropriate, or tests such as CRD-C 612-80, Test Method for Water Retentivity of Grout Mixtures, can be used to predict the amount of fluid separation to be expected from a waste/solidifier mix (U.S. Army 1949).

Processing parameters such as mix ratios, mix times, set times, and conditions of treated waste curing have to be examined in each specific waste solidification project. The detail of work involved approaches that used in designing concrete mixes. Much of the pilot testing can be patterned after concrete design procedures (U.S. Army 1949), but it is largely trial and error because of the wide variety of waste types and reagent properties. For instance, fly ash, which is a most common reagent, varies in sorption and pozzolanic activity depending upon the coal source and firing conditions in the furance, and its age and moisture content. Wastes will also vary between batches and even between the top and bottom of a single drum.

All solidification or absorption procedures result in some increase in waste volume. The volume increase can be seriously underestimated if too few measurements of additive requirements are made or if the moisture content of the absorbent or additive is greater in field specimens than in laboratory materials. Pilot tests with large, typical samples of additives usually provide more reliable estimates of additive volumes than laboratory bench studies, especially if care is taken to characterize additives (bulk density, moisture content, reactivity, etc.).

There is no substitute for a pilot study to evaluate a solidification program and develop production techniques in large-scale solidification projects. Pilot studies also provide large samples of material required for more accurate, realistic testing, and permit reconciliation of the complications with equipment and material handling. Pilot studies can also be used to train equipment operators on the characteristics of the waste and the solidified product. Although quite expensive and time-consuming, pilot studies can reduce the possibility of a major accident, reduce work stoppages, and increase product consistency and process reliability. Pilot studies pay for themselves many times over in large-scale projects.

REFERENCES

U.S. Army. 1949. Concrete Handbook. U.S. Army Engineer Waterways Experiment Station, Vicksburg, Mississippi, Loose-leaf revised quarterly.

Hatayma, H. K., et al. 1981. Hazardous Waste Compatibility Protocol. California Department of Health Services, Berkeley, California, Report on Grant R 804692010, U.S. Environmental Protection Agency, Cincinnati, Ohio.

6. Full-Scale Treatment Operations

6.1 Project Planning

Planning for the application of stabilization/solidification technology
at a particular remedial action site is divided into two distinct stages as
described in Section 1 (see Figure 1-1). The first planning stage considers
the specific treatment technology and reagents best suited to the particular
waste, including factors such as waste physical and chemical characteristics,
reagent cost and availability, and environmental desirability; this phase has
been considered in detail in the first five sections of this handbook. The
second phase, which is covered in this and succeeding sections, is concerned
with the overall operational and engineering plans for the actual completion
of the project at the specific site--i.e., the treatment scenario. Specific
aspects of this stage concern the development of equipment requirements,
construction sequencing, and cost estimation for the stabilization/
solidification portion of the remedial action project.

The development and selection of the solidification/stabilization operations plan for a particular remedial action site are dependent on several
factors such as the nature of the waste material, the quantity of the waste
material, the location of the site, the physical characteristics of the site,
and the solidification process to be utilized. When the solidification program is being developed, the primary goal is to create optimum efficiency
which is constrained by both short- and long-term environmental and public
health considerations.

This section identifies four alternative scenarios as applied to the
solidification/stabilization of hazardous wastes at remedial action sites and
examines their technical feasibility and comparative costs. The treatment
here is primarily concerned with the evaluation of equipment and project sequencing rather than with process chemistry. For purposes of this section,
it is presupposed that the waste solidification/stabilization process has
been selected and optimized, and that the site is geographically and geologi-
cally suitable for onsite disposal. The additional cost of transport and
offsite disposal of the final product may be incurred if onsite disposal is
not possible, but this possibility should not affect the validity of these
discussions.

Onsite solidification/stabilization programs can be classified according
to the manner in which the reagents are added to and mixed with the materials
being treated. Four onsite solidification/stabilization alternatives are

examined in this document: in-drum mixing, in-situ mixing, mobile plant mixing, and area mixing. Modifications to these basic operational techniques are identified and discussed where appropriate. The selection of an appropriate solidification/stabilization technique is based on an analysis of waste, reagent, and site-specific factors. As a result, only generalized criteria can be developed as applied to conditions expected at any given remedial action site.

In-drum mixing is best suited for application to highly toxic wastes that are present in relatively small quantities. This technique may also be applicable in cases where the waste is stored in drums of sufficient integrity to allow rehandling. In-drum mixing is typically the highest-cost alternative when compared with in-situ, mobile plant, and area mixing scenarios. Quality control also presents serious problems in small batch mixing operations; complete mixing is difficult to achieve, and variations in the waste between drums can cause variations in the characteristics of the final product.

In-situ mixing is primarily suitable for closure of liquid or slurry holding ponds. In-situ mixing is most applicable for the addition of large volumes of low reactivity, solid chemicals. The present state of technology limits application of in-situ mixing to the treatment of low solids content slurries or sludges. Where applicable, in-situ mixing is usually the lowest cost alternative. Quality control associated with in-situ mixing is limited with present technology.

Mobile mixing plants can be adapted for application to liquids, slurries, and solids. This technique is most suitable for application at sites with relatively large quantities of waste materials to be treated. It gives best results in terms of quality control. Mobile plant mixing is applicable at sites where the waste holding area is too large to permit effective in-situ mixing of the wastes or where the wastes must be moved to their final disposal area.

Area mixing consists of spreading the waste and treatment reagents in alternating layers at the final disposal site and mixing in place. This technique is applicable to those sites where high-solids-content slurries or where contaminated soils or solids must be treated. Area mixing requires that the waste materials be handled by construction equipment (i.e., dumptrucks, backhoes, etc.) and is not applicable to the treatment of liquids. Area mixing is land-area intensive, as it requires relatively large land areas to carry out the process. Area mixing presents the greatest possibility for fugitive dust, organic vapor, and odor generation. Area mixing ranks below in-drum and plant mixing in terms of quality control.

6.2 Cost Analysis and Comparison

The cost analyses in this chapter are by necessity general and based on generic techniques and equipment. They are included not as definitive numbers but as illustrations of the kinds of considerations which go into such

analyses. They also give a feel for the applicability of the different procedures which are discussed. We wish to emphasize that specific site and/or
waste characteristics can change these estimates by severalfold.

To increase the usefulness of comparisons the cost calculations are
based upon factors and assumptions which are consistent for the different
alternatives. Further, they illustrate the relative proportion that each
cost subcategory contributes to the overall cost of the process and then
allow estimates of the effect of substitution of alternate reagents or
equipment on the total process cost. The treatment reagents chosen for all
alternatives, Portland cement and sodium silicate, are not universally
applicable as might be implied by their inclusion in all alternatives; but
they are used in all examples because they make the comparisons valid and
because their cost is typically about average or slightly higher than other
reagents. Discussion and comparisons with other treatment reagents are
included in the summary (Section 6.7.2).

Labor costs shown in the illustrations are uniform throughout and include 25% fringe benefits. Reagents are priced at onsite costs as shown in
Appendix A. All equipment is charged at a daily rate of 0.5% of market value
which includes all fuel, interest, maintenance, and depreciation (3-year
base); this rate is unrealistic in some cases, but it serves well for comparison purposes. The equipment rental rates thus calculated are in line with
those quoted by industrial sources (see Appendix B).

6.3   In-Drum Mixing Alternative

The disposal of drums containing toxic and hazardous liquids and sludges
in landfills or open outdoor storage areas has been a common practice in the
United States. Many of the problems with uncontrolled disposal sites can, in
part, be linked to inadequate drum disposal activities. Typically, these
drums are 55 gallons (208 liters) in size although other sizes may also be
encountered. In-drum solidification is an attempt to utilize onsite assets
(i.e. drums) as both mixing vessel and container for the solidified waste
materials.

Handling of the drums of materials onsite and offsite before and after
solidification/stabilization is a major consideration in this alternative.
Related problems of selection and implementation of equipment and methods for
handling drums must be independently determined. Factors that influence the
selection of drum handling equipment or methods include worker safety, site-
specific variables, environmental protection, and costs. An EPA (1983)
manual reviews the applicability, advantages and disadvantages of equipment,
and methodologies for handling drums. The manual addresses detecting and
locating drums, determining drum integrity, excavation and onsite transfer of
drums, recontainerization and consolidation, and storage and shipping.

In-drum mixing can use existing or new drums. Where drum integrity
allows, the reagents are added directly to the drum in which the waste has
been previously stored. Drum reuse has the advantage that maximum use is

made of onsite assets, and drum crushing and disposal considerations are
eliminated, with subsequent cost savings. However, in-drum mixing is often
precluded because the poor condition of the drums or the need for head space
in the drum does not allow for addition of the solidification/stabilization
reagents and resulting expansion of the treated wastes. Typical head space
requirements range between 50% and 30% of drum volume. Thus, if all drums
have sufficient integrity for use, 0.5 to 1 additional drum is required for
each drum of existing wastes.

Most drums found at abandoned waste sites have only a bung hole in a
solid top. These drums pose a special problem because the opening is too
small to insert bulk reagents or an adequate mixing apparatus. Testing of
the composition of the contents or of their homogeneity is also difficult.
The most common procedure used to overcome these problems is to redrum the
contents in new or used, open-topped drums at which time the contents can be
visually inspected for uniformity or phase separation. A second alternative
is to cut a larger opening in the drum top for access. Although this procedure is cost-effective with drums which are in good condition, the added labor and equipment cost and exposure of employees lessens the benefits of the
latter method. Care should be taken to use a nonsparking cutting apparatus
(e.g., one of bronze), as the head space may contain explosive gas mixtures.

If new drums are required, the cost of the in-drum mixing option is substantially increased. Although the labor cost increases because of implied
redrumming requirements, the primary increase in cost is that of the drums.
The cost of drums (July 1983) ranged between $10 and $60 per drum depending
on the supplier and transportation costs.

6.3.1 Project Sequencing

Project sequencing for in-drum solidification can be divided into seven
steps:

a. The contents of each drum to be treated must be evaluated and/or
    identified. Particular care must be taken to ensure compatibility
    with the proposed solidification/stabilization process and the
    wastes. Each drum should be marked with appropriate identifying
    information. Costs associated with this testing are _not_ included in
    this analysis and can be substantial.

b. The condition of each drum should also be evaluated. Drums that are
    in sufficiently good condition for reuse should be marked. Head
    space in each drum should be noted on the exterior of the drum, and
    materials should be redrummed as required to accommodate head space
    and drum condition requirements.

c. A materials handling location should be prepared. Chemical storage
    and mixing equipment should be centrally located. A concrete pad or
    gravel surface should be prepared to ensure an adequate materials-
    handling facility for all weather conditions. Consideration should

be given to materials flow, including incoming empty drums, incoming
drums containing waste materials, and outgoing product drums. For
large sites, multiple materials handling locations may be cost-
effective.

d. Solidification/stabilization chemicals should be added to and mixed
with the wastes being treated.

e. The drums of mixed materials should be placed in a secure area and
allowed to set or cure until stable enough for safe handling.

f. After curing, any remaining head space should be filled with inert
material and the top replaced.

g. The drums should be removed for final disposal.

6.3.2 Equipment Requirements

Equipment requirements for the in-drum mixing process include: onsite
chemical storage system, chemical batching system, mixing system, and drum
handling system. Prior to actual solidification, a temporary enclosure for
the equipment should be erected. The mixing equipment should be installed on
a prepared surface that will facilitate the cleanup of spills and ensure ease
of daily cleanup. Requirements for the mixing area depend on the size of the
remedial action process and the nature of the wastes being treated. The enclosure serves to protect personnel from the elements and provides a controlled environment to minimize airborne hazards.

Mixing equipment for in-drum solidification includes the change-can
mixer and the top-entering propeller. Figure 6-1 illustrates in-drum solidification using the top-entering propeller.

6.3.3 Costs

In-drum mixing has the highest per unit cost of the four solidification/
stabilization techniques examined (Table 6-1). The total cost of cement-
silicate solidification using the equipment, labor, and assumptions listed
below is over $50 per drum holding only 40 gal of waste. Since reagent costs
are only a small part (about 12%) of the total cost per drum, using smaller
amounts of cheaper reagent would not greatly affect the overall cost (see
Section 6.7.1). Labor, equipment rental, and used drums each account for
between 15% and 25% of the total cost, for a total of about 60% of the total
(not including the 30% for profit and overhead). The high labor and equipment costs result from the very low throughput of the system--only 4.5 drums
per hour, which is less than 1 cu yd. Increasing this throughput would produce an appreciable reduction in treatment costs since over half of the cost
is sensitive to the production rate. The cost of treating 500,000 gal of
waste using this system is about $750,000--about $1.50 per gallon. Economic

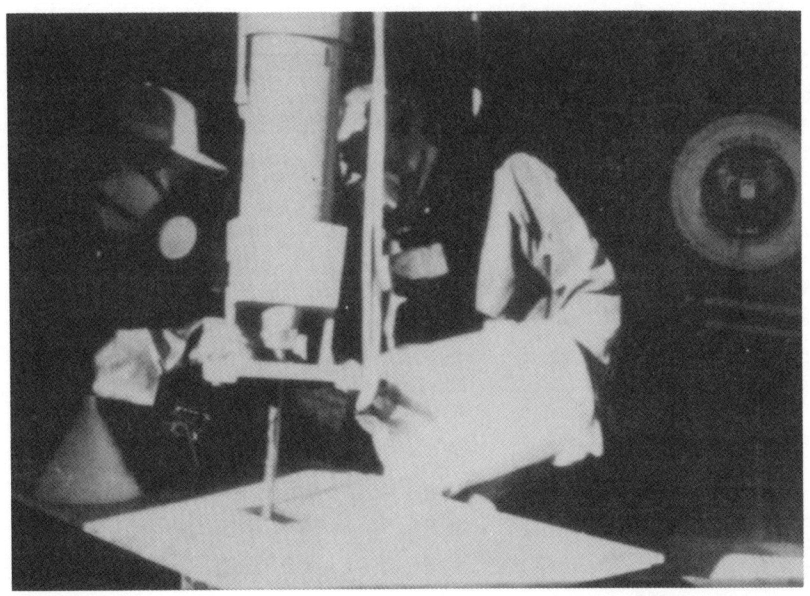

Figure 6-1. In-drum mixing using a top-entering propeller mixer.

considerations alone limit this treatment system to small amounts of very toxic wastes; it cannot compete with the dedrumming and bulk treatment of compatible wastes as done in other alternative techniques to be discussed. Even for wastes already contained in re-usable drums, the total cost would decrease only by about 20% to about $207/ton ($40.90/drum) in this example. Costs of initial classification, screening, and handling from remote site locations and to the point of final disposal and final disposal are not included.

The procedure for estimating the cost of in-drum solidification/ stabilization is summarized in Table 6-1 and detailed as follows:

a. <u>Assumptions</u>.

   (1) Solidification/stabilization process selected using Type I Portland cement (30%) and sodium silicate (2%).

   (2) Specific weight of waste to be solidified/stabilized is 85 lb/cu ft.

   (3) Approximately 40 gal of untreated waste can be placed in a drum and leave enough head space for reagent addition.

   (4) Processing rate averages 4.5 drums per hour.

TABLE 6-1.   COST ESTIMATES FOR THE IN-DRUM TREATMENT ALTERNATIVE

Note: Stabilization/solidification with 30% (w/w) Portland cement (Type I) and 2% sodium silicate of 40 gal of waste (85 lb/cu ft) in 55-gal drums at 4.5 drums per hour throughput.

TREATMENT REAGENTS:

  30% Portland cement = 137 lb/drum × ($0.0275/lb) = $3.77/drum
   2% sodium silicate =   9 lb/drum × ($0.10/lb)   = $0.90/drum

  Total cost for 12,500 drums: $58,275 $ 4.67/drum

LABOR COST FOR TREATMENT:

  1 ea Project supervisor = $27.50/hr = $6.11/drum
  2 ea Laborers @ $12.50   = 25.00/hr = 5.55/drum

  Total labor cost for 12,500 drums: $145,750 $11.66/drum

MATERIALS: Used, reconditioned drums: 12,500 for $137,500 $11.00/drum

EQUIPMENT RENTAL:

	Capacity	Value	Per hour	Per drum
Chemical storage silo	2,000 cu yd	$20,000	$13.15	$2.92
Change-can mixer	5 cu yd	15,000	9.90	2.20
Forklift	1 ton	14,250	9.40	2.09
Chemical feed system	100 lb/min	8,700	5.70	1.27

  Total rental for 12,500 drums: $106,000 $ 8.48/drum

MOBILIZATION-DEMOBILIZATION AND CLEANUP: 10% add-on = $44,750 $ 3.58/drum

TOTAL COST OF TREATMENT: 12,500 drums for $492,275 $39.36/drum

    PROFIT AND OVERHEAD (@ 30% of cost): $147,682 $11.81/drum

TOTAL CONTRACTED PRICE PER DRUM: $51.17/drum

TOTAL CONTRACTED PRICE FOR 12,500 DRUMS
  (500,000 gal or 2,850 tons of waste):

$639,957 or $224.29/ton

    (5)   Onsite cost of reagents is approximately $0.0275 per pound ($55 per ton) for Portland cement and $0.10 per pound ($200 per ton) for sodium silicate.

    (6)   Onsite labor dedicated to the solidification process includes two general laborers at $12.50 per hour, and one project supervisor at $27.50/hr.

    (7)   Reconditioned drums costing $11.00 each are used. Note that new drums can cost up to $40.00 each.

b.   <u>Chemical requirements per drum</u>.

    (1)   Portland cement at 30% by weight:

$$\frac{(40 \text{ gal/drum}) * (85 \text{ lb/cu ft}) * (0.30)}{(7.48 \text{ gal/cu ft})} = 137 \text{ lb/drum}$$

    (2)   Sodium silicate at 2% by weight:

$$\frac{(40 \text{ gal/drum}) * (85 \text{ lb/cu ft}) * (0.02)}{(7.48 \text{ gal/cu ft})} = 9 \text{ lb/drum}$$

c.   <u>Equipment rental and operation cost</u>. Equipment rental and operation costs are computed for a 2,000-cu-yd chemical storage silo ($20,000), a 5-cu-yd change-can mixer ($15,000), a 1-ton forklift ($14,250), and a 100-lb/min chemical feed system ($8,700).

d.   <u>Allowance for profit and overhead</u>. Profit and overhead allowances for this type work (based on construction company rates) range between 20 and 40%. Since this is assumed to be a high-risk operation, assume 30% profit and overhead.

e.   <u>Costs not included</u>. Note that the above cost includes the solidification/stabilization process and handling immediately before and after mixing. The following costs, which may be substantial, are not included: Identification and evaluation of drum contents, evaluation of drums, transport of drums to treatment area and of solidified/stabilized material to the final disposal site, and site preparation and closure activities.

f.   <u>Summary of in-drum mixing</u>. As seen in Table 6-1, the estimated actual cost of stabilization/solidification including profit and overhead is around $51 per drum ($244/cu yd or $258/ton). Of this, only about 10% is for the treatment reagents, while 30% goes for labor (including mobilization-demobilization), 21.5% for reconditioned drums, and about 16.5% for equipment. Since only about half of the cost of treatment is fixed per unit of waste (drums and reagents), the unit price is quite sensitive to production rate. Doubling the rate from 4.5 drums per hour to 9 drums per hour with the same equipment essentially lowers the unit treatment cost by about 25% to around $148/ton ($29.32/drum) or total cost with profit and overhead to $192.50/ton ($38.11/drum). If original drums are usable, the total cost of treatment will drop another 25%.

6.3.4 Safety and Environment

In-drum solidification/stabilization can provide the safest and most environmentally controlled work environment. Equipment can be purchased and installed to meet all Occupational Safety and Health Administration (OSHA) standards. A variety of standard accessories including dust hoods, dust shields, and vacuum hoods are available for the change-can mixer (these items are not included in costs shown in Table 6-1). In addition, the equipment can be easily operated by personnel in protective clothing. Typical protective clothing will include rubber gloves, safety glasses, hard hat, and dust mask or respirators. Equipment operation can also be accomplished in full air pack. Note that if full air pack protective equipment is required, a 50% to 60% reduction in productive capacity can be anticipated.

6.3.5 Modifications

In-drum or in-container solidification has been used extensively in the disposal of low-level radioactive waste materials. Specialized in-drum mixing equipment has been developed for this application. Particular attention has been given to the safety-related aspects of such equipment. Special drum fill-heads and remote monitoring systems have been developed to allow the drum to be filled, the reagents to be added, the contents mixed, and the drum sealed by operators isolated from the waste. Because of the high cost of these systems, they have not been widely used for the treatment of toxic and hazardous waste materials. They may have applicability to the solidification/stabilization of extremely toxic or hazardous wastes.

Another product of the nuclear industry is the prepackaged spill solidification kits. These systems are designed for the cleanup of small spills and include the mixing drum, premeasured solidification reagents, and disposable mixing blades. Kits come in a variety of available sizes complete with instructions. The user must supply a driver for the mixing blades. Figure 6-2 illustrates a typical spill cleanup drum solidification system. A 1981 price quote for a 55-gal drum system (40-gal maximum waste volume) was $600 per pallet of four drums f.o.b. plant of manufacture. These systems are considered for specialty purposes only and are not economically practicable for large-scale sites.

A final modification of the in-drum solidification scenario is the bulking of drummed liquids, solidification of the bulked liquids in a mobile or portable plant (Section 6.5), and repackaging of the solidified wastes in salvaged or new drums. This modification may be appropriate at sites with significant numbers of broken or leaking drums containing compatible wastes,

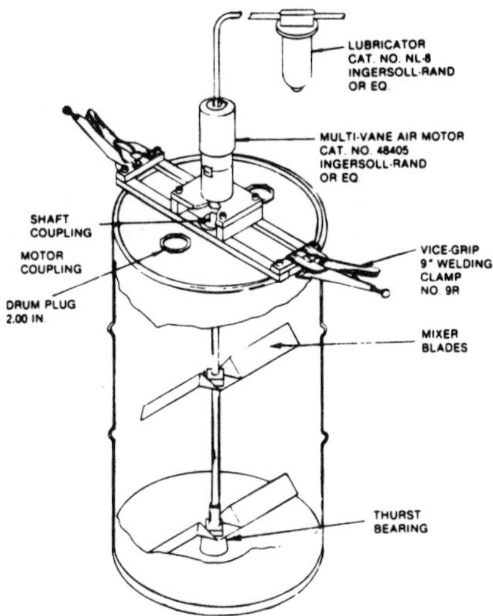

Figure 6-2. Typical spill cleanup system
(Courtesy Delaware Custom Materials).

or as a method to reduce the unit cost by increasing the production rate and
simplifying the equipment required.

6.4   In-Situ Mixing Alternative

The simplest solidification/stabilization alternative examined in this
study is in-situ mixing which incorporates the use of common construction
machinery (typically a backhoe or pull-shovel) to accomplish the mixing process. Where large lagoons are being treated, clamshells and/or draglines have
also been utilized. This technique is suitable for application to liquids or
light flowable sludges having a high liquid content. The technique is suited
more to those solidification/stabilization processes incorporating the addition of large amounts of bulk powdery solids (kiln dust, fly ash, etc.) to the
waste materials. In those cases where small amounts of admixture (fluidizers,
plasticizers, retardants, etc.) are to be added, the mixing efficiency of
available in-situ processes is not uniform. Data are not currently available
on the mixing efficiency of the in-situ processes when applied to large-scale
field projects.

6.4.1 Project Sequencing

Two in-situ solidification/stabilization alternatives are developed. In the first, the existing lagoon is used as both the mixing vessel and the final site for disposal of the treated wastes so that the waste materials are not removed from the existing lagoon or holding pond. In the second, the waste material is removed from the holding pond and placed in specially prepared mixing pits. After mixing, the treated wastes are either removed from the mixing pits to a prepared disposal site or are left in the mixing pits which become the final disposal site.

Under the first alternative, the existing holding lagoons are used as the final disposal site. The reagents are added to the lagoon by pneumatic or mechanical means. Pneumatic addition uses blowers to distribute the reagents over the surface of the lagoon. Mechanical addition incorporates the use of dump trucks, front-end loaders, or clamshells to mechanically add the required reagents. Mixing of the reagents is accomplished with a backhoe, clamshell, or dragline. The selection of mixing equipment is based on the size of the lagoon being treated and general site topography. Lagoons less than about 30 ft (10 m) in radius (or effective radius in the case of rectangular or odd-shaped lagoons) are amenable to backhoe mixing. Larger lagoons would require the use of a clamshell or dragline to ensure an adequate reach for mixing the contents in the middle of the lagoon.

The second alternative involves the preparation of special, onsite mixing pits. The waste material is transferred from the holding lagoon to the mixing pit. Pumps can be used to transfer liquids and light sludges whereas clamshells and trucks can be used to transfer heavy sludges. Reagents are added using the same methods described in the first alternative. Since the mixing pit can be constructed to a specified size, mixing is generally accomplished with a backhoe. After thorough mixing, the material is allowed to gel, or set, for the required amount of time. The solidified/stabilized material is then either capped in place (in the mixing pit) or removed to a prepared onsite disposal facility.

6.4.2 Equipment Requirements

Equipment required for in-situ solidification/stabilization varies with the specific site. Generally, an average site would require equipment in the following categories: dump trucks, front-end loader, excavator or backhoe, and onsite chemical storage and handling facilities. The size and amount of equipment depend on the location and topography of the remedial action site as well as the quantity of material to be treated. Figure 6-3 illustrates in-situ mixing using a backhoe.

Figure 6-3.   In-situ mixing with a backhoe at a large site.
(Courtesy Albert H. Halff Associates)

6.4.3 Costs

The cost of in-situ solidification/stabilization techniques is based primarily on the production rate achieved by the equipment mix selected for the specific remedial action project. Field data for the cost of in-situ mixing alternatives applied to remedial action sites are not available. However, production rates were determined for two RCRA sites using the backhoe-mixing pit technique. A daily (8-hr shift) production rate ranging from 1,000 to 1,200 cu yd (approximate 1000 cu m) of wastes solidified/stabilized was reported under the following conditions:

a. Construction of an earthen mixing basin (5 to 10 ft deep, 40 to 50 ft in diameter).

b. Introduction of liquid wastes received in bulk tankers or from de-drumming of liquids received in drums.

c. Addition of 40% to 60% (by volume) of fresh kiln dust, mechanically added with a front-end loader.

d. Mixing with backhoe (Caterpillar 225) until solidification/ stabilization process begins.

    e. Setting or gelling for 24 to 48 hr in the pit.

    f. Removal of solidified/stabilized material from pit with front-end loader or backhoe and spreading in secure landfill with dozer.

Another RCRA site using a similar scheme and equipment, except for the substitution of permanent concrete mixing pits (4 pits, 100 ft × 20 ft × 10 ft), reported daily (8-hr shift) production rates of 2,000 cu yd. This second RCRA site also had the capability of pneumatically adding bulk solid reagents.

The cost of solidification/stabilization at these RCRA sites was reported to range between $10 and $20 per cubic yard of waste material treated. The primary variable was the amount of kiln dust required for a specific waste. This factor affected chemical costs, material handling costs, and mixing labor costs.

The daily production rate for the backhoe mixing technique depends on the material being handled, size and quantity of equipment being used on a particular project, site conditions, quantity of material being treated, and quantity of reagent being added. Production rates for a remedial action site are expected to be somewhat less than those associated with a permanent installation. An in-situ treatment scheme incorporating one backhoe (Caterpillar 225 or equivalent) is anticipated to have a daily (8-hr shift) production rate ranging between 750 and 1,500 cu yd.

The procedure for estimating the cost of in-situ solidification/ stabilization is presented below and summarized in Table 6-2:

    a. <u>Assumptions</u>.

        (1) Approximately 500,000 gal of waste liquids and light sludges is to be solidified in situ using cement and sodium silicate. Mixing will be accomplished with a backhoe (Caterpillar 225 or equivalent). Wastes to be treated are contained in a rectangular-shaped lagoon approximately 120 ft × 60 ft × 10 ft.

        (2) Bench-scale studies indicate that the reagent must be added on a weight-to-weight ratio of 30% cement and 2% sodium silicate.

        (3) Waste and reagents will be mixed in the lagoon and left in place.

        (4) Onsite cost of cement is $55.00 per ton; sodium silicate is $200 per ton.

        (5) The remedial action site is located 200 miles from the nearest equipment.

TABLE 6-2. COST ESTIMATES FOR THE IN-SITU TREATMENT ALTERNATIVE

Note: Stabilization/solidification with 30% (w/w) Portland cement and 2% sodium silicate of a pumpable waste (85 lb/cu ft) from bulk tankers or drums mixed with a backhoe in an 8-ft-deep, 40-ft-diameter earthen mixing basin, and removed after 24 to 48 hr setting time. Total waste 500,000 gal (2,475 cu yd or 2,850 tons) and production rate is 800 cu yd per 8-hr shift (4 days required).

TREATMENT REAGENTS:

    30% Portland cement = 855 tons × ($55/ton) = $47,025
     2% sodium silicate = 57 tons × ($200/ton) = $11,400

    Total cost of treatment reagents: $58,425 $20.50/ton

LABOR COST FOR TREATMENT:

    1 ea Project supervisor   = $27.50/hr × 32 hr = $ 880
    2 ea Heavy equip. operators @ $22. = $44.00/hr × 32 hr = 1,408
    1 ea Laborer   = $12.50/hr × 32 hr =   400

    Total labor cost: = $2,688 $ 0.94/ton

    Expenses: @ $75/day for 4 men 4 days   = $1,200 $ 0.42/ton

EQUIPMENT RENTAL:

	Capacity	Value	Per hour	Per 6 days
Backhoe	(1.5 cu yd)	$95,000	$62/hr	= $2,976
Front-end loader	(1 cu yd)	29,000	$20/hr	= 960

    Total rental cost: $3,936 $ 1.38/ton

MOBILIZATION-DEMOBILIZATION AND CLEANUP:

    Labor and expenses for 3 days: $2,016 + $900   = $2,916
    Transportation: 200 mile/trip × 4 trips × $2/mile = 1,600

    Total $4,516 $ 1.58/ton

TOTAL COST OF TREATMENT: 500,000 gal   = $70,765 $24.83/ton

    PROFIT AND OVERHEAD: (@ 30% of cost) = $21,230 $ 7.45/ton

TOTAL CONTRACTED PRICE: 500,000 gal   = $91,995 $32.28/ton

b. <u>Mobilization and demobilization costs</u>. Mobilization costs are those incurred in preparing the equipment for shipment, transporting it to the site, and setting it up for mixing. Demobilization includes cleanup of the equipment and site and transportation back to origin. Mobilization-demobilization will take about 1 day each. Transportation costs are those associated with actually transporting the equipment to the site. For this example, it is assumed that local equipment rental is not available. Two tractor trailer loads will be required. The estimated cost of heavy equipment transport is $2.00 per load mile.

c. <u>Project duration</u>.

   (1) Based on field experience, a daily production rate (8-hr shift) is estimated to be 700 cu yd/day of wastes mixed.

   (2) Required project time is calculated as follows:

500,000 gal $\div$ 7.48 gal/cu ft $\div$ 27 cu ft/cu yd $\div$ 700 cu yd/day = 3.54 days (use 4 days).

   (3) Since processing will be accomplished at a remote site, personnel will be reimbursed for onsite expenses. Assume an expense rate of $75.00 per man per day.

d. <u>Summary of in-situ costs</u>. The in-situ treatment alternative is the fastest and least expensive of those discussed in this section. The speed and economy are largely due to the reduction in the amount of handling of the waste mass. Other than for mixing, the wastes are usually moved only once, or if not hazardous, they are often not even removed from the original waste lagoon but mixed and left in place. The method lends itself best to liquid or low-solids sludges which are easily mixed. However, heavy sludges can be mixed with heavy equipment like draglines or clamshells, but with less uniformity in the treated product. The low labor and equipment requirements result in the highest proportion of the cost (63%) going for the treatment reagent. Thus the cost of the method is quite sensitive to reagent cost and proportion. Major limitations of the method are the low amount of mixing attained and the inability to control accurately the proportion of reagent to waste which can result in a nonuniform, unevenly mixed final product. This can be overcome to some extent by using excess reagent to decrease zones of low reagent content, but this increases cost and treated product bulk.

Two new pieces of equipment which are designed and used specifically for in-situ mixing have recently been introduced; they are shown in Figures 6-4 and 6-5. These items pneumatically meter and inject the reagent directly into the waste mass at the lower end of the cylinders, which are used to stir and mix the wastes. One design (Figure 6-5) has augers at the ends of the

Figure 6-4. In-situ mixing by direct reagent injection (Courtesy ENRECO, Inc.).

Figure 6-5. In-situ mixing equipment (Courtesy American Resources Corporation).

cylinders which can be used to dig into underlying soils or sludges which may also be contaminated and incorporate them into the total waste mass.

6.4.4   Safety and Environment

In-situ mixing is the most difficult alternative in terms of control of safety and environmental considerations. Since the entire process is open to the atmosphere, anticipated problems include the generation of odors, vapors, and fugitive dust. In addition to the standard safety precautions associated with the operation of construction equipment, a strict program for minimizing exposure of personnel and equipment to the materials being treated should be implemented. Equipment should be decontaminated on a daily basis and the wash water should be collected for treatment or solidification.

Standard personnel protective procedures should be implemented as necessary, depending on the waste being handled. Reduction in production efficiency can be anticipated to be a function of the degree of protective apparatus required. Level A protection is expected to reduce production by up to 75 percent.

The ability to control adequately the in-situ mixing process is a subject of concern. Quantitative measurement of the degree of mixing produced by in-situ processes is not available. Most in-situ mixing operations are found at RCRA waste disposal sites where the mixed waste and solidification reagents are removed to a landfill after gelling. The rehandling of the processed materials allows some quality control of the adequacy of waste-reagent mixing. This additional level of quality of control may be lacking in the field environment unless the materials are rehandled and transported to a separate disposal area. Assurance of adequate quality control requires significant levels of experienced, onsite inspection and supervision.

6.4.5   Modifications

The chemical addition and mixing techniques currently used for in-situ solidification have been adopted from the construction industry and as such are relatively unsophisticated. Major modifications to in-situ solidification/stabilization include the development of reagent addition or mixing equipment that allows better control of the process. Equipment specifically designed for in-situ solidification/stabilization operations at pits, ponds, and lagoons is currently being used and marketed commercially. The equipment combines the injection of fly ash or kiln dust into the wastes by use of an injection head using a hydraulic/pneumatic system with the mixing of the materials by the injection head (Figures 6-4 and 6-5). The fly ash or kiln dust is added to the basin material at a predetermined rate until the consistency of the mix is sufficient for setting to occur within 1 to 3 days. An air compressor is used in conjunction with the injector head which is installed on a boom on a tracked vehicle. A hydraulic pump provides the drive for hydraulic motors on the injection head.

In one configuration (Figure 6-4) the hydraulic pump is mounted on the rear of a tracked vehicle for convenience of operation and to counterbalance the injection head boom when the boom is fully extended. Fly ash is delivered to the multibarreled injection head via a compressed air system. Besides providing a delivery system, the pneumatic system also prevents back flow of the basin material into the submerged ends of the barrels. Hydraulically driven augers in the lower section of the barrel force the fly ash out of the barrel into the basin contents. As fly ash is forced from the barrels into the waste, the boom simultaneously moves the injection head back and forth (in the plane of the boom) as well as up and down. This motion provides mixing of the fly ash and basin contents. Approximately 1,000 cu yd of waste material can be solidified per day. This equipment is best applicable to basins deeper than approximately 4 ft. In shallower basins, the necessary pneumatic pressure on the fly ash delivery to the injection heads causes loss of fly ash to the air at the basin surface which results in a burst of fly ash dust. Basins deeper than 4 ft require a larger injection head system and appropriately heavier duty equipment. New adaptations of this equipment which overcome these difficulties have been introduced (see Figure 6-5).

6.5   Mobile Plant Mixing Alternative

Plant mixing refers to those systems which incorporate mobile or fixed units to handle, meter, and mix the solidification/stabilization reagents and the wastes being treated. In this alternative, the wastes being treated are physically removed from their location, mechanically mixed with the solidification/stabilization reagents, and then redeposited in a prepared disposal site. Plant mixing is primarily oriented towards the treatment of pumpable liquids and high-liquid-content sludges; however, special equipment adaptations have been utilized to handle sludges with high solids contents and contaminated soils. A schematic of a typical plant mixing scenario is illustrated in Figure 6-6. Two plant mixing examples will be discussed—one used with pumpable wastes and one with high solids content wastes which must be handled with construction equipment.

6.5.1   Project Sequencing

Many plant mixing systems include all required solidification/ stabilization equipment in one trailer- or truck-mounted unit, whereas others are transported in modular form and are put together at the remedial action site.

Basic project sequencing for plant mixing is as follows:

a.   Prepare site for installation of the mobile system. This step includes any necessary utility hookup such as electricity. Some mobile systems have on-board power generation systems and require no onsite power connections.

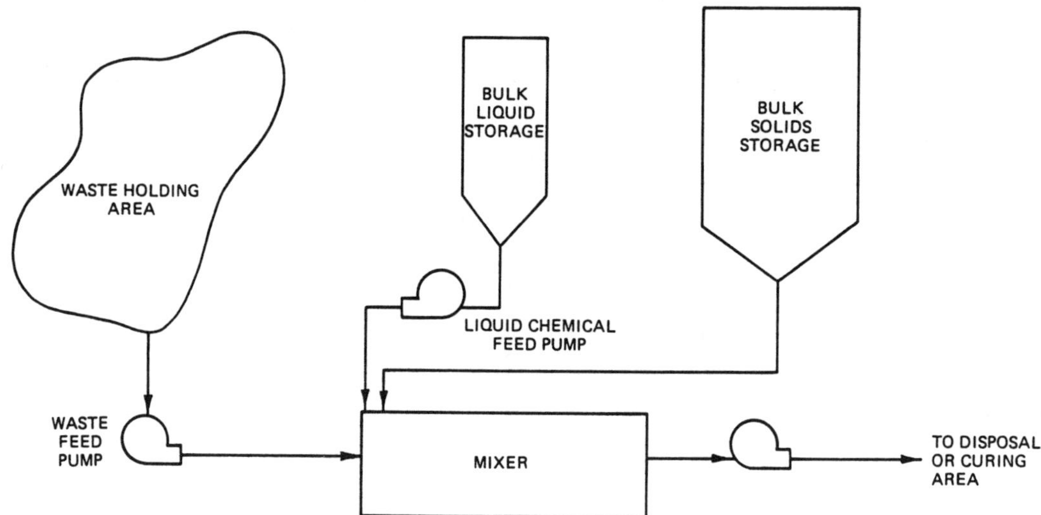

Figure 6-6.   Schematic of plant mixing scenario.

b. Prepare final disposal area for solidified/stabilized wastes.

c. Install raw and treated waste handling systems. These usually in-
    clude centrifugal or diaphragm pumps with electrical or gasoline-
    powered drivers, but they may be simple construction equipment for
    high-solids wastes.

d. Transport the portable system to the remedial action site and erect
    equipment, interfacing with utilities.

e. Initiate solidification/stabilization process and monitor as
    required.

6.5.2 Equipment Requirements

    Mobile, trailer-mounted plants may come complete with chemical storage
hoppers, chemical feed equipment, mixing equipment, and waste handling equipment. Some mobile plants have on-board power generation facilities; however,
more commonly, an onsite power hookup or separate power generation system is
required. Although the basic concept for the systems illustrated is identical, significant variations exist in the details of construction of each.
Variations found in those systems examined during this study were in the
mounting configuration (trailer or closed van), in the types of chemical feed
systems, in the types of mixing apparatus, and in the setup requirements.

Both weight- and volume-based chemical feed systems are used.
Volumetric-based systems are utilized exclusively for the addition of liquid
reagents, whereas the addition of bulk solid reagents may be controlled with
either weighing conveyor systems, batch weighing systems, or volumetric screw
feeder systems. Flow of the waste is controlled by the capacity of the
transfer pumps used to transfer the wastes from the holding area to the mixing vessel.

A variety of mixing systems have been successfully used on the mobile
plants currently in use. These include ribbon blenders and single and double
shaft rotor mixers. The type of mixer utilized appears to have little effect
on the quality of the final product, but production efficiency may be
affected. Illustrations and photographs of currently available mobile mixing
plants in operation are presented in Figures 6-7 through 6-11.

The design of mobile plants has been oriented toward the treatment of
liquids and light slurries. Materials handling is most often accomplished
using pumps. The capacity of the typical mobile plant ranges from 60,000 to
150,000 gal per 24-hr day of waste material treated. The controlling factor
in determining capacity is generally the handling characteristics of the
waste materials being treated. Thus the capacity of the same equipment will
vary significantly from job to job. The size of the equipment applicable to
mobile plants is limited by weight, length, and width restrictions associated
with over-the-road transportation requirements.

Modular plant systems consist of separate pieces of equipment that can
be tailored more closely to fit specific site requirements. Whereas mobile
plants are usually self-contained on one van or trailer, modular plants are
usually delivered to the site on several trailers. Typical modular plant
installations are illustrated in Figures 6-12 through 6-14.

The typical modular plant will include equipment modules for: onsite
chemical storage, usually a silo; chemical feed system, usually a weight
batching system; a mixing system of a type dependent on the waste materials
being treated; a raw waste handling system of a type also dependent on the
waste material; and a final product handling system.

The modular system illustrated in Figure 6-12 is designed primarily to
handle liquids and light flowable sludges up to 30% solids content. Mixing
is accomplished in a 1-1/2-cu yd ribbon blender. Waste materials can be
charged to the ribbon blender using pumps, a clamshell, or a front-end
loader. Mixing time is approximately 1-1/2 to 2 min depending on the material being handled. Solidified/stabilized material is discharged at the base
of the ribbon blender and removed by front-end loader. Material can be
transported to the final disposal site by dump truck.

The modular system illustrated in Figure 6-14 is designed to handle
heavy materials such as contaminated soils and low moisture content sludges.
In this particular application, the waste materials were slurried in order to
ensure reaction with the solidification/stabilization agents. A unique
aspect of this system was the use of concrete transit mixers to mix the

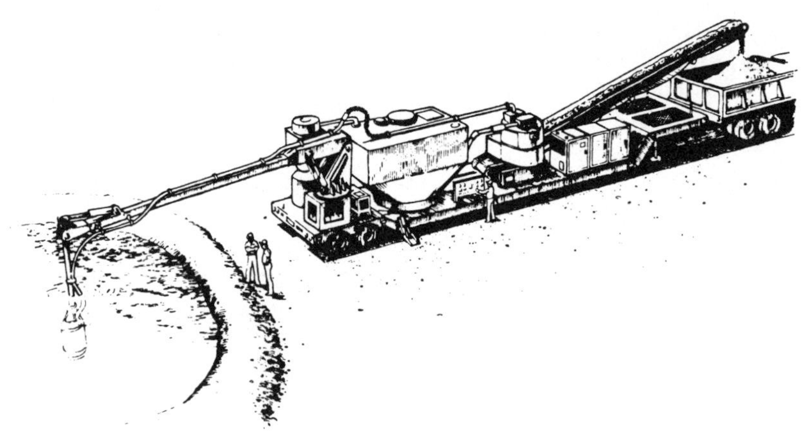

Figure 6-7. Schematic of a trailer-mounted mobile mixing plant
(Courtesy Beardsley & Piper).

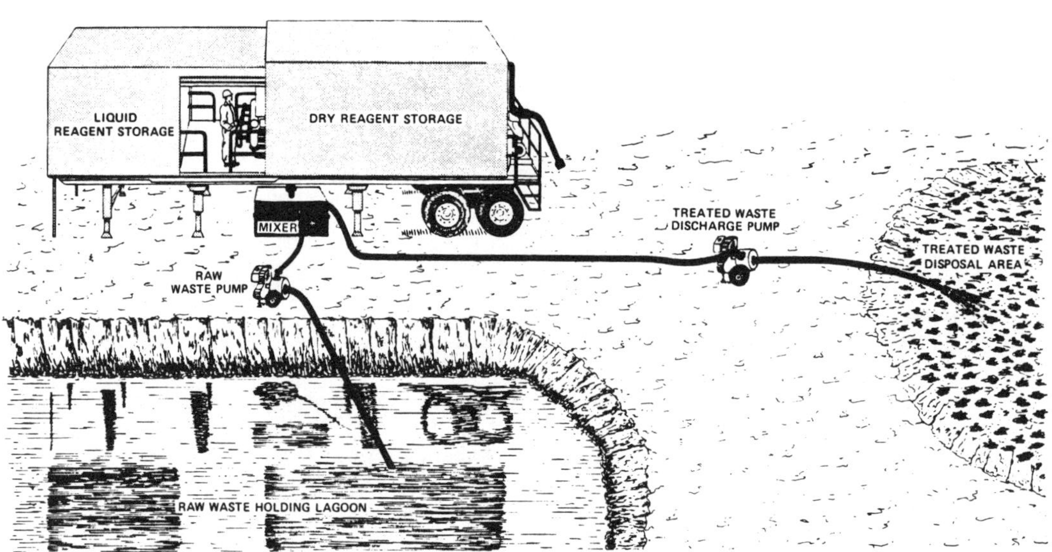

Figure 6-8. Schematic of a van-mounted mobile mixing plant
(Courtesy Chemfix).

Figure 6-9. Open mobile mixing plant (Courtesy Beardsley & Piper).

Figure 6-10. Enclosed mobile mixing plant (Courtesy Chemfix).

Figure 6-11. Drum handling mobile mixing plant (Courtesy Solid Tek).

Figure 6-12. Small modular mixing plant (Courtesy Solid Tek).

Figure 6-13. Large modular mixing plant (Courtesy IU Conversion).

Figure 6-14. Modular mixing plant for heavy slurries
(Courtesy Solid Tek).

wastes with the reagents. Reagents and waste materials were batched into the transit mixer. Mixing was accomplished while in transit to the final disposal site.

6.5.3 Costs

The costs of using mobile or portable mixing plants for a particular project are dependent on the process selected (reagent to be added) and the waste material being handled. These factors are the primary variables in determining the production rate on a particular project. Project costs include both fixed costs (i.e. transportation to and from the site and setup costs) and variable costs (i.e. chemicals and processing labor), which depend on the quantity and type of material treated.

General ranges of costs for application of both mobile and portable mixing plants to remedial action sites were provided by the owners of equipment discussed previously. Costs ranged between $20.00 and $75.00 per cubic yard of material treated. These costs included handling of the waste materials from their existing holding area to an onsite disposal site. Costs presented do not include the further handling of the material at the disposal area or capping and landscaping of the disposal area.

6.5.3.1 Mobile Mixing Plant for Pumpable Wastes.

The procedure for estimating the cost of mobile plant solidification/stabilization of a pumpable waste is presented below and summarized in Table 6-3.

a. <u>Assumptions</u>.

   (1) Approximately 500,000 gal of waste liquids and pumpable sludges in an open lagoon are to be solidified using a two reagent process consisting of Portland cement and sodium silicate.

   (2) Bench-scale studies indicate that the reagents need to be added in weight-to-weight ratios of 30% Portland cement and 2% sodium silicate.

   (3) An onsite disposal area is available.

   (4) The onsite cost of the reagents is $55.00 per ton for Portland cement and $0.10 per pound for liquid sodium silicate.

   (5) The remedial action site is located 200 miles from the nearest mobile unit.

TABLE 6-3. COST ESTIMATES FOR THE MOBILE PLANT MIXING
ALTERNATIVE FOR PUMPABLE WASTES

Note: Stabilization/solidification with 30% (w/w) Portland cement and 2%
sodium silicate of 500,000 gal (2,850 tons) of pumpable sludge (85 lb/cu ft)
in a mobile mixing plant with daily throughput of 250 cu yd (10 days
required). Onsite disposal available.

TREATMENT REAGENTS:

  30% Portland cement = 855 tons × ($55/ton) = $47,025
   2% sodium silicate = 57 tons × ($200/ton) = $11,400

  Total costs of treatment reagents: $58,425 $20.50/ton

LABOR COST FOR TREATMENT
  1 ea Project supervisor = $27.50/hr × 80 hr = $ 2,200
  2 ea Technicians @ $18.50 = $37.00/hr × 80 hr =   2,960
  2 ea Laborers   @ $12.50 = 25.00/hr × 80 hr =   2,000

  Total labor cost: = $ 7,160 $ 2.51/ton

  Expenses: @$75/day for 5 men 10 days   = $ 3,750 $ 1.32/ton

EQUIPMENT RENTAL:

	Capacity	Value	Per hour	Per 10 days
2 ea Trash pumps	(6 in.)	$31,000	$20/hr =	$1,600
1 ea Mobile plant		180,000	120/hr =	9,600

  Total rental cost: $11,200 $ 3.93/ton

MOBILIZATION-DEMOBILIZATION AND CLEANUP:

  Labor and expenses for 3 days: $2,148 + $1,125   = $3,273
  Transportation: 200 mile/trip × 2 trips × $2/mile =   800

  Total $4,073 $ 1.43/ton

TOTAL COST OF TREATMENT: 500,000 gal = $84,608 $29.69/ton

    PROFIT AND OVERHEAD (@ 30% of cost) = $25,382 $ 8.91/ton

TOTAL CONTRACTED PRICE: 500,000 gal = $110,000 $38.60/ton

b. <u>Mobilization and demobilization cost</u>. Mobilization costs are those costs incurred in preparing the equipment for shipment, transporting the equipment, and setting the equipment up for actual waste processing; these costs include labor costs and transportation costs. Demobilization includes cleanup of site and equipment and transportation back to origin. These activities are expected to take about 3 days total. Transportation is the cost of actually transporting the equipment. The estimated cost of transporting the equipment is $2.00 per load mile. Assuming that there will be the equivalent of two tractor trailer loads, we obtain a $4.00 per mile cost.

c. <u>Project duration</u>. Total processing time is based on the estimated production rate of the mobile unit. For the material to be processed, a production rate of 250 cu yd per day (8-hr shift) is assumed so that 10 working days is necessary to treat the entire lagoon (2,475 cu yd). This includes only the solidification activity.

d. <u>Cost summary for plant mixing of pumpable wastes</u>. Plant mixing techniques used with pumpable wastes are the least expensive of the alternatives developed here, except for the in-situ mixing scenario. The efficiency is largely due to the economical movement of the materials by pumps rather than by loading and trucking. For applicable wastes, this method permits precise reagent addition and complete and uniform mixing, both of which are lacking in the in-situ methodology at this time. This tighter control of mixing proportions and duration gives the ability to precisely tailor the reagent addition for maximum efficiency and effectiveness. The use of less reagent to attain adequate stabilization results in less final product to be disposed of which often makes this method quite competitive with in-situ methodology.

6.5.3.2 Modular Mixing Plant for Unpumpable Wastes.

The procedure for estimating the cost of in-situ solidification/ stabilization for a high solids waste is presented below and summarized in Table 6-4:

a. <u>Assumptions</u>.

(1) Approximately 500,000 gal of nonpumpable, high solids sludge is to be solidified using a two reagent process consisting of Portland cement and liquid sodium silicate.

(2) Bench-scale studies indicate that the reagents must be added in weight-to-weight ratios of 30% Portland cement and 2% sodium silicate.

(3) Onsite equipment will include a mobile plant that has a silo for cement storage, a weight batcher for control of the cement feed, a ribbon blender for mixing, a front-end loader for

TABLE 6-4.   COST ESTIMATES FOR THE MODULAR PLANT MIXING ALTERNATIVE
FOR UNPUMPABLE OR SOLID WASTES

Note: Stabilization/solidification with 30% (w/w) Portland cement and 2%
sodium silicate of 500,000 gal (2,850 tons) of unpumpable sludge or solid
waste (85 lb/cu ft) in a mobile mixing plant with daily throughput of
180 cu yd (14 days required). Onsite disposal available.

TREATMENT REAGENTS:

30% Portland cement = 855 tons × ($55/ton) = $47,025
 2% sodium silicate = 57 tons × ($200/ton) = $11,400

Total costs for treatment reagents: $58,425 $20.50/ton

LABOR COST FOR TREATMENT

1 ea Project supervisor   = $27.50/hr × 112 hr = $ 3,080
1 ea Technician   @ $18.50 = $37.00/hr × 112 hr =   2,072
2 ea Truck drivers @ $15.00 = 30.00/hr × 112 hr =   3,360
2 ea Laborers   @ $12.50 = 25.00/hr × 112 hr =   4,928

Total labor cost: = $13,440 $ 4.72/ton

Expenses: @ $75/day for 6 men 14 days   = $ 6,300 $ 2.21/ton

EQUIPMENT RENTAL:

	Capacity	Value	Per hour	Per 14 days
1 ea Mobile plant		$125,000	$82.25 =	$ 9,212
1 ea Front-end loader	2 yd	44,000	29.40 =	3,293
2 ea Dump trucks	12 yd	54,000	33.60 =	3,987
1 ea Backhoe	1.2 yd	68,000	44.70 =	5,006

Total rental cost: $21,498 $ 7.54/ton

MOBILIZATION-DEMOBILIZATION AND CLEANUP:

Labor and expenses for 4 days: $3,840 + $1,800   = $5,640
Transportation: 200 mile/trip × 2 trips × $2/mile =   800
Total $6,440 $ 2.26/ton

TOTAL COST OF TREATMENT: 500,000 gal = $106,103 $37.23/ton
    PROFIT AND OVERHEAD (@ 30% of cost) = $31,831 $11.17/ton
TOTAL CONTRACTED PRICE: 500,000 gal = $137,934 $48.40/ton

materials handling, two dump trucks to transport the raw and treated wastes, and a backhoe to load raw waste into the dump truck.

(4) An onsite disposal area is available.

(5) The onsite cost of the reagents is $55.00 per ton for Portland cement and $0.10 per pound for the sodium silicate.

(6) The remedial action site is located 200 miles from the nearest portable unit.

b. <u>Mobilization and demobilization costs</u>. Mobilization costs are those incurred in preparing the equipment for shipment, transporting the equipment, and setting the equipment up for actual waste processing. Mobilization costs include labor, equipment, and transportation costs. Demobilization costs include site and equipment cleanup and transportation of equipment back to its source. These activities are expected to take 4 days to complete. Transportation is the cost of actually transporting the equipment. There are two loads to be transported at a cost of $2.00 per load mile.

c. <u>Project duration</u>. Estimated production rate for the modular mixing plant and peripheral equipment is 180 cu yd per day. Thus about 14 working days are required to process the 2,475 cu yd (500,000 gal). This includes only the solidification activity.

d. <u>Cost summary for plant mixing of unpumpable wastes</u>. The more expensive and time-consuming handling and transportation of high solids waste which cannot be moved by pumps, or transport distances which make pumping impractical, increase the cost of the plant mixing alternative. Labor and equipment costs are about doubled in the example given here (Table 6-4) over the more easily handled, pumpable wastes so that this is the most expensive of the bulk-handling treatment options. This method does retain the precision of reagent dosing and mixing uniformity so that some efficiencies can be gained by producing treated waste with a lower proportion of treatment reagents, and therefore less total volume for disposal. This method is often the method of choice for highly toxic or hazardous wastes since the mixing process is under close control.

6.5.4 Safety and Environment

Special safety and environmental concerns associated with plant mixing include the generation of odors, organic vapors, and fugitive dust. Under normal conditions, the process is open to the atmosphere and thus presents a greater potential for problems than pumping liquid wastes or in-drum mixing. Equipment moving around the site should be decontaminated daily. Stationary processing equipment should be cleaned as operational requirements necessitate and decontaminated after project completion.

Standard personnel protective measures should be implemented as necessary, depending on the waste being handled. The reduction in production efficiency can be anticipated as a direct function of the level of protective apparatus required. Level A protection is anticipated to reduce production rate by 50 to 75 percent.

Quality control for plant mixing scenarios is expected to be better than that associated with area and in-situ mixing and similar to that obtainable with in-drum mixing. The material handling and rehandling requirements give better control of the chemical addition and mixing process; however, they also provide added potential for offsite contamination.

The solidification equipment proposed above should incorporate several fail-safe design features. First, the motor for the mixer is located outside the solidification area and contains a hand crank. This permits emptying of the mixer should the process be stopped in mid-stream due to motor failure or loss of electrical power. Maintenance on the motor can also be performed without entering a contaminated area. Second, the system flush is controlled through a flush module mounted outside the solidification area, again for maintenance purposes. The flush water is kept under pneumatic pressure at all times so that it is available even during loss of electrical power. Capped containers are inspected and tested for external contamination and decontaminated if necessary. The container is labeled and stored for shipment to the final disposal area.

6.5.5 Modifications

Both mobile and modular mixing systems have been developed for the solidification of low-level radioactive waste materials usually associated with the nuclear power industry. These facilities are similar in concept to the mobile plants that have been developed for the treatment of hazardous wastes; however, the attention given to operator safety is significantly greater than that associated with the hazardous waste plants. The primary concern is shielding of the operator and decontamination of equipment that has come in contact with the waste materials. Use of remote and automatic control systems is stressed in the nuclear environment. The emphasis on safety generally raises the cost per unit of waste treated with these systems significantly above that typically found in the treatment of hazardous wastes. The number and kinds of modifications of mobile and modular treatment facilities are as numerous as the vending companies which offer their services, as can be seen in the illustrations (Figures 6-7 to 6-14). They vary in size from large, semipermanent installations at very large sites which can treat 500 to 1,000 cu yd per day, to very small, portable units which treat 10 to 50 cu yd per day. Mixing, storage, and measuring facilities also are sized and changed to optimize the equipment for the specific job and level of hazard encountered.

A modification of the plant mixing alternative is the use of the plant to add and mix the reagents with the waste materials and then package the treated materials in drums. This modification incorporates the bulk materials handling features of plant mixing with the secure containerization features of in-drum mixing. If containerization is required, this procedure offers significant labor saving over the in-drum scenario. These savings, however, are substantially offset by the cost of the drums. Field experience indicates that approximately 300 drums per 8-hr shift could be handled using a typical portable mixing plant. Costs for this modification are anticipated to range between $30.00 and $50.00 per drum ($0.55 to $0.91 per gallon). Figure 6-15 illustrates a portable plant being used for this purpose.

Figure 6-15. Portable plant mixing followed by drum encapsulation (Courtesy Solid Tek).

6.6 Area Mixing or Layering Alternative

Area mixing, or in-place layering, provides an economical method for stabilization/solidification of homogeneous and nonhomogeneous waste liquids and sludges. The system avoids the use of conventional, stationary mixing equipment. The waste is placed in layers over the disposal area in lifts of from 2 in. to 24 in., depending upon its consistency and handling ability. The waste is then overlaid with a layer of treatment reagents which have been selected for the specific waste being treated. Once the two lifts are placed, a mechanized vehicle lifts and turns the layer much like a roto-tiller, using multiple passes. The resulting mixture is left to air dry and/or is compacted in-place using standard earth compaction equipment. Additional layers are then constructed over the lift in an identical manner until the final height of material has been attained. Typically the final lift is covered with earth, seeded, and maintained as the final cap. Alternatively, after mixing, the treated waste can be removed to a final disposal area using standard earth-moving equipment; but this may leave a very large area to clean up if hazardous wastes are being treated.

6.6.1 Project Sequencing

Project sequencing for area mixing is patterned after the construction techniques used for the soil cement or lime stabilization of roadway subbase materials. A typical project incorporates the following steps:

a. Select and prepare the onsite disposal area.

b. Excavate the untreated material from the holding lagoon and transport it to the disposal area.

c. Spread the untreated material in a lift of desired thickness on the disposal area using standard construction techniques (Figure 6-16).

d. Spread the solidification/stabilization reagents over the material in the required amount (Figure 6-17).

e. Mix the materials using a high-speed rotary mixer such as a pulvimixer. This equipment, illustrated in Figure 6-18, works in a manner similar to a large rototiller and can mix layers up to 24 in. in depth.

f. Compact the mixed material as required with standard roadway compaction equipment.

g. Repeat steps b through f until all material has been treated or until the designed depth of material has been attained.

6.6.2 Equipment Requirements.

Equipment requirements are based on the nature and quantity of waste material to be treated. An additional consideration is the location, topography, and size of the remedial action site. Minimum equipment requirements would include a backhoe, clamshell, or front-end loader to excavate the material from the holding lagoon; one or two dump trucks to haul the material to the disposal site; a motor grader, excavator, or dozer to spread the material in lifts; a high speed rotary mixer; a dry-chemical spreader; and a pneumatic-tired roller or vibratory compactor.

Depending on the size of the project, additional equipment could be efficiently added. Production rates will be a function of equipment size, mix, and quantity. Production rates ranging from 400 to 500 cu yd per day were obtained with the following equipment mix: two 10-yard dump trucks, two excavators, two chemical spreaders, two high-speed rotary mixers, two compactors, and one motor grader.

Figure 6-16. Spreading untreated material for area mixing (Courtesy Soil Recovery).

Figure 6-17. Adding stabilization/solidification reagent for area mixing (Courtesy Soil Recovery).

Figure 6-18. Mixing waste materials with stabilization/
solidification reagents in area mixing (Courtesy Soil
Recovery).

6.6.3 Costs

The procedure for estimating the cost of area mixing for solidification/
stabilization of an applicable waste is presented below and summarized in
Table 6-5.

a. <u>Assumptions</u>.

   (1) Approximately 500,000 gal (2,850 tons or 2,575 cu yd) of non-
        pumpable, high solids sludge is to be solidified using 30%
        cement and 2% sodium silicate.

   (2) The waste sludge is handleable using construction equipment
        such as a front-end loader and will support the spreading and
        mixing equipment when layered on the disposal area. Sludges
        often must be pretreated in situ with an absorbent such as fly
        ash to produce such a handleable product. Pumping lower solids
        sludges onto the disposal site to dry to a manageable solids
        content is feasible, but the additional time required and the
        low lift height attainable by this method often makes this
        option infeasible.

   (3) Onsite cost of cement is $55.00 per ton; sodium silicate is
        $200 per ton.

   (4) The waste site is 200 miles from the nearest equipment.

TABLE 6-5.   COST ESTIMATES FOR THE AREA MIXING (OR LAYERING) ALTERNATIVE

Note: Stabilization/solidification with 30% (w/w) Portland cement and 2% sodium silicate of 500,000 gal (2,850 tons) of high solids waste (85 lb/cu ft) in 12-in. lifts of waste to which a reagent layer is added and mixed with a high speed rotary mixer. Daily capacity is 250 cu yd (10 days required). Onsite disposal available.

TREATMENT REAGENTS:

```
30% Portland cement = 855 tons × ($55/ton)  = $47,025
 2% sodium silicate =  57 tons × ($200/ton) = $11,400

  Total cost of treatment reagents:             $58,425          $20.50/ton
```

LABOR COST FOR TREATMENT

```
1 ea Project supervisor        = $27.50/hr × 80 hr = $ 2,200
3 ea Heavy eq. operators @ $22 =  66.00/hr × 80 hr =   5,280
3 ea Truck drivers       @ $15 =  45.00/hr × 80 hr =   3,600
1 ea Laborer                   =  12.50/hr × 80 hr =   1,000

  Total labor cost:                              = $12,080      $ 4.24/ton

Expenses:  @ $75/day for 8 men 10 days           = $ 6,000      $ 2.11/ton
```

EQUIPMENT RENTAL:

	Capacity	Value	Per hour	Per 10 days
1 ea Front-end loader	2 yd	$44,000	$29.40 =	$ 2,352
1 ea Dump truck	12 yd	27,000	17.80 =	1,424
1 ea Chem. spreader	8 ton	22,500	14.80 =	1,184
1 ea Rotary mixer	12 ft	36,000	23.70 =	1,896
1 ea Roller compactor	14 ton	28,000	18.75 =	1,500
1 ea Motor grader	14 ton	61,500	40.63 =	3,250

```
  Total rental cost:                               $11,606    $ 4.07/ton
```

MOBILIZATION-DEMOBILIZATION AND CLEANUP:

```
Labor and expenses for 1 day:  $1208 + $600       = $1,808
Transportation:  200 mile/trip × 4 trips × $2/mile =  1,600

  Total                                             $3,408    $ 1.20/ton
```

```
TOTAL COST OF TREATMENT:     500,000 gal = $ 91,519            $32.11/ton

    PROFIT AND OVERHEAD (@ 30% of cost) = $ 27,456            $ 9.63/ton

TOTAL CONTRACTED PRICE:      500,000 gal = $118,975            $41.75/ton
```

     (5) Sufficient land area is available at the site for the complete treatment process.

b. <u>Mobilization and demobilization costs</u>. Transportation of equipment to the site is estimated to require four trips of 200 miles with flat-bed trucks @$2.00 per mile, for $1,600 total. Other than transportation costs, area mixing requires little equipment setup or break-down at the waste site since only standard construction equipment is required. One day should be sufficient for equipment cleanup. Unusual preparation of the disposal site (such as grading uneven terrain or installing leachate collection systems or final cover) is not included in these costs.

c. <u>Project duration</u>. The daily production rate, considering loading, transporting, spreading, mixing and compacting operations, is estimated to be about 250 cu yd per day when using a single loader and dump truck, and an eight-man crew (see 6.6.2, above). Therefore, approximately 10 days would be required to complete the 2,575 cu yd of waste. Some efficiencies might be realized by using a larger crew with more or larger equipment.

d. <u>Summary of area mixing costs</u>. Project costs are dependent upon the quantity of material treated, the distance to the disposal site, the amount and size of equipment used, and the type of reagents selected. Cost estimates for area mixing of 500,000 gal of waste are summarized in Table 6-5 in a form comparable with that used for the other alternatives. Total cost of treatment in this example is about $32 per ton (~$28 per cu yd) of which about 65% is for treatment reagents and about 20% each for labor and equipment.

Costs shown include disposal site preparations, excavation of waste material, transportation to treatment and disposal area, treatment reagents, and mixing and compaction of the treated product. Not included are any pretreatment costs, land cost (which may be quite high), capping and revegetation of the site, treatment of any decanted liquid, or removal of the waste to a final disposal site, if necessary. Total costs reported for actual remedial site stabilization projects including all of the above-listed parameters have run from $95 to $105 per cu yd (110 to $120 per ton).

6.6.4 Safety and Environment

Special safety and environmental concerns associated with the plant area mixing scenario are similar to those associated with in-situ and plant mixing. Of primary concern is the generation of fugitive dust, release of organic vapors, release of odors, and decontamination of equipment. Each of these areas of concern should be addressed in detail in the overall remedial action plan.

To date, the use of the area mixing scenario has generally been limited to the treatment of oil sludges and other semisolid wastes with relatively

low associated hazard levels so that little emphasis has been given to associated safety and environmental concerns. The potential for offsite release
of contaminants, particularly fugitive dust and vapor releases, should re-
ceive additional scrutiny should this scenario be adopted.

6.6.5 Modifications

Major modifications have not been identified. Modifications are
expected to be limited to the types of solidification reagents used in the
process and the types of equipment used to handle the waste materials and
solidification reagents.

6.7 Summary

The number of waste processing, handling, and mixing technologies is as
varied as the number of treatment reagent-waste formulations. Waste and site
characteristics, and reagent cost and availability are the major factors
which must be weighed in project planning to ascertain the most cost-
efficient and reliable containment strategy. This section has discussed a
representative sampling of possible stabilization/solidification scenarios,
all of which are currently available commercially. This should give the
reader a good understanding of the wide diversity of applicable technology
now in use. A formal decision process outline as recommended for remedial
action alternatives is discussed in an EPA Guidance Manual (U.S. EPA 1983).

6.7.1 Comparison of Treatment Alternative Costs

Attributes of the four stabilization/solidification alternatives discussed in this section are summarized in Table 6-6. Similar assumptions were
used in all of the alternative cost estimates, as were production rates from
actual equipment now in use at remedial action sites. It is emphasized that
these estimates are for comparison purposes only and cannot be extended to
specific wastes and/or sites, as cost and reliability of all processing technologies are quite waste- and site-specific.

In-drum mixing is by far the most expensive and takes the greatest
amount of production time due obviously to the very small quantities processed in each batch. Mixing done inside the drum is reasonably complete but
difficulties are often encountered in the corners, especially if the complete
top of the drum cannot be removed. In-drum mixing is most applicable to
sites which have a wide variety of incompatible and highly toxic wastes which
occur in individual drums. Since each drum must be analyzed individually (an
expense not included in the estimates), customized formulations of reagents
and mixing times can be determined for each drum or waste type. The cost of
reagent is a small fraction of the whole (generally less than 10%), while
labor and equipment make up about half of the total cost. If sufficient

TABLE 6-6.   SUMMARY COMPARISON OF RELATIVE COSTS FOR STABILIZATION/
SOLIDIFICATION ALTERNATIVES

Parameter	In-drum	In-situ	Plant Mixing Pumpable	Plant Mixing Unpumpable	Area mixing
NOTE: In all cases, 500,000 gal (2,850 tons) of waste was treated with 30% Portland cement and 2% sodium silicate with onsite disposal; costs include only those operations necessary for treatment. All costs are per ton of waste treated. Data taken from Tables 6-1 through 6-5.					
Metering and mixing efficiency	Good	Fair	Excellent	Excellent	Good
Processing days required	374	4	10	14	10
Cost/ton					
Reagent	$ 20.50 (9%)*	$20.50 (63%)	$20.50 (53%)	$20.50 (42%)	$20.50 (49%)
Labor and per diem	51.07 (23%)	1.36 (4%)	3.83 (10%)	6.93 (14%)	$ 6.35 (15%)
Equipment rental	37.14 (17%)	1.38 (4%)	3.93 (10%)	7.54 (16%)	4.07 (10%)
Used drums @ $11/drum	48.18 (21%)	–	–	–	–
Mobilization-demobilization	15.68 (7%)	1.58 (5%)	1.43 (4%)	2.26 (5%)	1.20 (3%)
Cost of treatment process	$172.57	24.83	29.69	37.23	32.11
Profit and overhead (30%)	51.72 (23%)	7.45 (23%)	8.91 (23%)	11.17 (23%)	9.63 (23%)
TOTAL COST/TON	$224.29	32.28	38.60	48.40	41.75

* % of total cost/ton for that alternative.

drums of identical or compatible wastes are found, it is much more economical
to bulk the wastes and use other mixing techniques, as this greatly decreases
cost and increases mixing efficiencies. This is also true when it is desired
to place the treated waste back into drums, either for ease of handling or for
increased, short-term containment; the output from bulk mixers usually can be
easily loaded directly into new drums or rinsed original drums.

The remaining bulk mixing alternatives are much more consistent in cost
and production rates, the two handling liquid or pumpable wastes being the
less expensive alternatives. All are quite sensitive to reagent cost since
it typically makes up from 40 to 65% of the total cost. The in-situ technique is the fastest and most economical of the bulk methods because the
wastes typically need to be handled only once, or not at all if they are to
be left in place, as is done with most nonhazardous wastes--only the reagent
is handled. Labor and equipment each make up less than 5% of the total
treatment cost. However, in-situ mixing is the least reliable because of
difficulties in accurate reagent measurement and in getting uniform and/or
complete mixing of wastes and treatment reagents. Also, in-situ mixing requires a liquid or a semisolid sludge. If the wastes are to be left in
place, the waste site must be dedicated as the final waste disposal area. In
some cases, liquid or sludge wastes are stabilized or solidified in situ so
that they can later be removed from the site using standard earth-moving
equipment.

Mobile or modular mixing plants, although giving excellent mixing and
relatively high production rates, require that both the untreated waste and
the treated product be handled. The cheapest and fastest material handling
technique is that in which the waste can be pumped directly from the waste
lagoon, mixed, and then pumped to the final disposal site. Pumpable waste
can be treated for about 15% less, in which case, labor and equipment cost
each make up only about 10% of the total treatment cost. Nonpumpable waste
requires more manpower and machinery for material handling and transport so
that labor and equipment costs each increase to around 15%. Plant mixing
scenarios are probably the most used alternatives for large amounts of bulk
or drummed waste which have a high degree of hazard, as the wastes are always
under control of the operators, and reagent dosing is the most accurate and
the mixing the most complete of any of the bulk processes.

In area mixing technology, the waste is usually moved only once to the
final disposal site where it is mixed and compacted in place. The waste can
be removed to another site if needed, but this lessens the other benefits of
the technique and leaves large areas to be cleaned up. Very large and standard construction equipment can be used for increased efficiency. Major disadvantages of this technique are that larger land areas are often necessary,
and mixing reagent dosing cannot be as accurately controlled.

6.7.2 How Using Different Treatment Reagents Affects Cost

For comparison purposes, all treatment alternatives were developed using
the 30% Portland cement and 2% sodium silicate formulation which is about

average in reagent cost. However, this formulation is not really universal as implied. It lends itself especially to in-drum and plant mixing techniques with their better mixing efficiencies, and to inorganic, aqueous sludges with toxic heavy metals. In-situ and area mixing techniques do not usually lend themselves to the addition of liquid reagents (although it has been done) or to formulations where uniform and/or extensive mixing are necessary. The higher unit cost of these reagents tends to limit their use to those techniques with good mixing efficiencies.

Table 6-7 compares the costs of the four alternatives using different amounts of other common treatment reagents with different delivered cost. In these examples it is assumed that the change in reagents will not affect equipment requirements or production rates. Total cost of each alternative and proportional cost of the reagent only are shown in each case.

Changing reagent costs from $34/ton to $0/ton has only a small effect on the total cost of in-drum mixing since it is labor- and equipment-intensive. In-situ mixing is the most sensitive to reagent cost, since it is by far the largest part of the total cost of this technique. Other bulk mixing techniques are also quite sensitive to reagent costs; as reagent costs decrease, the proportional differences among the four increase, but their ranking remains the same. The sensitivity of total treatment cost to delivered reagent price is well illustrated in these calculations.

Reagent costs for the other waste products such as fly ash, cement or lime kiln dust, or furnace slag are highly variable. The major component of their cost is usually transportation to the site. The reagent used is typically based upon the nearest source of suitable pozzolanic materials and not through preference of one over the others. As these waste materials have been incorporated into waste treatment systems, they have come to have appreciable value in some areas.

TABLE 6-7.   COMPARISON OF TREATMENT COSTS WITH DIFFERENT REAGENTS

Reagent type, amount, and cost	In-drum	In-situ	Plant Mixing		Area mixing
			Pumpable	Unpumpable	
1. 80% fly ash (Type F) @ $30/ton, 20% lime @ $50/ton Total reagent cost/ton of waste = $34					
Reagent cost	12.5%	68%	60%	52%	57%
Total cost/ton	$237.06	$49.89	$56.15	$65.95	$59.30
2. 30% Portland cement @ $55/ton, 2% sodium silicate @ $200/ton Total reagent cost/ton of waste = $20.50					
Reagent cost	9%	63%	53%	42%	49%
Total cost/ton	$224.29	$23.28	$38.60	$48.40	$41.75
3. 50% fly ash (Type C) @ $20/ton Total reagent cost/ton of waste = $10					
Reagent cost	4%	54%	40%	29%	36%
Total cost/ton	$209.90	$18.63	$24.95	$34.75	$28.04
4. Free reagent (including delivery)					
Reagent cost	0%	0%	0%	0%	0%
Total cost/ton	$198.57	$5.63	$11.95	$21.75	$15.10

NOTE: Data are from Table 6-6. They have been recalculated for different reagent cost, but for the same equipment, project duration, and mobilization costs. All reagent proportions are in weight of reagent per weight of waste.

REFERENCE

U.S. EPA. 1985. Guidance on Feasibility Studies under CERCLA. EPA-540/
G-85-003, Office of Emergency and Remedial Response, U.S. Environmental
Protection Agency, Washington, D.C., 103 pp.

BIBLIOGRAPHY

Carson, A. B. 1961. General Excavation Methods. F. W. Dodge Corporation,
New York, New York.

Caterpillar Tractor Co. 1981. Handbook of Earthmoving. Caterpillar Tractor
Company, Peoria, Illinois.

Caterpillar Tractor Co. 1982. Caterpillar Performance Handbook. Caterpillar
Tractor Company, Peoria, Illinois.

Gallagher, G. A. 1981. Health and Safety Program for Hazardous Waste Site
Investigation. New England Section of the Association of Engineering
Geologists, Boston, Massachusetts.

Perry, R. H. 1973. Chemical Engineers' Handbook. McGraw-Hill Book Company,
New York, New York.

Peurifoy, R. L. 1956. Construction Planning Equipment and Methods. McGraw-
Hill Book Company, New York, New York.

Peurifoy, R. L. 1975. Estimating Construction Costs. McGraw-Hill Book
Company, New York, New York.

Robert Snow Means Company, Inc. 1983. Site Work Cost Data. Construction
Consultants & Publishers, Kingston, Massachusetts.

Terex. 1981. Production and Cost Estimating of Material Movement with
Earthmoving Equipment. Terex Corporation.

U.S. Army Corps of Engineers. 1980. General Safety Requirements Manual.
Washington, D.C.

U.S. Army Corps of Engineers. 1983. Preliminary Guidelines for Selection
and Design of Remedial Systems for Uncontrolled Hazardous Waste Sites.
Washington, D.C.

U.S. EPA. 1980. Closure of Waste Surface Impoundments. SW-873. U.S. Environmental Protection Agency, Office of Waste Management,
Washington, D.C.

U.S. EPA. 1981. Interim Standard Operating Safety Procedures. U.S. Environmental Protection Agency, Washington, D.C.

U.S. EPA. 1982. Inplace Closure of Hazardous Waste Surface Impoundments. U.S. Environmental Protection Agency. Municipal Environmental Research Laboratory, Cincinnati, Ohio.

U.S. EPA. 1985a. Drum Handling Practices at Hazardous Waste Sites (Draft). U.S. Environmental Protection Agency, Municipal Environmental Research Laboratory, Office of Research and Development, Cincinnati, Ohio.

U.S. EPA. 1985b. Remedial Action at Waste Disposal Sites (Revised). EPA-625/6-85-006, Municipal Environmental Research Laboratory, U.S. Environmental Protection Agency, Cincinnati, Ohio.

U.S. EPA. 1985c. Guidance on Remedial Investigations under CERCLA. EPA-540/G-85-002. Office of Emergency and Remedial Response. U.S. Environmental Protection Agency, Washington, D.C.

U.S. EPA. 1985d. Remedial Action at Waste Disposal Sites. EPA-625/6-85-006. U.S. Environmental Protection Agency, Washington, D.C.

7. Quality Control, Safety, and Environmental Considerations for Waste Treatment

The waste stabilization and solidification processes are similar to any chemical treatment operation in that the product must be periodically tested to ensure that the physical integrity and containment characteristics are adequate. The treated waste must be sampled in such a way that representative material is obtained and tested using reliable screening tests to verify performance.

7.1 Sampling of Treated Wastes

Stabilization and solidification systems which are batch operations bear some similarity to batch cement blending systems. Approaches similar to those for fresh concrete can be employed for fluid waste, whereas cured material can be sampled using sampling techniques employed with hardened concrete. Standard ASTM method C 172-71, Standard Method of Sampling Fresh Concrete (U.S. Army 1949; CRD-C4-71) outlines procedures to be used in taking samples from stationary and truck mixers, paving machines, and agitating and non-agitating concrete transports. Standard method CRD-C 620-80 outlines techniques for sampling grouts from mixers, pumps, and discharge lines (U.S. Army 1949).

For solidified or hardened concrete, techniques such as those recommended in ASTM C 823-75 (U.S. Army 1949) or in Abdun-Nur (1978) can be used. In general, careful visual inspection and selected sampling can be used to augment purely random approaches. The objective of any waste testing program is to ensure complete treatment of all materials so that nonrandom testing in areas of poorly performing waste (for example, materials that fail to solidify or have excessive weep water) is justified. If waste treatment is a batch operation, each successive batch should be tested. Some solidification systems that are used with flue gas cleaning wastes have similar problems with regard to producing a consistent set. Interim ponding systems where the treated sludge is allowed to cure for 30 days have been developed to ensure that treatment is complete before disposal. This approach requires double handling of the treated wastes, but it ensures that unsatisfactory materials can be retrieved for reprocessing (Duvel et al. 1978).

7.2   Testing of Stabilized and Solidified Wastes

Early testing of any product that cures slowly presents problems in that
the ability to predict the final properties of the cured material from short-
term tests is generally poor (Arni 1978). This problem has been thoroughly ·
studied with regard to early strength development in concrete, and no generally satisfactory testing and prediction system has evolved.

In a waste treatment system where the treated material must be placed in
a land disposal area shortly after treatment, it is necessary to develop
testing that will ensure waste containment in a minimum period of time. This
testing can take the form of early strength testing (24-hr compressive
strength) and leach testing of cured, ground material (where strength is not
a primary consideration). Details of the types of testing that can be used
for these purposes are given in Sections 3 and 4 of this report.

7.3   Safety and Environment

In this handbook the solidification/stabilization process is considered
to be a subset of the remedial action plan as a whole. As such, it may be
assumed that the environmental and safety aspects of the solidification/
stabilization process will be addressed in development of the overall remedial action plan. A brief summary of the major safety and environmental
aspects of a solidification/stabilization project is presented in the following paragraphs. Detailed safety and environmental guidance may be found
in the following publications:

a.   Chemical Manufacturer's Association, Inc. 1982. Hazardous Waste
     Site Management Plan, Washington, D.C.

b.   Environmental Protection Agency. 1981. Hazardous Materials Inci-
     dent Response Operations: Training Manual. National Training and
     Operational Technology Center, Cincinnati, Ohio.

c.   Environmental Protection Agency. 1981. Technical Methods for
     Investigating Sites Containing Hazardous Substances Training Pro-
     gram. Technical Monograph Nos. 2, 3, and 12.

d.   Environmental Protection Agency. 1985. Remedial Action at Waste
     Disposal Sites. EPA 625/6-85-006, Office of Emergency and Remedial
     Response, Washington, D.C.

e.   Meluold, R. W., S. C. Gibson, and M. D. Rogers. 1981. Safety Pro-
     tection for Hazardous Materials Cleaning: Management of Uncon-
     trolled Hazardous Waste Sites. American Society of Civil Engineers,
     New York, New York.

    f. U.S. Army Corps of Engineers. 1984. Preliminary Guidelines for
       Selection of Remedial Actions for Hazardous Waste Sites.
       EM 1110-2-505 (Draft), Washington, D.C.

7.3.1 Safety

Safety concerns associated with solidification/stabilization of hazardous wastes are primarily related to the protection of onsite personnel. These concerns can be addressed through development of a Personnel Protection Program (PPP). At a minimum, the PPP should include the following elements:

a. Medical Surveillance Plan.

b. Industrial Hygiene Support Plan.

c. Employee Training Plan.

d. Entry Control Plan.

e. Respiratory Protection Plan.

f. Eye Protection Plan.

g. Skin Protection Plan.

h. Personnel and Equipment Decontamination Plan.

i. Emergency Response Plan.

j. Record Keeping and Reporting Plan.

The detailed requirements of the PPP must be developed on a site-specific basis. Obviously, the more hazardous the waste, the more rigorous must be the PPP.

Good management and work practices, as well as legal requirements, emphasize the need for placing top priority on the health and safety of the worker. Various legal and regulatory requirements establish the minimum guidelines for the development and implementation of a comprehensive health and safety program. The Occupational Safety and Health Administration (OSHA) has established regulations designed to decrease accidents associated with the construction site. Many of these requirements are also applicable to the solidification/stabilization process itself. The regulations may be found in Title 29 of the Code of Federal Regulations. Examples of the specific parts and subparts most likely to apply to the solidification/stabilization scenarios are listed in Table 7-1. Compliance with applicable OSHA regulations should be a mandatory requirement of the PPP. In addition, the EPA has referenced various policies and mandatory requirements for occupational health and safety. A listing of pertinent documents is presented in Table 7-2.

TABLE 7-1.   CITATIONS FOR CURRENT OSHA REGULATIONS LIKELY TO BE
APPLICABLE AT LAND-BASED DISPOSAL SITES

<u>29 CFR Part 1926</u>

Subpart D   Occupational Health and Environmental Controls
(Sections 1926.50 through 1926.57)

Subpart E   Personal Protection
(Sections 1926.100 through 1926.107)

Subpart F   Fire Protection
(Sections 1926.150 through 1926.155)

Subpart G   Signs and Signals
(Sections 1926.200 through 1926.203)

Subpart L   Ladders and Scaffolding
(Sections 1926.450 through 1926.452)

Subpart O   Mechanical Handling Equipment
(Sections 1926.600 through 1926.606)

Subpart P   Excavation and Trenching
(Sections 1926.650 through 1926.653)

Subpart S   Tunnels and Shafts
(Sections 1926.800 through 1926.804)

Subpart U   Blasting and Explosives
(Sections 1926.900 through 1926.914)

<u>29 CFR Part 1910</u>

Subpart Z   Toxic and Hazardous Substances
(Sections 1910.1000 through 1910.1046)

TABLE 7-2.   POLICIES APPLICABLE TO REMEDIAL ACTIONS

EM 385-1-1, Safety and Health Requirements Manual.

29 CFR 1910, Parts 16, 94, 96, 106, 109, 111, 134, 151, Occupational Health and Safety Standards.

Executive Order 12196, Section 1-201, Sec. (k), Occupational Health and Safety Programs for Federal Employees.

29 CFR 1960.20 (1), Occupational Safety and Health for the Federal Employee.

EPA Occupational Health and Safety Manual, Chapter 7 (1).

EPA Training and Development Manual, Chapter 3, Par. 7 (b).

Occupational Health and Safety Act of 1971, PL 91-596, Sec. 6.

EPA Order on Respiratory Protection (Proposed).

49 CFR, Parts 100-177, Transportation of Hazardous Materials.

EPA Order 1000.18, Transportation of Hazardous Materials.

EPA Order 3100.1, Uniforms, Protective Clothing, and Protective Equipment.

7.3.2   Environment

Environmental concerns during the remedial action project are primarily related to waste containment, to retention of the environment in its natural state to the greatest extent possible, and to the enhancement of site appearance in its final condition. Environmental protection as applied to the remedial action as a whole generally includes consideration of air, water, and land resources. As specifically applied to the solidification/stabilization processes, environmental considerations include the elimination of the spread of contamination through minimization of organic vapor and/or fugitive dust generation, decontamination of personnel and equipment, and prevention and control of spills.

7.3.2.1   Organic Vapor and Dust Generation

Depending on the nature of the wastes found at a site and the solidification/stabilization reagent selected, the possibility exists for a release of volatile organic compounds which may have an adverse impact on

public health. Objectives of the remedial action project must include minimizing the release of organic vapors and monitoring onsite and offsite to
measure concentrations and types of vapors that may be released. The potential for volatile organic vapor generation should be addressed during the
bench or pilot study phase (Section 5 of this Handbook) of the
solidification/ stabilization scenario selection process. Other than
elimination or minimization of the generation of organic vapors by a judi-
cious selection process, few technical options are available for control of
vapors. The general approach has been limited to the monitoring of organic
vapors. Both onsite and site-perimeter monitoring are recommended. Area-
type monitoring should be conducted on a periodic basis to determine whether
contaminants are migrating out of the contaminated area.

Migration of contaminants through transport of airborne particulates
(fugitive dust) could present a significant health and environmental hazard
during remedial action activities. Such hazards are particularly likely with
large-scale solidification/stabilization scenarios such as in-situ mixing and
area mixing. Fugitive dust that could cause a hazard or nuisance to others
must be eliminated.

The meteorological conditions at the site will strongly influence the
potential for this fugitive dust problems. Hot, dry, windy conditions produce the greatest potential for entrainment and transport of contaminants.
The solidification/stabilization reagent and application scenario, as well as
the waste being treated, will also affect the amount of fugitive dust
formation.

Techniques that can be used during the solidification/stabilization
process to mitigate airborne particulate transport include the following:

    a. Minimizing the rehandling of waste materials.

    b. Erecting portable wind screens.

    c. Applying surface stabilizers or dust palliatives.

    d. Using portable surface covers on the work area during periods of
       inactivity.

    e. Constructing temporary enclosures around the solidification/
       stabilization processing area.

7.3.2.2 Equipment and Personnel Decontamination

Although a maximum effort is made to prevent contamination of personnel and equipment, such contamination will inevitably occur as a result of
contact with the wastes being treated. Contamination may occur in a number
of ways, including the following:

a. Contacting vapors, gases, mists, or particulates in the air.

b. Being splashed by materials while sampling, opening containers, or conducting the solidification/stabilization process.

c. Walking through puddles of liquids or on contaminated materials.

d. Using contaminated instruments or equipment.

To prevent the spread of contaminants, methods for reducing contamination and decontamination procedures must be developed before the initiation of site operations. Decontamination consists of physically removing the contaminants and/or changing their chemical nature to innocuous substances. The nature and extent of the required decontamination process depends on a number of factors, the most important of which is the type of contaminants being solidified. This topic is treated further in Section 8, Cleanup and Closure.

7.3.2.3 Spill Control

Another important environmental concern is preventing the spread of contamination through spills. A continuous effort should be made to prevent any spillage of contaminated materials during the solidification/ stabilization process. A spill control program should as a minimum provide all physical controls possible in areas where spills are likely to occur and proceed in a deliberate and controlled fashion in handling all hazardous materials. Activities presenting the highest probability of material spillage include the transfer of liquid or solid material to a staging area, handling of deteriorated drums of liquid waste, and staging of liquid waste. During solidification/stabilization operations, preventing spills is the responsibility of all workmen at the site.

REFERENCES

Abdun-Nur, E. A. 1978. Techniques, Procedures, and Practices of Sampling of
Concrete and Concrete-Making Materials. In: Significance of Tests and Properties of Concrete and Concrete-Making Materials, ASTM Publ 169B, American
Society for Testing and Materials (ASTM), Philadelphia, Pennsylvania.
pp. 5-23.

Arni, H. T. 1978. Statistical Considerations in Sampling and Testing. In:
Significance of Tests and Properties of Concrete and Concrete-Making Mate-
rials, ASTM Publ. 169B, American Society for Testing and Materials (ASTM),
Philadelphia, Pennsylvania. pp. 24-43.

Duvel, W. A., Jr., et al. 1978. State-of-the-Art of FGD Sludge Fixation.
Publication FP-671, Vol. 3, Electric Power Research Institute, Palo Alto,
California, not paginated.

U.S. Army. 1949. Handbook for Concrete and Cement, Vols. I and II.
U.S. Army Engineer Waterways Experiment Station, Vicksburg, Mississippi,
unpaginated, loose leaf.

8. Cleanup and Closure

After completion of waste treatment and the final placement of the stabilized and solidified waste, it will be necessary to ensure that all equipment is adequately cleaned to prevent material from moving offsite and that the plans for monitoring are implemented in a timely fashion. Programs for decontaminating equipment are generally part of the safety planning involved in the site activity. The postclosure monitoring program is developed as part of the master plan for site closure. Examples of cleanup and closure activities at actual remedial sites are found in EPA (1984).

8.1 Cleanup of Equipment

Stabilization and solidification require extensive mixing and materials handling equipment. Decontamination of equipment may require high-pressure washing systems and manual scraping. Most mixers are cleaned by putting clean material in the mixer and cycling through several mixing operations.

Discarded equipment and cleaning water must be treated as a contaminated waste and be disposed of in an EPA-approved manner. Where residual contamination of equipment is suspected, a swabbing or rinsing procedure and chemical analysis of swabs and rinse water can be used to confirm the effectiveness of the cleaning procedure.

8.2 Site Monitoring

A monitoring system is routinely established at any remedial action site before, during, and after cleanup operations. This system ensures that no adverse impact to air, surface water, or ground water occurs during the remedial activities. These monitoring activities would normally continue after site closure to evaluate the effects of remediation and to act as an early warning system for possible breakdown of liners or other containment structures (EPA 1985a).

If the remedial program involves leaving stabilized or solidified wastes onsite, the monitoring should be designed to ensure that the treated wastes do not become a new source of air or water pollution. The solidified wastes are designed to provide the needed waste containment; therefore, placement of monitoring wells directly under or adjacent to the solidified waste should be considered in developing the postclosure monitoring plan.

Even structural concrete can break down from exposure and weathering; therefore, the possibility of solidified materials disintegrating or chemical stabilization systems being defeated in natural weathering processes must be considered in monitoring. For example, sulfate-rich ground water can cause swelling and disintegration of Portland-cement/fly-ash-solidified waste, or leaching by rainwater can remove buffering materials in a stabilized waste and allow the pH to drop and metals to be taken into solution in contacting water. If the breakdown of the treated waste is a possible problem, the monitoring program should include the coring and retrieval of solidified waste for leaching tests. Test holes in the wastes can also be filled with clean water, and in-situ leaching rates can be determined.

8.3 Capping of Solidified Wastes

Most solidified wastes are not designed for constant exposure to weathering. Freezing and thawing and wetting and drying can cause the material to fragment badly (Bartos and Palermo 1977). A cap that is thick enough to ensure that the solidified material maintains uniform moisture and is not subjected to freezing is necessary to ensure that the waste does not deteriorate. The cap also should minimize the percolation of water into the waste.

Details on the design of closures are given in Brown and Associates (1982) and Wyss et al. (1980). Selection of soils and vegetation for capping landfills is discussed in detail in Lutton et al. (1979) and for solid hazardous waste in Lutton (1982) and U.S. EPA (1985b).

A program for the periodic inspection and maintenance of the waste caps is generally part of a remedial site master plan.

REFERENCES

Bartos, M. J., Jr., and M. R. Palermo. 1977. Physical and Engineering Properties of Hazardous Industrial Wastes and Sludges. EPA-600/2-77-139, U.S. Environmental Protection Agency, Cincinnati, Ohio. 89 pp.

Brown, K. W., and Associates. 1982. Inplace Closure of Hazardous Waste Surface Impoundments. Draft Report, U.S. Environmental Protection Agency Contract 68-03-2943. 92 pp.

Lutton, R. J., G. L. Regan, and L. W. Jones. 1979. Design and Construction of Covers for Solid Waste Landfills. EPA 600/2-79-165, U.S. Environmental Protection Agency, Cincinnati, Ohio. 274 pp.

Lutton, R. J. 1982. Evaluating Cover Systems for Solid and Hazardous Waste. SW-872 (NTIS-PB81-181505). U.S. Environmental Protection Agency, Washington, D.C.

U.S. EPA. 1984. Case Studies 1-23: Remedial Response at Hazardous Waste Sites. EPA-540/2-84-0026, Office of Emergency and Remedial Response, U.S. Environmental Protection Agency, Washington, D.C. 637 pp.

U.S. EPA. 1985a. Remedial Action at Waste Disposal Sites (Revised). EPA-625/6-85-006, U.S. Environmental Protection Agency, Cincinnati, Ohio. 497 pp.

U.S. EPA. 1985b. Covers for Uncontrolled Hazardous Waste Sites. EPA-540/2-85-002. U.S. Environmental Protection Agency, Cincinnati, Ohio.

Wyss, A. W., et al. 1980. Closure of Hazardous Waste Surface Impoundments. EPA-530/SW-873, U.S. Environmental Protection Agency, Cincinnati, Ohio. 100 pp.

Appendix A—Acquisition and Costs of Reagents

One of the items of concern which is associated with all onsite solidification/stabilization alternatives is the ability to obtain the necessary process chemicals and transport them to the proposed project site at reasonable cost. Onsite cost of the required chemicals is a major portion of overall project costs. The cost of chemicals associated with an onsite solidification/stabilization project includes the purchase price from the manufacturer, transportation cost from the point of manufacture to the point of use, cost of onsite storage and handling of the chemicals, and the quantity of chemicals required for a particular project.

A.1 Purchase Price

The purchase price of chemicals is usually the most significant cost associated with the total cost of chemicals for an onsite solidification/ stabilization project. Generally, prices are quoted as free on board (f.o.b.) at the manufacturer's plant. The price for chemicals varies from day to day and is a function of a variety of factors including the cost of raw materials and manufacturing at a particular plant location, the current demand for the product as reflected by general economic conditions, the quantity of chemicals to be purchased, the nature of the shipment (e.g. bulk versus bag for cement), and the reactivity of the material.

The major chemicals or materials used in the solidification/ stabilization of hazardous wastes are products associated with the construction industry. For this reason, the cost of these materials is strongly related to construction activity. An example is the availability and cost of Portland cement. Increased construction activity results in increased demand which tends to drive prices up. Likewise, decreased construction activity has the opposite effect. Note that this effect is also noticeable in the secondary materials, i.e. cement-kiln dust and lime-kiln dust.

The results of an April 1983 survey of chemical costs for materials commonly used for solidification/stabilization are presented in Table A-1. These costs represent telephone quotes for the materials f.o.b. at the points of manufacture. A wide range of prices can be noted. This range represents geographic differences in material costs. Also note that these prices are probably depressed because of the recent slump in major construction activity. These prices are presented for comparison purposes only.

TABLE A-1. TYPICAL COSTS OF CHEMICALS USED FOR STABILIZATION/
SOLIDIFICATION (APRIL 1983)

Chemical	Units	Cost Range	
Portland cement	\$/ton* (bulk)	\$40	– \$65
Portland cement	\$/ton (bag)	70	– 85
Quick lime (CaO)	\$/ton (bulk)	45	– 55
Hydrated lime ($Ca(OH)_2$)	\$/ton (bulk)	45	– 55
Hydrated lime ($Ca(OH)_2$)	\$/ton (bag)	60	– 75
Cement kiln dust	\$/ton	5	– 25
Waste quick lime	\$/ton	4	– 10
Fly ash	\$/ton	0	– 40
Gypsum	\$/ton	0	– 35
Sodium silicate	\$/pound	0.05	– 0.20
Concrete admixtures	\$/gallon	1.50	– 9.00

* Customary units are used because price quotations are made in these units.
All prices f.o.b. at point of manufacture.

A.2 Transportation Costs

The cost of transporting chemicals from the point of manufacture to the
point of use is generally the second most costly item associated with the
total onsite chemical cost for a solidification/stabilization project. In
those cases where waste materials (kiln dust) are used as the solidification/
stabilization agent, the cost of transportation may actually exceed the cost
of the material itself. The materials associated with solidification/
stabilization are commonly shipped by rail or truck. For application at
remedial action sites, truck haulage has the particular advantage of geographic flexibility, which limits consideration of rail transportation. Therefore, for purposes of this discussion, the costs of chemical transportation
to the project site are based on haulage by trucks.

The cost of chemical transportation is primarily a function of the characteristics of the material being handled (specific weight, liquid versus
solid, etc.), the quantity of material being transported, the nature of packaging (bulk versus container), the distance over which the material is transported, and the type of carrier performing the transport services.

Because of the quantity of materials required at a typical remedial
action site, bulk transport is generally the method used for obtaining the
required chemicals. Truck types used for the movement of bulk materials are
essentially limited to two: dump trucks (open top with tarpaulin cover)
and/or tank-type trucks. Dump trucks are very commonplace and are used for

hauling a variety of materials over relatively short distances. Tank-type trucks are often used in the transport of lime and cement products. The tank-type truck is fully enclosed and is loaded and unloaded pneumatically. The time required to unload the tank-type truck is considerably longer than the dump truck; however, the material is not exposed to the weather, which is a definite advantage. Each type of truck is capable of transporting payloads in the 40,000- to 50,000-pound (20,000- to 25,000-kg) range. The actual pay-load capacity depends primarily on the specific weight of the material being transported. The tank-type truck is the primary type of carrier employed for the transportation of materials associated with onsite solidification/ stabilization projects.

Transportation rates are generally established as a tariff in the case of common carriers, or they are negotiated between the carrier and the manu-facturer in the case of contract carriers. For planning purposes, it is easier to develop costs based on common carrier tariffs. Note, however, that these tariffs can vary significantly within a region and certainly across the Nation. The basis for a tariff may vary between carriers in such areas as minimum load and distance traveled. At the planning stage, it is somewhat difficult to compare tariffs directly. In any event, the chemical manufac-turer or supplier generally arranges transportation to the site.

Figure A-1 presents typical transportation costs of major chemicals associated with solidification/stabilization technology. The costs presented include the cost of transporting the material from the place of manufacture to the project site. The manufacturer pays loading costs. In the case of bulk shipments, the rate includes the cost of unloading. In the case where packaged materials (lime or cement in bags) are transported, the person to whom the materials are shipped is usually responsible for unloading services (i.e. forklifts, etc.). Bag shipments are usually palletized for easy off-loading.

The basic transportation cost will generally include a free time to effect unloading. Typical free time ranges from 1-1/2 to 3 hr. Should un-loading fail to be accomplished in this time frame, demurrage will be charged. These demurrage rates are highly variable and are a function of the demand for transportation services. Typical demurrage rates range from $20 to $50 per hour.

A.3 Onsite Chemical Handling

When compared with the purchase costs and transportation costs, the on-site handling costs of solidification/stabilization chemicals are usually minimal. Onsite handling costs incorporate those costs relating to the stor-age and handling of the chemicals between the time of delivery and the mixing of chemicals with the wastes being treated. The costs of onsite chemical handling are a function of the method of materials delivery (containers or bulk), the nature and quantity of materials being handled, the method of storage, and the method used to mix the chemicals with the waste being treated. Many of these factors are interrelated and difficult to define.

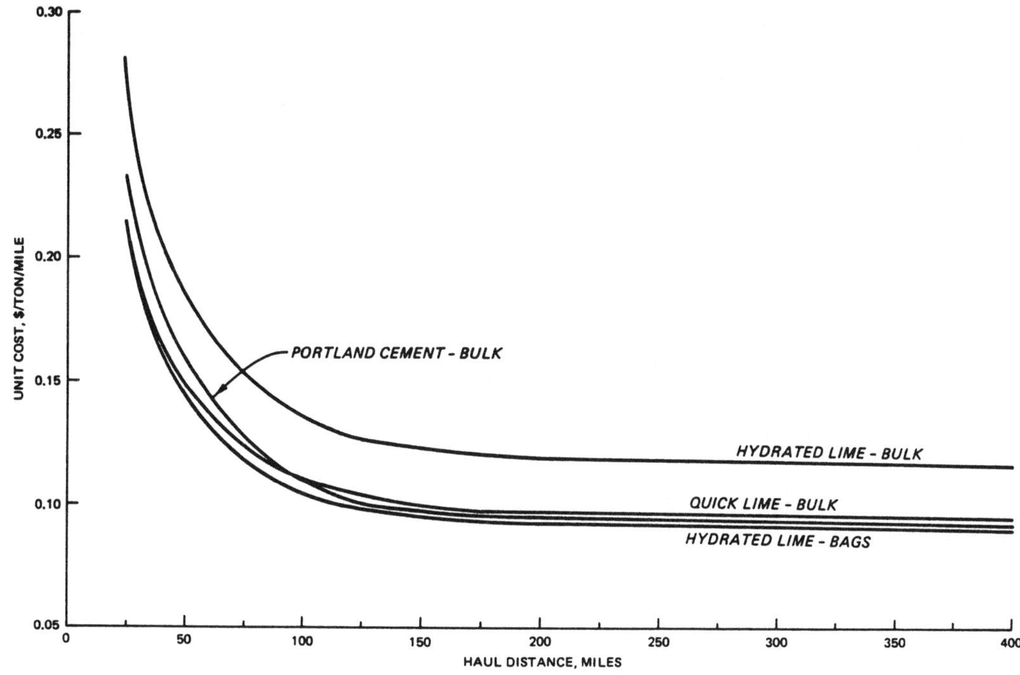

Figure A-1. Typical chemical transportation costs.

Total costs for onsite chemical handling are expected to range from $0.10 per ton of chemical handled for automated conveyor or pneumatic systems to as high as $0.50 per ton of added chemical for manual addition methods.

A.4   Quantity and Cost of Chemicals Required

The quantity of chemicals required on a specific remedial action project is the driving force behind all other costs associated with the total onsite chemical costs. The cost of chemicals can represent up to 95 percent of the total cost of an onsite solidification/stabilization remedial action project. The quantity of reagents required to ensure adequate performance of a particular process are usually determined through pilot- or laboratory-scale studies. Reagent requirements can be determined on the basis of volume of reagent per volume of waste, or weight of reagent to weight of waste. For pilot or laboratory studies, it is often easy to determine requirements on a weight/ weight basis. Results are usually expressed on a percentage basis (i.e., 20 percent by weight Portland cement to be added).

In the field, it is often more convenient to measure the quantities of wastes on a volume basis such as gallons or cubic yards to be treated. The relationship between volume and weight is expressed as a specific weight, usually in units such as pounds per cubic foot or pounds per cubic yard or

metric equivalents. The specific weight for materials may vary depending on
the condition of the material (i.e., natural state, disturbed state, compacted state, etc.). Specific weights for typical materials are presented in
Table A-2. Once the volume of waste material to be treated is determined
from field surveys, the total weight of material to be treated can be determined by multiplying the volume by the estimated (or measured) specific
weight.

Once the total weight of waste materials to be treated is determined,
the total quantity of reagents required can be determined using the results
of the pilot- or laboratory-scale studies. The weight of reagents required
is simply the reagent percent by weight obtained from the pilot or laboratory
study multiplied by the total weight of onsite material.

TABLE A-2.   SPECIFIC WEIGHTS FOR COMMON MATERIALS AT REMEDIAL
ACTION SITES

Material	Weight in bank (lb/BCY)*	Percent swell	Swell factor	Loose weight (lb/LCY)
Ashes, hard coal	700-1,000	8	0.93	650-930
Ashes, soft coal, ordinary	1,080-1,215	8	0.93	1,000-1,130
Ashes, soft coal w/clinkers	1,000-1,515	8	0.93	930-1,410
Cement	2,970	20	0.83	2,465
Clay, natural bed	3,400	22	0.82	2,800
Clay, dry	3,100	23	0.81	2,510
Clay, wet	3,500	25	0.80	2,800
Clay with gravel, dry	2,800	18	0.85	2,380
Clay with gravel, wet	3,100	18	0.85	2,640
Earth, top soil	2,350-2,550	43	0.70	1,650-1,790
Earth, dry	2,450-2,600	43	0.70	1,720-1,820
Earth, moist	2,700-3,000	33	0.75	2,030-2,250
Earth, compacted	3,000	25	0.80	2,400
Earth, w/sand and gravel	3,100	11	0.90	2,790
Gypsum, fractured	5,300	75	0.57	3,020
Gypsum, crushed	4,700	75	0.57	2,680
Kaolin	2,800	30	0.77	2,160
Lime	--	--	--	1,400
Lime, slaked	--	--	--	800-1,500
Limestone, blasted	4,200	6,765	0.570.60	2,400-2,520
Limestone, loose, crushed	--	--	--	2,600-2,700
Mud, dry (close)	2,160-2,970	20	0.83	1,790-2,470
Mud, wet (moderately packed)	2,970-3,510	20	0.83	2,470-2,910
Peat, dry	800-1,300	80	0.56	450-730
Peat, wet	1,600-1,800	80	0.56	900-1,010
Sand, dry	2,450	12	0.89	2,180

(Continued)

* BCY = Bank cubic yards, all specifications are in customary units.
  LCY = Loose cubic yards

TABLE A-2.   (Concluded)

Material	Weight in bank (lb/BCY)	Percent swell	Swell factor	Loose weight (lb/LCY)
Sand, dry, fine	2,700	12	0.89	2,400
Sand, damp	3,200	12	0.89	2,850
Sand, wet	3,500	14	0.88	3,080
Sand and gravel, dry	3,300	12	0.89	2,940
Sand and gravel, wet	3,700	11	0.90	3,330
Slag, sand	1,670	12	0.89	1,490
Slag, solid	4,320-4,830	33	0.75	3,240-3,620
Slag, crushed	--	--	--	1,900
Slag, furnace, granulated	1,600	12	0.89	1,420

Appendix B—Typical Stabilization/Solidification Equipment

Many of the solidification/stabilization alternatives use similar equipment and/or groups of equipment. The processing equipment used for the solidification/stabilization of hazardous materials at remedial action sites has generally been adapted from the materials processing and construction industries. The equipment or groups of equipment used for the various treatment programs identified in this study have been fabricated from readily available, off-the-shelf equipment modules. The discussion that follows provides information on the technical attributes, available capacities, and costs associated with each identified equipment module.

The equipment that has been adapted for use in solidification operations is divided into four basic categories: chemical storage, materials handling, materials mixing, and materials control. A variety of equipment modules are available under each category. The more common types of equipment modules identified during site visits of operating facilities conducted as a part of this study are the primary focus of this discussion. No attempt has been made to review all available equipment to optimize equipment sizes and mixes.

The cost information presented is based on the purchase cost or rental cost of equipment modules. The costs presented have a July 1983 base year and result from interviews with equipment manufacturers. Note, however, that most of the identified equipment modules are readily available in the used or rental equipment market at substantial cost savings. In addition, the type of equipment generally utilized is designed for portability. As a result, it can be moved from site to site with minimal loss of productive capacity. Thus once it is purchased, the equipment could be amortized over several projects at substantial savings when calculated on a basis of per-unit cost of waste treated.

B.1 Chemical Storage Facilities

Onsite facilities may be required for the storage of both dry and liquid chemicals. The nature and size of required storage facilities are a function of the solidification/stabilization process selected (the types of chemicals required), the quantity of chemicals required, and the method of chemical shipment (bulk or container). Chemical deliveries should be programmed to minimize onsite storage requirements and ensure their continuous availability at the site.

The majority of remedial action projects are assumed to be large enough to justify the bulk purchase of chemicals; however, some specialty chemicals used in the various solidification/stabilization processes may be purchased in smaller, containerized quantities. Therefore, consideration must be given to the protection of both bulk and containerized chemicals during the planning phase.

B.1.1 Dry Chemical Storage

On a volume or weight basis, the major dry chemicals used in a solidification/stabilization process will normally be either Portland cement, quick lime, hydrated lime, fly ash, gypsum, cement-kiln dust, or lime-kiln dust. The quality of these materials, measured by their reactivity, is subject to severe degradation by exposure to moisture from precipitation or excessive humidity. Storage can be provided in one of four ways: open storage, storage with fabric or membrane covers, storage in a warehouse environment, or closed bins and silos.

Open storage can be utilized for short periods of time during appropriate weather conditions for the less reactive dry reagents. Open storage of the more reactive, dry reagents, such as Portland cement and quick lime, would not be appropriate. For example, small amounts of the less reactive dry reagents (e.g., weathered kiln dust) could be stored in the open pending use in an in-situ mixing program without significant loss of reactivity. Fugitive dust may be a severe problem when using this storage option in dry, windy climates. Long-term open storage of dry reagents is not a recommended option. A zero-cost, not including losses of material, may be given to the open storage option.

Storage under a fabric or membrane cover is more appropriate than open storage for low-reactivity materials such as kiln dust, fly ash, or gypsum. Short-term storage in this manner should not result in significant deterioration in these materials. Fugitive dust, however, may still be a significant problem when this method of storage is used. The cost of storage with fabric or membrane covers is estimated to range between $2.00 and $4.00 per square foot of storage area provided. The majority of this cost is involved in the cost of the fabric or membrane covering. This category of storage is not appropriate for high-reactivity reagents such as quick lime, hydrated lime, or Portland cement.

Covered storage in a warehouse environment provides an alternative for onsite storage. Unheated warehouse storage can be provided for a cost ranging between $8.00 and $10.00 per square foot. Fugitive dust control and access to the materials may present problems. The high reactivity chemicals still may suffer degradation from humidity effects.

Covered storage for the bulk solid materials associated with solidification/stabilization processes is often provided in the form of metal storage silos similar to those used for the storage of Portland cement. Storage capacities ranging from 1,000 to 5,000 cu ft are readily available. The estimated installed cost for these dry chemical storage silos is presented in Figure B-1.

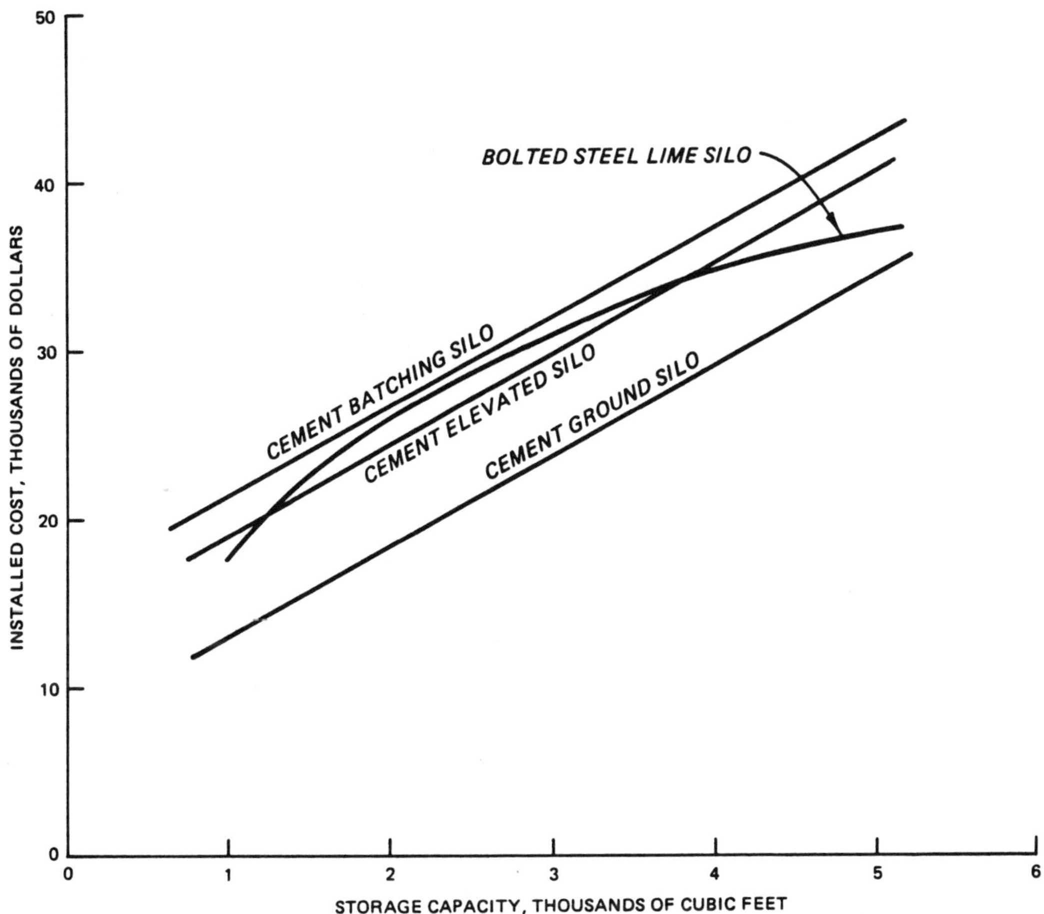

Figure B-1.   Installed cost of dry chemical storage.

The required size for a storage silo is a function of the rate of chemical usage and the anticipated chemical delivery schedule. The minimum size silo should be capable of holding at least the quantity of material in a bulk tank truck (approximately 500 cu ft). Material suppliers should be consulted to determine delivery schedules, minimum order quantities, and delivery times.

Containerized dry chemicals (generally bags or drums) can be stored in open storage or covered storage. Some specialty chemicals may require protection from extreme cold or heat. Appropriate covered storage should be provided for such materials. Heated warehouse space can be provided for approximately $10.00 to $12.00 per square foot.

Rather than providing for the construction of onsite storage facilities, it may be desirable to use the bulk transport trailer for onsite storage. The cost for long-term use of the bulk transport trailers for such use is subject to extreme variation. Business conditions may preclude the use of this option because of the demand for bulk transportation services and resultant high demurrage rates for bulk trailers. Materials transporters should be consulted during the project planning phase.

B.1.2 Liquid Chemical Storage

Liquid reagents may be received in containers (generally drums) and in bulk form. Containerized liquid reagents may be placed in open storage or covered storage. Although less susceptible to degradation caused by moisture (because of the nature of the shipping container), liquid reagents may be more sensitive to temperature extremes. Changes in both degradation and handling characteristics may result from exposure to temperature extremes. Open and covered storage has been discussed under dry chemical storage above. Similar storage facilities can be provided for liquid chemical storage.

Bulk liquid storage is provided in tanks. Typically, horizontal and/or vertical tanks may be provided. Tanks may be equipped with heating coils to ensure the maintenance of handling characteristics when exposed to low temperatures. The estimated installed costs of various tank storage facilities are presented in Figure B-2. As in the case of dry chemical storage, the proper planning of chemical delivery schedules can be used to minimize onsite storage requirements.

B.2 Materials Handling Equipment

One of the most important factors in the application of solidification/stabilization technology to waste at a remedial action site is the form or nature of the wastes to be processed. The forms that wastes may take include:

    a.   Liquids from lagoons, settling ponds, drums, and the container.

    b.   Sludges from lagoons, settling ponds, and leaking drums or other containers.

    c.   Contaminated soils caused by leaking containers or direct dumping of liquids and sludges on the soil.

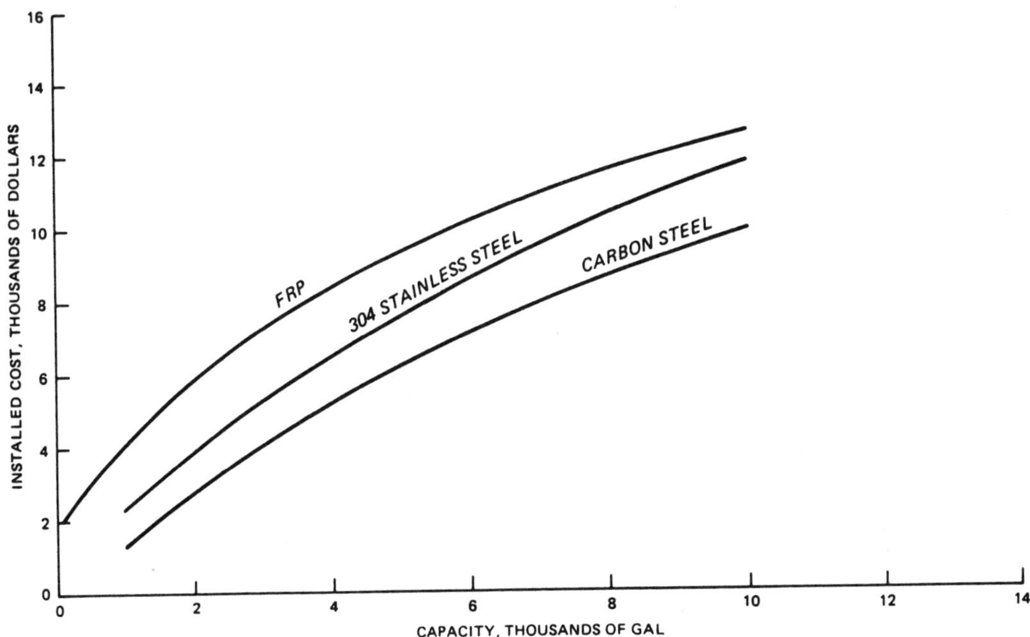

Figure B-2. Installed cost of liquid reagent storage
(FRP = fiber reinforced plastic).

d.   Pasty solids from breached and/or intact containers.

e.   Solids in drums or in other containers or from open contaminated
     sites.

Materials handling equipment selected for a particular remedial action
project will depend on the forms of waste to be handled. Selection of
equipment for materials handling is a function of the physical characteristics of the waste material being handled (percent solids, viscosity, etc.),
the packaging of the waste materials (drums, lagoons, open area, etc.), the
quantity of waste materials being handled, and the physical characteristics
of the solidified/stabilized wastes. It is desirable to transport liquids
and high-moisture-content sludges with pumps. Some low-moisture-content
sludges can be handled with special pumps. Low-moisture-content and/or
viscous sludges may be handled with earth-moving equipment such as clamshells, backhoes, and dump trucks. Contaminated soils are handled with
earth-moving equipment. Material conveying systems can also be utilized for
low-moisture-content sludges and contaminated soils. Care must be taken to
ensure compatibility between the material to be handled and the equipment
selected to do the handling.

B.2.1 Pumps

Either centrifugal or diaphragm pumps may be used for the bulk transfer of liquids and high-liquid-content sludges. Centrifugal pumps have the advantage of higher capacities, whereas diaphragm pumps are capable of handling higher-solids-content materials, but generally have higher maintenance costs. Centrifugal pumps used for handling materials to be solidified or stabilized are generally referred to as self-priming, centrifugal trash pumps (Hicks 1971). Size ranges from 5 cm to 15 cm (2 in. to 6 in.) are commonly available with pumping capacities, based on pumping water, ranging between 95 and 5,100 ℓ/min at heads of up to 56 m. Capacity reductions may be significant for high-solids-content materials. Both motor- and engine-driven pumps are available on frame and trailer-mounted systems. A trailer-mounted, gasoline-engine-driven pump is illustrated in Figure B-3.

Figure B-3. Trailer-mounted centrifugal pump (Courtesy Gorman Rupp Company).

Self-priming trash pumps are generally limited to handling waste materials with a solids content less than 40 percent. Recent developments in centrifugal pumping systems, incorporating chopper pumps and floating platforms have produced systems capable of efficiently handling slurries containing up to 60% solids. Commonly available sizes range from 7 cm to 15 cm (3 in. to 6 in.) with pumping capacities, based on pumping water ranging between 1,000 and 5,200 ℓ/min at heads up to 44 m. As in the case with the self-priming, centrifugal pumps, capacity reductions are significant when pumping sludges with high solids content. Since the pump impeller on the floating system is in contact with the waste slurry, the floating system can handle a higher-solids-content slurry. Figure B-4 illustrates a typical floating pump system.

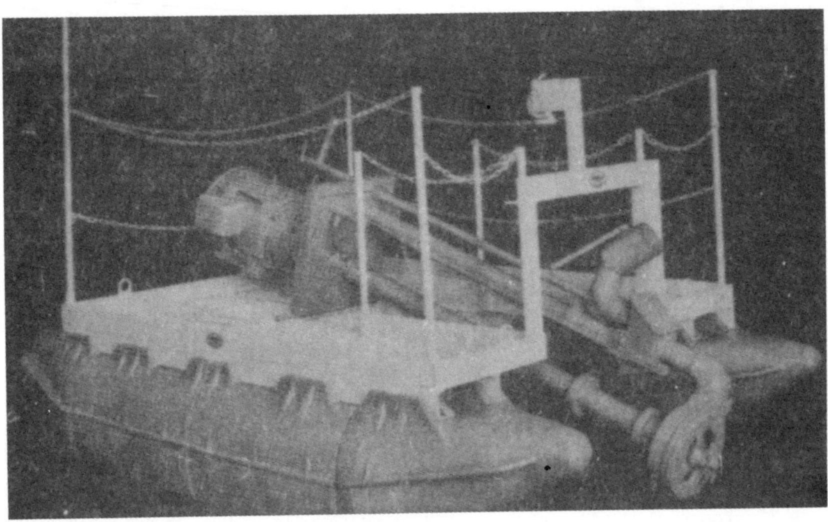

Figure B-4. Typical floating centrifugal pump (Courtesy
Vaughan Pump Company).

Diaphragm pumps can be utilized on more viscous material with higher
solids content; however, capacities and head are generally limited, and
maintenance costs are higher. Commonly available diaphragm pump sizes range
from 40 to 570 ℓ/min at heads up to 8 m. Both electric and engine-driven
diaphragm pumps are available.

Figure B-5 presents the purchase costs for self-priming centrifugal
trash pumps, floating centrifugal pumps, and diaphragm pumps.

Waste materials that have been mixed with solidification/stabilization
reagents can also be transported with pumping systems. In addition to the
systems described above, concrete pumps have been used to transport treated
waste materials. Available capacities range from 40 to 120 cu yd/hr. Concrete pumps can handle very high solids content slurries; however, the high
cost of these systems has prohibited their wide-scale use. Figure B-6 presents the purchase cost of available units.

B.2.2 Construction Equipment

In those cases where waste material which is not amenable to pumping is
to be handled, reliance has been placed on the use of conventional excavation
and earth-moving equipment. Typically, equipment used for the solidification/
stabilization of waste materials will include backhoes or all-purpose

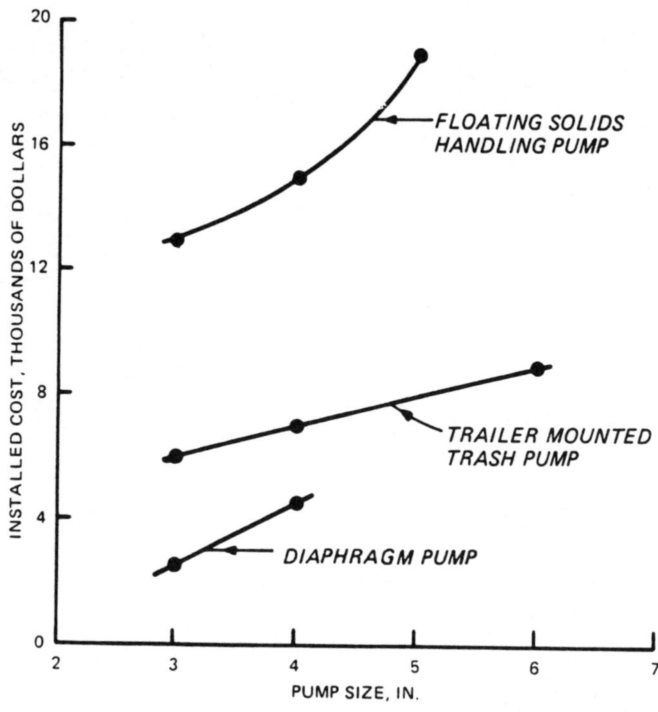

Figure B-5. Typical costs for pumping systems.

excavators, clamshells, or draglines; front-end loaders; and dump trucks.
Figure B-7 illustrates a backhoe-dump-truck operation for removal of contaminated soils. Other types of equipment including graders, dozers, compactors,
etc. may be used in the overall remedial action project; however, this discussion is limited to consideration of materials handling associated with the
solidification/stabilization process.

The required materials handling equipment is available in a wide range
of sizes. The selection of quantity, types, and size of equipment is primarily a function of the quantity of materials to be handled and the working
area available.

Production rates for construction equipment used on remedial action
projects may vary significantly. Estimates of production rates are beyond
the scope of this study; however, a number of excellent references are
readily available to assist the project planner in preparing production and
cost estimates (Terex 1981; Caterpillar Tractor Co. 1981, 1982). In addition, direct consultation will often be provided by the equipment
manufacturer.

Estimation of the production rates expected on a particular job requires
careful preparation, a thorough knowledge of the material to be handled, and
a complete understanding of equipment capabilities. Factors to be considered

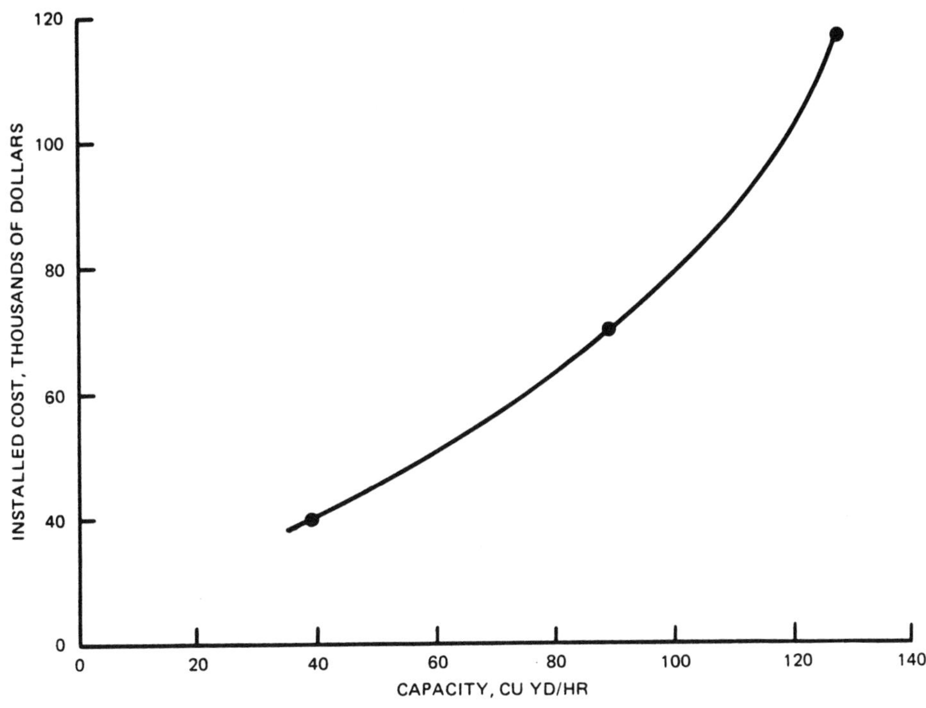

Figure B-6.   Typical costs for trailer-mounted concrete pumps.

in preparation of the estimate include (1) cycle time of materials-moving components; (2) job efficiency factors; (3) material weights, swell factors, and handling characteristics; and (4) vehicle payloads.

The cycle time in construction activities is defined as the time for a machine or group of machines to complete one cycle (i.e., load, haul, dump, return, spot, and delay). Each of these components affects the total cycle time and is controlled by a number of factors. Loading factors include: size and type of loading equipment, nature of material being handled, capacity of hauling equipment, and skill of the operator. Haul factors include: capability of hauling unit, hauling distance, haul road conditions, and grades. Dumping, or unloading, factors include: destinations of material (i.e. fill, stockpile, mixer, etc.), conditions of unloading area, maneuverability of the hauling unit, and nature of the material. Return factors include: capability of the hauling unit, return distance, haul road condition, and grades. Spot factors include maneuverability of the hauling unit, maneuver area available, type of loading machine, and location of the loading equipment. Delay factors include time spent waiting on the loading unit and time spent waiting to unload.

Figure B-7.   Backhoe-dump-truck operation for removal of contaminated soils (Courtesy Albert H. Halff Associates).

Job efficiency factors are used to estimate the sustained or average
materials handling capability over a long period of time. Job efficiency is
influenced by such factors as operator skill, repair time, personnel delays,
and job layout (Caterpillar Tractor Co. 1982). Since many of these factors
are difficult to quantify, estimates of job efficiency are very complex.
Typical job efficiency factors are presented in Table B-1. Note that a 75%
efficiency (45 min/hr) is estimated for a job with good working conditions
and good management. Job efficiency factors as low as 25% may be anticipated
for some remedial action projects due to safety factors and nonoptimum working conditions.

Weight and handling characteristics of materials being moved are also
important factors in determining production rates. Materials handled with
construction equipment on typical remedial action projects are low-moisture-
content sludges with difficult handling characteristics. Specific weights of
the materials in-place are expected to vary between 700 and 1,400 kg/cu m.
The materials may also be subject to swelling and/or hardening in the loading
equipment.

Payloads for the loading and hauling equipment must be determined from
the manufacturer or his representative. Again, it must be noted that payload
data are developed based on earth and rock loading and hauling capabilities.
Few if any data are available on handling of the waste materials that are
candidates for solidification/stabilization.

TABLE B-1. TYPICAL JOB EFFICIENCY FACTORS

Job condition	Management condition			
	Excellent	Good	Fair	Poor
Excellent	0.84	0.81	0.76	0.70
Good	0.78	0.75	0.75	0.65
Fair	0.72	0.69	0.69	0.60
Poor	0.63	0.61	0.61	0.52

Once production estimates have been developed, the onsite requirement for each piece of construction equipment can be estimated. With this time estimate, the job cost of each item of equipment can be estimated. Equipment can be either purchased or rented. Of course, purchased equipment can be amortized over more than one project. For planning purposes, the normal procedure is to estimate costs based on equipment rental rates. Table B-2 provides information on the rental rates for various items of construction equipment anticipated to be required on a typical remedial action project.

B.2.3 Conveyors

Belt conveyors, or stackers, can be used to transport materials with soil-like properties (i.e., contaminated soils or the solidified/stabilized waste material). Belt conveyors are not suitable for the transport of liquids, high-moisture-content sludges, or viscous materials. Portable conveying systems from 15 m to 70 m in length with 60-cm through 90-cm belt widths are readily available. Capacities range from 300 to 700 tons/hr. Estimated costs for an installed, portable conveyor system are presented in Figure B-8. Figure B-9 illustrates a typical portable conveyor system.

B.2.4 Drum Handling

Waste to be solidified or stabilized is often stored in drums. Efficient drum handling has been one of the most difficult problems in materials handling associated with remedial action projects. Appropriate procedures have been defined in the publication entitled "Drum Handling Practices at Hazardous Waste Sites" (EPA 1982).

TABLE B-2. APPROXIMATE RENTAL RATES FOR CONSTRUCTION EQUIPMENT USED
FOR STABILIZATION/SOLIDIFICATION PROJECTS

Equipment	Per Month (176 hr)	Per Week (40 hr)	Per Day (8 hr)	Approx. Purchase Price
Compactors - self-propelled				
Vibratory plates	$ 1,450.00	$ 485.00	$ 140.00	$ 25,500.00
3 wheel steel (14 ton)*	1,600.00	535.00	150.00	28,000.00
Tandem (14 ton)	1,600.00	535.00	150.00	28,500.00
Rubber tired (11 ton)	1,600.00	535.00	150.00	24,000.00
Vibratory drum (10 ton)	2,495.00	800.00	235.00	41,000.00
Graders				
14 ton	3,300.00	1,100.00	325.00	61,500.00
19 ton	4,650.00	1,550.00	450.00	83,500.00
25 ton	6,350.00	2,100.00	600.00	110,000.00
Front-end loaders				
1 cu yd	1,550.00	525.00	150.00	28,500.00
2 cu yd	2,400.00	825.00	235.00	44,000.00
4 cy yd	5,250.00	1,750.00	500.00	93,000.00
5 cu yd	6,250.00	2,100.00	600.00	112,000.00
Crawler tractors				
140 hp	2,950.00	1,000.00	290.00	52,500.00
300 hp	5,700.00	1,900.00	550.00	102,000.00
400 hp	8,650.00	2,900.00	835.00	156,000.00
Wheel tractors				
180 hp	3,750.00	1,250.00	360.00	66,200.00
300 hp	6,200.00	2,050.00	600.00	113,500.00
420 hp	7,250.00	2,400.00	685.00	128,000.00
Hydraulic pull shovel				
1-1/2 cu yd	5,100.00	1,700.00	495.00	95,000.00
2 cu yd	6,350.00	2,100.00	610.00	115,000.00
3 cu yd	7,950.00	2,650.00	770.00	143,000.00
All purpose excavators				
1/2 cu yd	3,875.00	1,275.00	380.00	71,600.00
3/4 cu yd	5,725.00	1,900.00	560.00	103,000.00
1-1/4 cu yd	7,450.00	2,500.00	715.00	133,000.00

(Continued)

* Ratings are in customary units.

TABLE B-2. (Concluded)

Equipment	Per Month (176 hr)	Per Week (40 hr)	Per Day (8 hr)	Approx. Purchase Price
Mechanical shovels				
2 cu yd	4,950.00	2,000.00	575.00	100,000.00
3-1/4 cu yd	10,000.00	3,300.00	900.00	170,000.00
4-1/4 cu yd	12,500.00	4,150.00	1,200.00	220,000.00
5-1/2 cu yd	17,500.00	5,950.00	1,700.00	320,000.00
Hydraulic crane				
10 tons	3,150.00	1,050.00	300.00	56,000.00
15 tons	3,250.00	1,100.00	310.00	59,000.00
18 tons	3,550.00	1,250.00	340.00	65,000.00
35 tons	6,400.00	2,150.00	620.00	120,000.00
Mechanical crane-crawler				
20 tons	3,850.00	1,250.00	375.00	71,500.00
30 tons	4,300.00	1,450.00	410.00	77,000.00
40 tons	6,000.00	2,000.00	575.00	110,000.00
50 tons	6,600.00	2,175.00	625.00	115,000.00
Truck crane				
25 tons	4,750.00	1,550.00	450.00	82,500.00
50 tons	7,000.00	2,350.00	675.00	125,000.00
Water pumps				
2-in. discharge	120.00	40.00	12.00	2,300.00
3-in. discharge	210.00	70.00	20.00	3,750.00
4-in. discharge	510.00	175.00	50.00	8,800.00
6-in. discharge	850.00	285.00	80.00	15,500.00
8-in. discharge	1,000.00	330.00	95.00	18,000.00

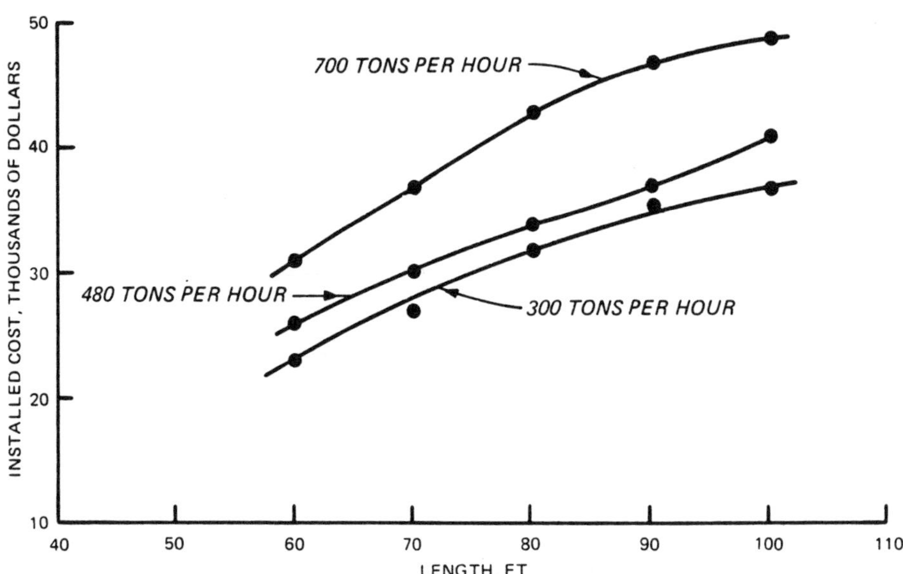

Figure B-8. Installed portable conveyor system costs.

Figure B-9. Typical portable conveyor system (Courtesy
The Vince Hagen Company).

B.3 Materials Mixing Equipment

Materials mixing equipment is used to blend reagents with the waste materials to accomplish the solidification/stabilization reaction.

B.3.1 Construction Equipment

Backhoes, clamshells, and draglines have been applied to the in-situ mixing of solidification/stabilization reagents with waste materials. Since this is not a "normal" use for this equipment, little detailed information is available concerning production rates and control of the mixing process (i.e., is mixing adequate or do pockets of unreacted waste material remain?). Backhoe mixing has been successfully applied at Resource Conservation and Recovery Act (RCRA) disposal sites; however, this is usually done in relatively small basins and the solidified/stabilized material is always re-handled. Thus adequate mixing is usually ensured.

The high-speed rotary mixer (Figure B-10) has been used to mix solidification/stabilization reagents with sludges and contaminated soils. The procedure for using this equipment places alternating layers of waste and treatment reagents. Data for application to the solidification/stabilization of waste materials are not available; however, based on highway construction experience, it is estimated that around 2,000 sq m (21,520 sq ft or about 1/2 acre) of surface per day could be mixed. Assuming a lift of 25 cm, 500 cu m/day of waste material could be mixed with the required reagents.

A variety of mixing and other types of materials handling equipment is available from the concrete and roadway materials industry. Products that can be readily adapted to the solidification/stabilization of hazardous wastes include materials storage, batching, and mixing equipment. Mobile, portable, and stationary equipment modules are readily available for all of these functions. Modules can be purchased and assembled to meet site-specific requirements. Equipment manufacturers provide consultative service to address specific materials handling requirements.

A typical adaptation of concrete technology is the use of a base stabilization plant for treating contaminated soils as illustrated in Figure B-11. Sizes for such plants range from 100 to 400 tons/hr and consist of materials storage, batching, and mixing facilities. Materials mixing is generally accomplished using a pug mill. The estimated cost of a base stabilization plant is illustrated in Figure B-12.

Other applications from concrete mixing technology include the use of concrete batch plants, central mixing facilities, and/or transit mixing trucks. These can be used for both apportioning and mixing solidification/stabilization reagents with the waste materials being treated. The costs of both mobile and modular batching plants are illustrated in Figure B-13.

Figure B-10. Typical high-speed rotary mixer (Courtesy
Albert H. Halff Associates).

Figure B-11. Typical base stabilization plant.

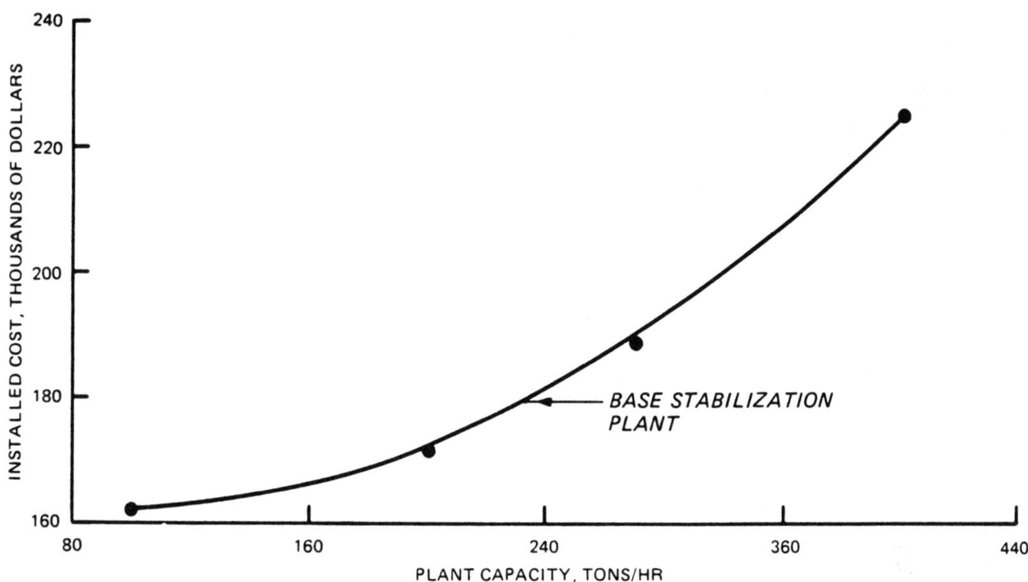

Figure B-12. Installed cost for base stabilization plant.

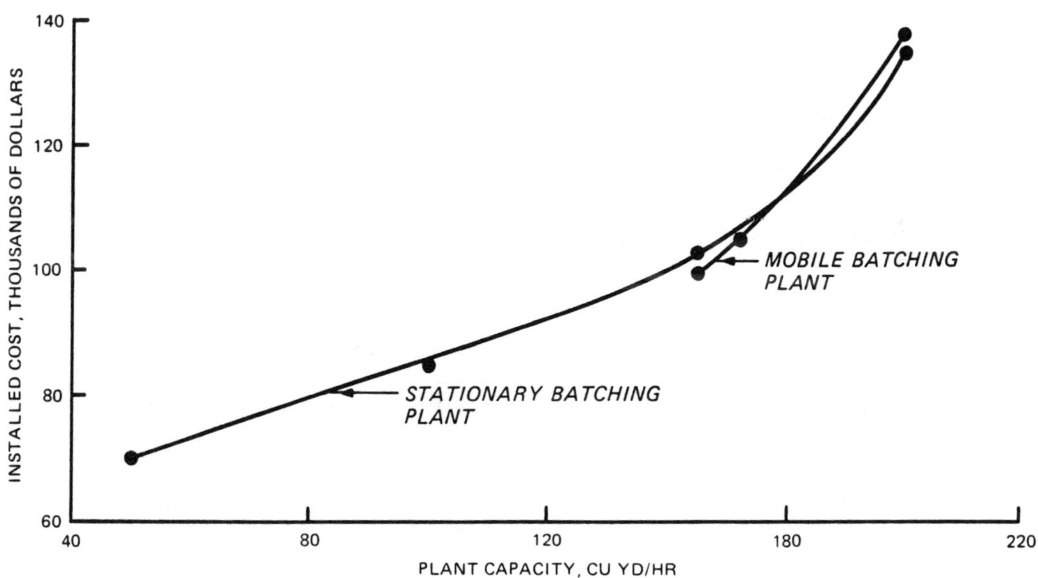

Figure B-13. Installed cost for mobile and modular concrete batching facility.

Materials mixing can be accomplished by central mixing equipment (tilting mixers) or in transit-mix trucks. Tilting mixers are available in sizes ranging from 6 to 12 cu yd per batch. The installed cost of a tilting mixer is presented in Figure B-14.

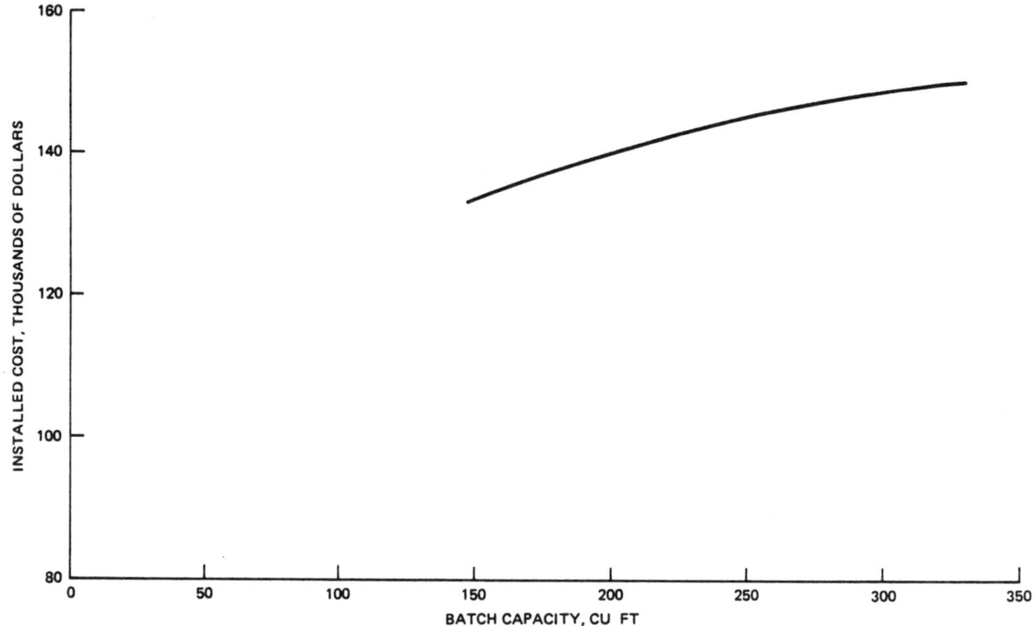

Figure B-14.   Installed cost for concrete tilting mixers.

Transit-mix trucks have been used to mix contaminated materials and solidification/stabilization reagents. Typically, the materials are batched in a mobile batch plant and mixed during transport to the final disposal area. Transit-mix trucks are available in capacities ranging from 6 to 12 cu m.

Although the concept for using modified equipment from the concrete industry has been developed, the equipment has not received widespread use because of the relatively high cost compared with equipment used in the scenarios developed in Sections 6.3 through 6.6 of this handbook. However, the use of concrete industry equipment should be included in alternative evaluations on a site-specific basis.

B.3.2 Process Mixing Equipment

A wide variety of process mixing equipment has been used, or is theoretically available for use in the mixing of reagents with waste materials to be solidified or stabilized. This equipment has been adapted from either the food or chemical processing industry. Basic parameters, which include mixing characteristics, available sizes, and costs for the more significant mixer types, are presented below. Additional information on the specific application of each is provided in Section 6.

The scientific design of mixing equipment is complex and usually requires detailed engineering study. Perry (1973) identifies properties of the materials to be mixed that affect the selection of appropriate mixing equipment: particle-size distribution; bulk density; true density; particle shape; surface characteristics; friability; state of agglomeration; moisture or liquid content of solids; density, viscosity, and surface tension; and temperature characteristics. Little if any scientific design has been applied to mixing required for solidification/stabilization processes. Most mixing equipment has been developed or modified by trial and error based on field experience. One reason for this is the wide range of materials that the typical system may be required to handle. The major types of mixing equipment for waste processing include the change-can mixer, ribbon blender, muller mixer, rotor mixer, and propeller mixer. Detailed engineering has not been performed to optimize the design of mixing equipment currently used for solidification/stabilization of hazardous waste.

B.3.2.1 Change-Can Mixer

The change-can mixer is a vertical batch mixer in which the container is separate from the frame of the machine. Capacities ranging from 0.5 ℓ to 1,100 ℓ are available. The most common size used in the solidification of hazardous wastes is the 200-ℓ drum. Figure B-15 illustrates a typical change-can mixer.

The change-can mixer is ideally suited for use in drum solidification/ stabilization of wastes. The mixing head may be raised from the can (drum) allowing the mixing blades to drain into the drum. If necessary, the blades may be wiped down or cleaned by rotating them in a solvent. When the can is removed, cleaning the blades and support is a rather simple process.

Mixing of can contents is achieved in one of two ways. First, the mixing unit assembly may rotate with a planetary motion so that the rotating blades sweep the entire circumference of the can. Second, the can is mounted on a rotating turntable so that all parts of the can will pass fixed scraper blades on the agitation blades at a point of minimum clearance. The mixing action is primarily in the horizontal, to and from the center of the can. Vertical mixing results from the shape of the blades.

As mixing progresses, the flow characteristics usually change. In order to achieve a minimum time cycle, variable speed or two-speed mixers are desirable. A slow speed at the start of mixing will reduce dusting or splashing.

The estimated costs for a change-can mixer installation are presented in Figure B-16.

Figure B-15. Typical change-
can mixer (Courtesy Charles
Ross & Son).

B.3.2.2 Ribbon Blender

A ribbon blender consists of a stationary shell and rotating horizontal
mixing elements (Figure B-17). To accommodate a wide variety of materials,
it is possible to modify such features as ribbon cross section, ribbon pitch,
the number of ribbons, and the clearance between ribbons and ribbons and
shell. The ribbon blender can be used for continuous or batch operations.
Installed costs for ribbon blenders of various sizes are presented in
Figure B-18.

B.3.2.3 Muller Mixer

The muller mixer consists of a stationary pan with rotating wheels and
plows (Figure B-19). The muller is typically used for batch operations; however, continuous-operation mullers are available. Installed cost for muller
mixing systems are presented in Figure B-20.

B.3.2.4 Rotor Mixers

Rotor mixers consist of shafts with paddles or screws contained in a
stationary trough. These mixers may be equipped with single or twin shaft

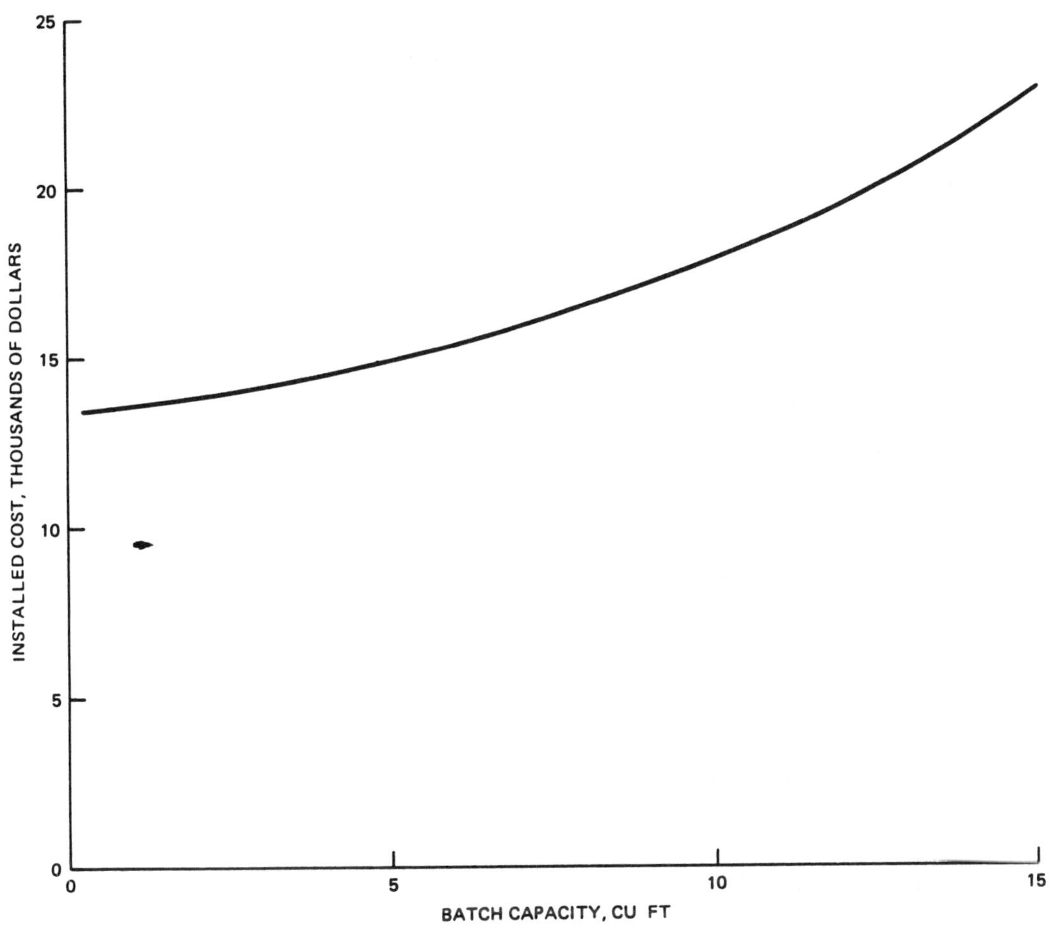

Figure B-16.   Installed cost for change-can mixers.

assemblies.   Figure B-21 illustrates a twin-shaft rotor mixer.   The installed cost for a twin-shaft rotor mixer is presented in Figure B-22.

B.3.2.5 Propeller Mixer

The top-entering propeller mixer consists of a driver, shaft, and propeller.   This mixer is lightweight and highly portable, and it can be easily changed from one drum to the next.   This mixer works by changing the mixer from drum to drum rather than by changing drums in the mixer.   The mixer is mounted on the drum with a clamp or special head frame.   Typical cost of the equipment is approximately $2,000.00.   Figure B-23 illustrates a typical propeller mixer.

Figure B-17. Typical ribbon blender (Courtesy Beardsley & Piper).

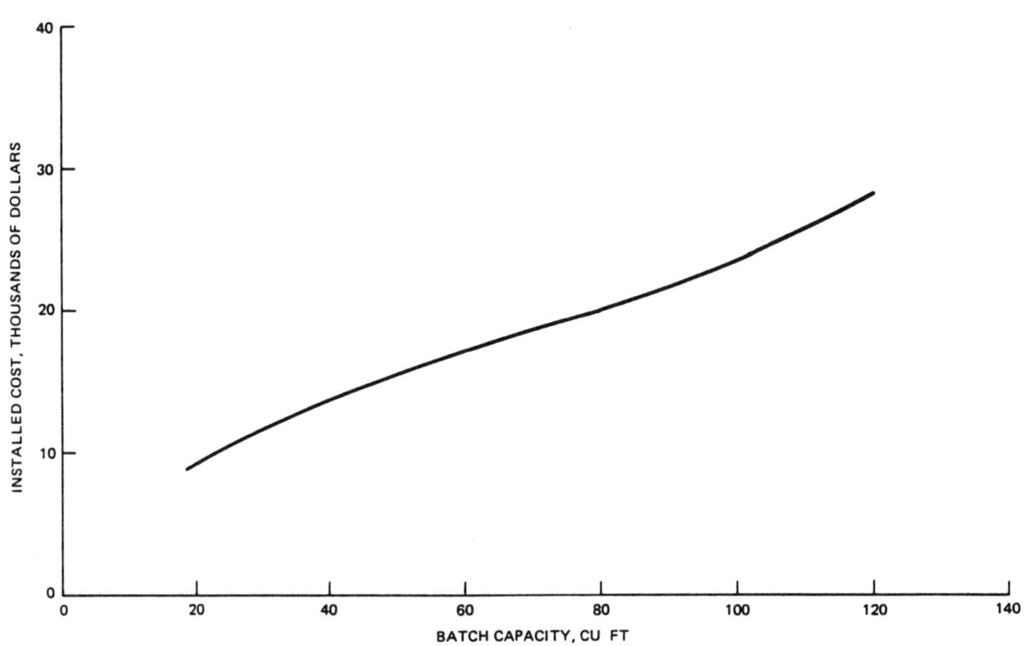

Figure B-18. Installed cost for ribbon blenders.

Figure B-19. Typical muller mixer
(Courtesy Beardsley & Piper).

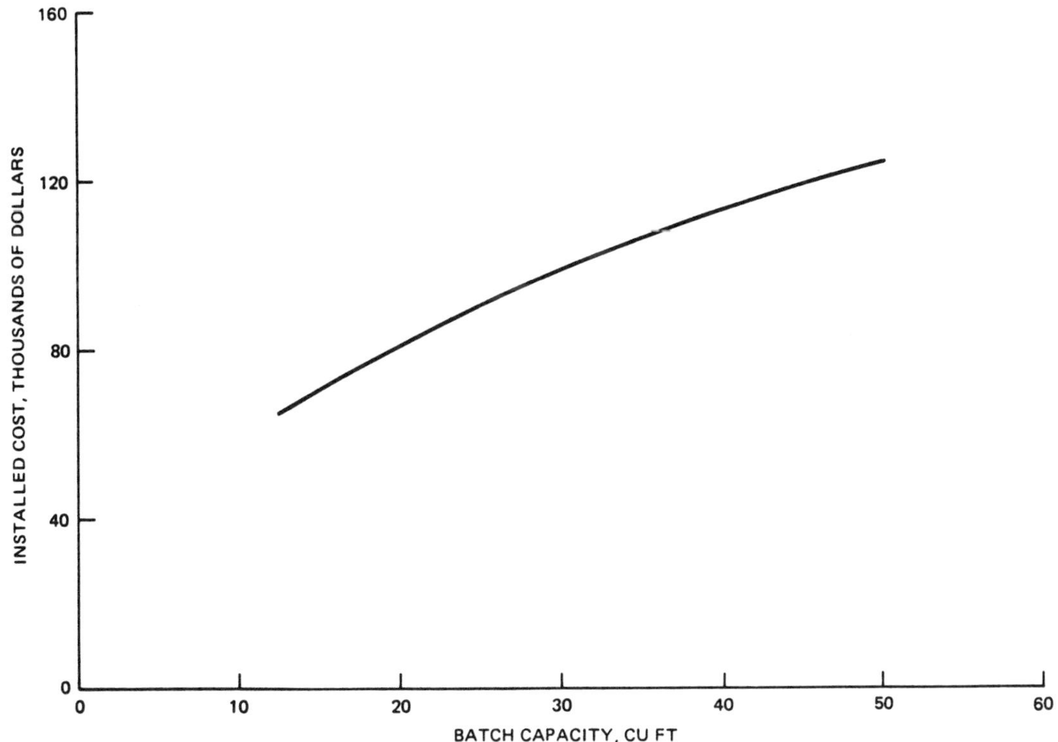

Figure B-20. Installed cost for muller mixers.

Figure B-21. Typical twin-shaft rotor mixer
(Courtesy Beardsley & Piper).

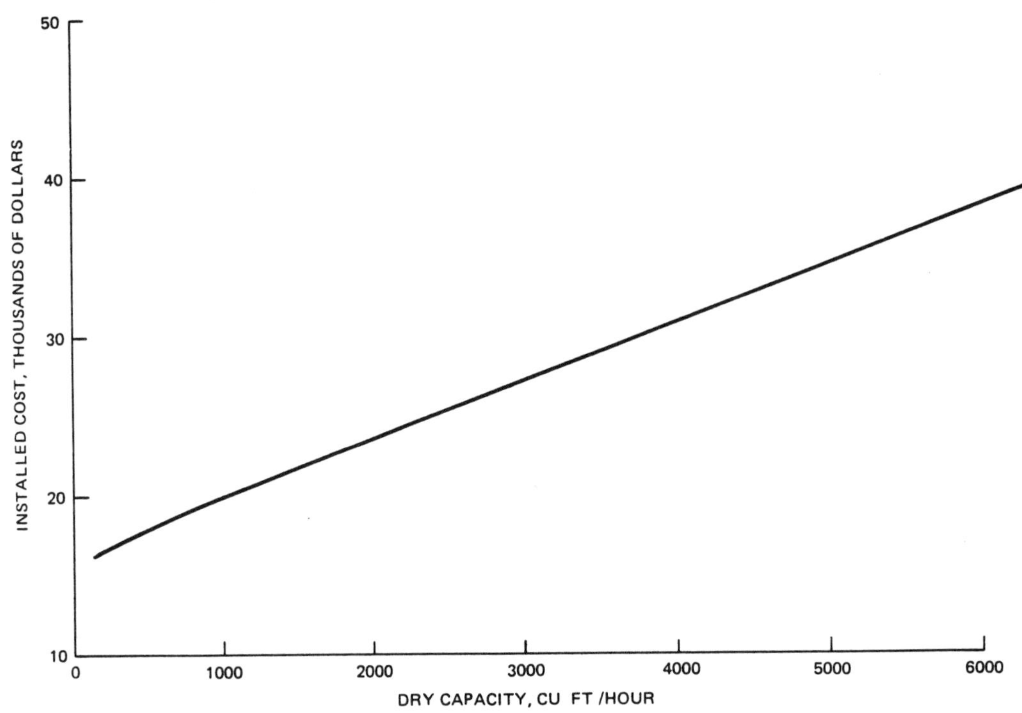

Figure B-22. Installed cost for rotor mixers.

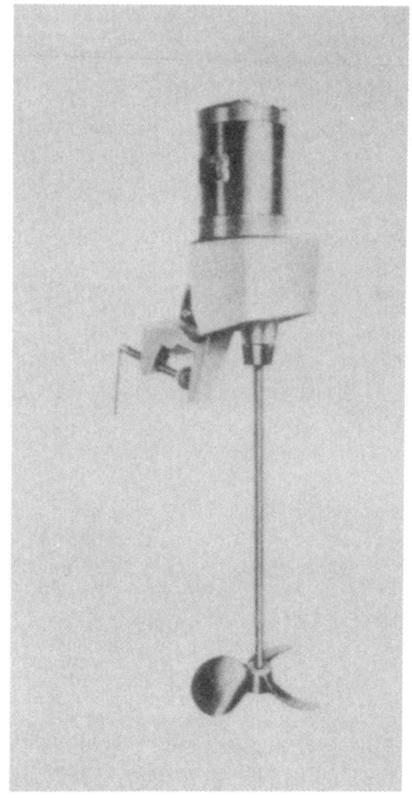

Figure B-23. Typical top-entering
propeller mixer (Courtesy Mixing
Equipment).

B.4 Materials Control Equipment

Solidification/stabilization processes require the addition of reagents
to waste materials in fixed, measured quantities, generally as determined from
pilot- or laboratory-scale studies. Adjustments are subsequently made as a
result of onsite experience with the particular waste being treated. The
control of materials, both the waste to be treated and the reagents to be
added, can be accomplished using methods based on either weight or volume.
In addition, either batch or continuous control systems are available. The
sophistication of the materials control technique selected for a particular
project can vary from simplistic systems incorporating manual feed to complex, fully automated equipment.

Materials control (i.e., the proper proportioning of waste materials and
solidification/stabilization reagents) is one key to the proper performance
of the treated waste materials. Numerous materials control systems are
available off-the-shelf. The most common types of equipment used for materials control purposes are discussed below.

B.4.1 Waste Materials Control

Waste materials control can be accomplished by either volume- or weight-
based methods. The type of control system selected will depend on the

materials handling equipment and the materials mixing equipment selected to
accomplish the solidification/stabilization process.

If the waste material is pumped and a continuous mixer is used, pump
curves can be consulted to determine the discharge under stated conditions.
Since manufacturers' pump curves are based on pumping clean water and the
typical remedial action project will handle sludges or high solids content
liquids, adjustment to the manufacturers' curves will be required. Calibration of pumps under field conditions may be required. For those systems
using pumps for waste material handling and batch-type mixing equipment, a
volumetric batching system can be employed. This system may consist of a
separate, level-controlled batch hopper, or the mixing vessel can simply be
filled to a predetermined level. Manual or automatic control can be used.

If the waste material is handled by construction equipment, material
control can be accomplished by volumetric measurement, or for granular material, aggregate weigh batches from the concrete batch plant industry can be
used. Figure B-24 illustrates a weigh batcher being used to meter waste
materials. Volumetric measurements can be used in the same manner as for
pumped wastes; however, feeding the measuring or mixing equipment will be
more difficult. A less sophisticated method of measurement is merely to
count the number of truckloads of material and make an estimate of the volume
of waste material on each based on the known truck capacity.

Figure B-24. Weigh batcher system for waste materials
control (Courtesy Solid Tek).

B.4.2 Solid Reagent Control

The control of solid reagents can be accomplished by either volumetric-
or weight-based methods. The type of equipment selected should be based on
the quantity of material to be fed to the waste and the solidification/
stabilization scenario selected.

The most common type of system for feeding dry solids is the screw
feeder (Figure B-25). The screw-type feeder is fairly rugged and well suited
for application in the field environment. The feed rate is controlled by
increasing or decreasing the speed of the screw. Assuming a constant bulk
density of material, the weight of material discharged from the screw feeder
can be accurately controlled. Screw feeders can be adapted for use with both
batch and continuous mixing systems.

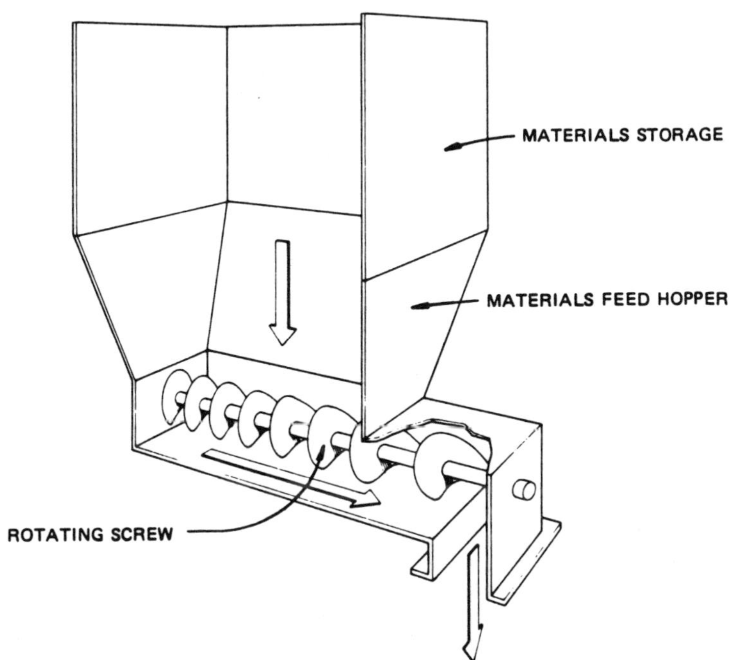

Figure B-25.   Typical screw feeder.

Batch and continuous-feed systems based on measurement of weight are
also available. These systems, although somewhat more delicate than screw
feeder systems, provide for more accurate materials control. Batch weighing
systems suitable for use with batch mixing systems usually consist of a containment vessel or hopper mounted on a scale or load cell. The entire assembly is usually mounted directly above the mixing unit. The material being
weighed is fed from a storage bin into the hopper. The flow of material is
controlled by signals from the load cell or scale. When the set point is

reached, the flow of material is stopped and the batch is ready for addition
to the mixer. Weighing accuracies within ±0.25% are available from batch
weighing systems. Figure B-26 illustrates a typical weigh feeder system.

Figure B-26. Typical weigh feeder system (Courtesy
Rexnord).

Continuous weighing involves a system that is sensitive to changes in
the weight of material on a continuous belt (Figure B-27). Typically, the
belt passes a weight-sensitive area (usually load cells) that measure and
total the weight of materials on the belt. A control signal is sent to a
gate controlling the flow of materials from a storage hopper to the belt.
Accuracies within ±1.0% are available from continuous weighing feeders.

B.4.3 Liquid Reagent Control

The control of liquid reagents is normally accomplished by volumetric
methods. Typically, liquid reagents will be proportioned with metering pumps
or flow-measuring systems sending a signal to a control valve. A popular
installation would include a turbine flow meter transmitter with output signal sent to digital or analog instruments for feed rate indication, totaling,
and flow control. Numerous other flow-measuring devices are also available,
including venturi meters, magnetic flow meters, orifice meters, etc. The
turbine flow meter seems, however, to offer greater sensitivity and control.

Figure B-27. Typical belt scale system (Courtesy Rexnord).

REFERENCES

Caterpillar Tractor Co.   1981.   Handbook of Earthmoving.   Caterpillar Tractor Co., Peoria, Illinois.

Caterpillar Tractor Co.   1982.   Caterpillar Performance Handbook.   Caterpillar Tractor Co., Peoria, Illinois.

Hicks, T. G., and T. W. Edwards.   1971.   Pump Application Engineering. McGraw-Hill Book Company, New York, New York.

Terex.   1981.   Production and Cost Estimating of Material Movement with Earthmoving Equipment.   Terex Corporation, Hudson, Ohio.

Perry, R. H.   1973.   Chemical Engineers' Handbook.   McGraw-Hill Book Co., New York, New York.

U.S. EPA.   1985.   Drum Handling Practices at Hazardous Waste Sites (Draft). Municipal Environmental Research Laboratory, U.S. Environmental Protection Agency, Cincinnati, Ohio.

Part III

The information in Part III is from *Immobilization Technology Seminar—Speaker Slide Copies and Supporting Information,* prepared by the U.S. Environmental Protection Agency, October 1989.

Immobilization Processes Overview

Mr. Carlton Wiles
USEPA/RREL
Cincinnati, Ohio

Mr. Edwin Barth
USEPA/RREL
Cincinnati, Ohio

Solidification/stabilization technology is being utilized as a treatment technology for Resource Conservation and Recovery Act RCRA listed waste and waste from uncontrolled hazardous waste sites. Several Best Demonstrated Available Technology (BDAT) levels for Resource Conservation and Recovery Act (RCRA) waste codes are based on solidification/stabilization technology. Vitrification technology is emerging as an alternative technology for hazardous waste. Approximately 25 percent of the Records of Decision (RODs) for Fiscal Year 1988 for the Superfund Program involved solidification/stabilization.

REFERENCES

Barth, E.F., Wiles, C., Technical and Regulatory Status of Solidification/Stabilization in the United States. Proceedings on the Application of U.S. Control Technology in Korea, Seoul, Korea (1989).

U.S. EPA Guide to the Disposal of Chemically Stabilized and Solidified Waste, SW-872 (September, 1980)

U.S. EPA Handbook for Stabilization/Solidification of Hazardous Waste, EPA 540/2-86/001 (June, 1986).

1. Technical and Regulatory Status of Stabilization/Solidification in the United States

E. F. Barth and C. C. Wiles
U.S. Environmental Protection Agency
Cincinnati, Ohio USA

1.0 INTRODUCTION

Solidification/stabilization (S/S) technology has been used for over 20 years to treat U.S. industrial waste and more recently for contaminated soils and municipal waste combustion residuals.

1.1 SOLIDIFICATION/STABILIZATION TECHNOLOGY

1.1.1 Definitions

Definitions for S/S technology vary depending upon the source. Other terms that have been used are immobilization and fixation. In very general terms, S/S, as it relates to managing hazardous waste, refers to a technology where one uses additives or processes to transform the waste into a more manageable form or less toxic form by physically and/or chemically immobilizing the waste constituents. It is important to understand the terminology being used in order to properly evaluate the technology for potential application.

1.1.2 Objectives

The broad objective of S/S technology is to contain a waste contaminant and prevent or minimize the release of the contaminant into the environment. In practice this broad objective may be realized by several mechanisms which include producing a solid; improving the handling characteristics of the waste; decreasing the surface area across which the transport of the contaminant may occur; and limiting the mobility of the contaminant when exposed to leaching fluids. The ideal objective is to chemically transform or bond the toxic contaminant into a non-toxic form. Realistically, chemical bonding of the binder to the contaminant does not routinely occur with available state-of-the-art inorganic S/S technologies. More vendors, however, are claiming that chemical bonding from additive reagents does take place, rather than or in addition to microencapsulation.

1.1.3 Binders and Binding Mechanisms

Binder systems can be placed into two broad categories, inorganic or organic. Most inorganic binding systems in use include varying combinations of hydraulic cements, lime, pozzolans, gypsum, and silicates. Organic binders used or experimented with include epoxy, polyesters, asphalt/bitumen, polyolefins (primarily polyethylene and polybutadiene), and urea formaldehyde. Combinations of inorganic and organic binder systems have been used. These include diatomaceous earth with cement and polystyrene; polyurethane and cement, polymer gels with silicate and lime cement and organic modified clays.

1.1.4 Process Types

S/S process types normally found in the U.S. are: in-drum processing, in-plant processing, mobile plant processing, and in-situ processing. Combinations of these process types may be used depending upon specific requirements at a given site. In the case of a contaminated soil area, in-situ and mobile type of processes are most often used.

1.2 Current Status of S/S in the U.S.

Although S/S has seen significant use for treating industrial waste in the U.S., there are technical and regulatory factors which may greatly affect its future use. U.S. EPA regulations and proposed regulations controlling the treatment and disposal of hazardous waste and contaminated soil and debris will be the major factor determining how much S/S technology will be used. The remaining portions of this paper will further discuss the U.S. EPA S/S programs and the factors important in determining how much S/S will be used in the future in the U.S.

2.0 THE U.S. EPA PROGRAM IN SOLIDIFICATION/STABILIZATION

The U.S. EPA programs in S/S include evaluating S/S as a best demonstrated available technology for treating hazardous waste, evaluating S/S for treating contaminated soil and debris, as well as municipal waste combustion residuals and conducting research to develop a more comprehensive scientific understanding of the technology.

2.1 RESEARCH

The current U.S. EPA research (Table I) has emphasized investigating interferences to S/S, investigating waste-binder interaction and waste disposition sites, and study of methods for predicting performance of S/S products.

2.1.1 Interfering Agents

Research on interfering agents will produce data on the effects that interfering inorganics (certain metals, sulfates, etc.) and organics (oil, grease, HCB, TCE, phenol, etc.) may have on generally used pozzolanic binder systems. The information will be useful in evaluating applications for delisting hazardous waste and for permits to treat hazardous wastes with S/S, particularly for those waste streams contaminated with organics. This information and that from research on factors critical to S/S will also aid decisions on potential waste pretreatment techniques for enhancing S/S performance. Data from physical and chemical tests are being analyzed to determine if a correlation exists between physical properties of the solidified waste form and its ability to resist stresses when exposed to leaching situations. An extensive literature search on interference compounds and interference testing has been completed (2,3).

2.1.2 Evaluating Test Methods

An effort with Environmental Canada is emphasizing evaluation of several leaching tests for determining the extent of toxic constituent binding. Protocols for examining physical properties are also being evaluated. This research includes actual waste and synthetic sludges solidified by vendors. Information on the performance of several different solidified products can be compared. This research will provide important data to compare current U.S. EPA regulatory leaching procedures with others beign tested. However, more research is needed to determine the long-term effectiveness of this treatment technology.

2.1.3 Morphological Studies

Electron scanning and x-ray diffraction microscopy techniques and solvent extractions are being used to investigate waste/binder interactions. The objective is to better understand S/S by identifying binder reaction phases where the waste from other research projects are being examined in efforts to correlate results of physical and chemical tests with performance of S/S products. Results from these specific studies have indicated that physical entrapment of inorganic metals is a predominant containment mechanism (4). However, some results are also indicating formation of altered or new crystal structures in some phases which appear to be chemically bonding some organics. Indications are that this type research could provide information useful in preparing binder formulations better able to treat a specific waste.

2.1.4 Air Emissions

Because of the nature of many solidification processes, uncontrolled air emissions are a potential problem to workers and the environment. Investigations are being conducted to determine the magnitude of these air emissions (5, 6). Processes being evaluated are Portland cement - fly ash and lime kiln dust - fly ash mixtures. As expected, mixing causes the greatest air emissions. Some additives such as lime result in exothermic reactions which increases the release of volatile compounds or may cause combustion. Capture and treatment of these emissions may be required to protect worker health and the environment, particularly in cases where the waste or contaminated soils contain volatile compounds.

2.2 EVALUATION OF S/S AS AN AVAILABLE TREATMENT TECHNOLOGY

2.2.1 Treatment of Hazardous Wastes

In the United States, the Hazardous and Solid Waste Amendments (HSWA), which amended the Resource Conservation and Recovery Act (RCRA) provide detailed procedures dictating how hazardous waste is defined, controlled, and managed. Wastes classified as hazardous under RCRA are often referred to as RCRA hazardous waste. Key provisions of HSWA are the ones which ban the land disposal of hazardous waste unless it is proven to be more protective of the environment and human health than other alternatives. The legislation requires that all hazardous waste be treated by the best demonstrated available treatment (BDAT) instead of and prior to land disposal. The U.S. EPA is required to determine and specify levels to which BDAT technologies can treat RCRA waste. S/S is one of several BDAT technologies being evaluated for non-waste waters (see Table II). In this program selected hazardous wastes are solidified/stabilized by portland cement, lime kiln dust, and lime/fly ash mixtures. Various ratios of waste to binder for each binder system is then evaluated by the Unconfined Compressive Strength Test (UCS) after a cure time of 7, 14, 21 and 28 days. Cured samples are then subjected to the U.S. EPA's Toxicity Characteristics Leaching Procedure (TCLP) extraction test. Leachate from the TCLP is analyzed for the pollutants of concern, to determine how effective S/S can be for treating the selected hazardous waste. Results will be important in determining how much S/S will be used to treat hazardous waste in the U.S.

2.2.2 Treatment of Contaminated Soils

The remediation of contaminated soils from uncontrolled dump sites, is controlled under legislation referred to as the Superfund Amendments and Reauthorization Act (SARA) which amended the Comprehensive Environmental Reclamation Compensation Liability Act (CERCLA). Under SARA provisions, permanent treatment of the contaminated soil and debris is being emphasized rather than the use of nontreatment containment systems such as covers, grout walls, and similar methods. Because of this, a program similar to the RCRA BDAT evaluations is being conducted for SARA remediation technologies including S/S. Mixtures of soils contaminated with selected chemicals were solidified/stabilized and tested to evaluate performance of S/S technology for treating contaminated soils. The test soil being used is a mixture of clay, sand, silt, topsoil and aggregate. As can be seen from Table III, solidification/stabilization was effective for reducing the leaching concentrations of arsenic, copper, lead, nickel, and zinc. No conclusion could be made from the chromium data since the initial concentrations were low.

2.3 SUPERFUND INNOVATIVE TECHNOLOGY EVALUATIONS (SITE)

The U.S. EPA SITE Program provides for demonstration and evaluation of innovative technologies to remediate Superfund sites. Currently six S/S processes are being evaluated (Table IV). The vendors are allowed access to the sites and pay for their operational and treatment expense. The U.S. EPA pays for site preparation work and sampling and analytical cost. Results will provide information on how well S/S can be expected to permanently treat contaminated soils. The evaluations will also help make better extrapolation of laboratory tests results to field conditions.

3.0 PROCESS SELECTION CONSIDERATIONS

3.1 IMPORTANT FACTORS

Factors important in the selection, design, implementation, and performance of processes and products are: waste characteristics (chemical and physical), processing requirements, S/S product management objective, regulatory requirements, and economics. These and other site-specific factors (i.e., location, condition, climate, hydrology, etc.) must be carefully considered to ensure acceptable performance.

3.1.1 Waste Characteristics

The chemical effects of some compounds can reduce the strength of the binder/waste mix, while some compounds can accelerate or retard the S/S curing rate. Temperature and humidity can also retard or accelerate curing. Size and shape of particles can affect the viscosity of the mix. The impact of strength on leachability has not been determined.

3.1.2 Process Type/Quality Assurance and Control

It is important to assess what process type and specific process requirements are required before selecting an S/S technology. For example, a waste-binder can be controlled and mixed more easily in a drum or in a plant process than in the in-situ solidification of a pit, pond, or lagoon. The performance of a process is related to the extent of mixing. More techniques are needed to determine mixing effectiveness. An effective real time Quality Assurance/Quality Control program may need to focus on indirect monitoring techniques (metering equipment, reagent quality and pilot test cells, operator experience, etc.) with then on-line sampling.

3.1.3 Treatment Objectives

The performance goals of the process must be established, for example, strength or leachate reduction required. The ultimate disposal condition must also be determined.

3.1.4 Regulatory Factors

Hazardous waste management regulations in the United States will be critical to the success of S/S. Processes can be altered to meet different performance criteria, which will become increasingly stringent, as regulations become more stringent. S/S will be competing with other treatment technologies to meet these regulatory criteria.

3.1.5. Costs

Costs will depend on site-specific conditions. Important are the waste's characteristics, type of process, disposal requirements and other special factors. What is the physical form and chemical make-up of the waste? Is pretreatment needed? Is transportation of raw materials and/or

are finished S/S products required? Which S/S process is needed? What
special health and safety requirements are needed? What is the quality
assurance/quality control cost involved? What regulatory criteria must be
met? Each of these factors must be considered. As regulatory criteria
become more demanding, the costs of acceptable solidification processes may
increase.

4.0 STUDIES NEEDED FOR S/S AS A TREATMENT TECHNOLOGY

For S/S technology to be effective in managing hazardous waste,
proposed processes must be properly selected, formulated, and used.
Improved knowledge of process selection considerations and the interactions
among the various candidate binders and waste types is critical to the
successful use as an acceptable treatment technology.

Studies are needed:

- to determine the long-term physical-chemical stability of S/S
  products when placed on the land;

- to determine how and under what conditions S/S products should be
  placed on the land to ensure long-term environmental protection;

- to more accurately predict and measure the performance of S/S
  processes and products;

- to provide a correlation between regulatory criteria and real
  world situations;

- to evaluate the effectiveness of protocols (e.g., leaching tests,
  durability test, etc.) to characterize S/S products, provide
  effective measurement techniques, and correlate results of such
  tests with performance in the field;

- to determine if micro-encapsulation is an effective technique
  without bonding;

- to determine the effectiveness of processes and equipment to
  effectively solidify/stabilize contaminated soil and/or lagoons;
  the effectiveness of mixing methods; and the resulting
  solidified/stabilized soil performance at varying soil depths;

- to determine the amounts of organic compounds that can be included
  in inorganic waste streams without requiring pretreatment before
  S/S;

- to determine how effectively S/S processes treat residuals from
  other alternative treatments.

5.0 SUMMARY

Solidification/stabilization is being evaluated by the U.S. EPA as a best demonstrated available technology for treating hazardous waste and contaminated soils and debris. Future use of the technology in the United States will depend on how well it performs compared to other available treatment processes. The evaluations and current research being conducted will provide some answers regarding performance, however, additional studies are required for a better scientific understanding of S/S. Whether or not S/S becomes an important technology for treating hazardous waste and contaminated soils in the U.S. ultimately depends upon regulatory requirements and the capability of the technology to meet these requirements. As performance criteria become more severe, S/S developers may need to improve their processes. The future technology direction is showing progress. In the case of RCRA waste, S/S may be the only acceptable method to treat selected inorganic waste and hazardous residues from incinerators and other treatment processes. In the case of contaminated soils and debris, S/S offers a relatively inexpensive method for treating large areas in situ. However, the capability of the technology to perform satisfactorily over long periods of time has yet to be determined.

REFERENCES

1. Wiles, C. C: A review of Solidification/Stabilization Technology.
    Journal of Hazardous Materials, Vol. 14, (1987)

2. Jones, L. W: Interference Mechanisms in Waste
    Solidification/Stabilization Processes, Final Report for U.S. EPA. IAG
    No. SW-219306080-01-0 (1988)

3. Cullinane, M. J: An Assessment of Materials that Interfere With
    Solidification/Stabilization Processes. Final Report for U.S. EPA. IAG
    No. DW-219306080-01-0 (1988)

4. Cartledge, F. K., et.al.: A Study of the Morphology and Microchemistry
    of Solidified/Stabilized Hazardous Waste Systems. Final Report for U.S.
    EPA #CR-812318 (1989)

5. Weitzman, L., Hammel, M., and Barth, E: Evaluation of
    Solidification/Stabilization as a BDAT for Contaminated Soils,
    Proceedings of Fourteenth Annual HWERL Symposium, Cincinnati, OH (1988)

6. Weitzman, L. et.al.: Volatile Organic Emissions from Stabilized
    Hazardous Waste. Final Report. U.S. EPA Contract #68-02-3994 (1989)

Table I.   CURRENT U.S. EPA SOLIDIFICATION/STABILIZATION
RESEARCH PROJECTS

Project Title	Objective
Evaluation of S/S for Treating Ash Residues, Hazardous Sludges, and Contaminated Soils	Evaluate efficacy for treating several waste types from each category
Evaluation of Factors Affecting Solidification/Stabilization Process	Determine effects of interfering agents on performance of S/S
Investigation of Comparative Test Methods for Solidified – Waste Characterization	Develop and evaluate methods for testing performance of solidification processes
Study of Morphology and Microchemistry of Solidified/Stabilized Waste	Investigate bonding mechanisms
Air Emissions from Waste Stabilization	Determine air emissions from S/S processes
Use of S/S in Treatment Train	Determine efficacy of S/S following incineration, low temperature desorption, and soil washing
Potential Use of Organophilic Clays and Other Organic Binders	Determine treatment potential of organic binders
Construction Quality Assurance/ Quality Control Parameters	Development of QA/QC procedures for real time field use

TABLE II.   EXAMPLE OF U.S. EPA RCRA HAZARDOUS WASTES FOR WHICH
S/S IS BEING EVALUATED AS A TREATMENT TECHNOLOGY

Waste Code	Description of Waste	Pollutant of Concern for S/S
K048-52	Dissolved air flotation (DAF) float from the petroleum refining industry	chromium, lead
K061	Emission control dust/sludge from the primary production of steel in electric furnaces	chromium, lead, cadmium
K046	Wastewater treatment sludges from manufacturing formulation and loading of lead-based initiating compounds	lead
F006	Metal finishing sludges	cadmium, chromium, lead, nickel, silver
F012, F019	Metal finishing sludges	cadmium, chromium lead, nickel, silver
K022	Distillation tar (treated)	chromium, nickel
K001	Wood preserving sludges (treated)	lead

TABLE III.   CONTAMINATED SOIL (SARM) TCLP RESULTS (mg/l)
FROM U.S. EPA SARA BDAT STUDY

Metal	Raw	Treated
As	6.4, 9.6	ND, ND
Cd	33.1, 35.3	ND, ND
Cr	No, .06	.07, .07
Cu	80.7, 10.0	.09, .17
Pb	19.9, 70.4	ND, .37
Ni	17.5, 26.8	ND, ND
Zn	359, 396	.69, .74

Source of data: Weitzman, Hammel, Barth (5)

TABLE IV.   SOLIDIFICATION/STABILIZATION PROCESSES BEING
EVALUATED IN THE U.S. EPA SITE PROGRAM

Chemfix Technologies, Inc.
Soliditech, Inc.
Silicate Technology, Inc.
Hazcon, Inc.
International Waste Technologies
Separation and Recovery Systems

OVERVIEW OF SEMINAR

- Introduction
- Descriptions of S/S technologies
- Description of vitrification technology
- Technology screening procedures
- Physical testing methods
- Chemical testing methods
- Field implementation procedures
- Quality assurance procedures
- Case histories

OVERVIEW OF INTRODUCTION SECTION

- Definitions
- Range of immobilization technologies
- Process descriptions
- Where being utilized
- Regulations

IMMOBILIZATION TECHNOLOGIES

Solidification   **Vitrification**   **Other**
Stabilization

Macroencapsulation
Microencapsulation

SOLIDIFICATION

VS.

STABILIZATION

HIERARCHY OF HAZARDOUS WASTE MANAGEMENT

- Waste minimization/reduction
- 3-R's
  - recovery
  - reuse
  - recycle
- Treatment
  - destruction
  - reduction
    mobility, toxicity, volume
- Storage

GOALS OF S/S PROCESSES

- Reduce pollutant mobility
- Decrease surface area (reduce loss or transfer of contained pollutants)
- Produce solid with no free liquid
- Improve handling and physical characteristics of waste

WHY CONSIDER S/S?

- Minimizes leachate rate from metal waste
- Lower cost than other immobilization technologies, especially if done in situ

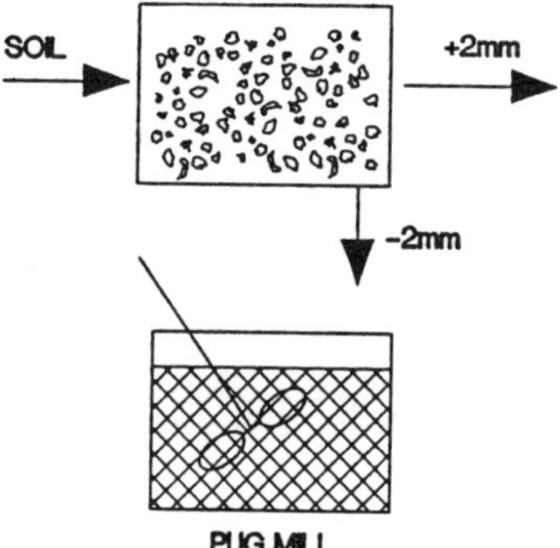

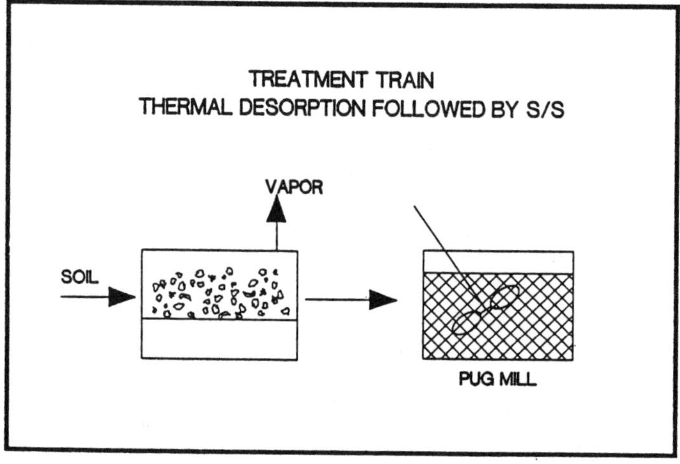

TREATMENT TRAIN
INCINERATION FOLLOWED BY S/S

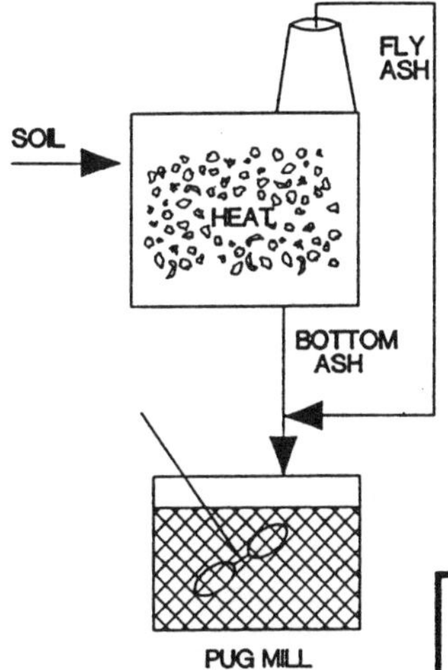

CURRENT STATUS OF S/S UTILIZATION IN U.S.A.

- 25% of Superfund sites in FY 1988
- Several RCRA waste codes for BDAT
- Being considered for MWC ashes

EXAMPLES OF U.S. EPA RCRA HAZARDOUS WASTES FOR WHICH S/S IS BEING EVALUATED AS A TREATMENT TECHNOLOGY

Waste Code	Description of Waste	Pollutant of Concern for S/S
K048-52	Dissolved air flotation (DAF) float from the petroleum refining industry	chromium, lead
K061	Emission control dust/sludge from the primary production of steel in electric furnaces	chromium, lead cadmium
K046'	Wastewater treatment sludges from manufacturing formulation and loading of lead-based initiating compounds	lead

---

EXAMPLES OF U.S. EPA RCRA HAZARDOUS
WASTES FOR WHICH S/S IS BEING
EVALUATED AS A TREATMENT TECHNOLOGY
(continued)

Waste Code	Description of Waste	Pollutant of Concern for S/S
F006	Metal finishing sludges	cadmium, chromium, lead, nickel, silver
F012, F019	Metal finishing sludges	cadmium, chromium, lead, nickel, silver
K022	Distillation tar (treated)	chromium, nickel
K001	Wood preserving sludges (treated)	lead

CURRENT STATUS OF S/S UTILIZATION FOR RADIOACTIVE SITES

- NRC has guidance for low level disposal

SITE PROGRAM DEMONSTRATIONS

HAZCON
IWT
SOLIDITECH
CHEMFIX
STC
SRS

INSTITUTIONAL CONSIDERATIONS

- Waste is neither destroyed nor altered unless volatilized
- Rate of release to groundwater is minimized, but "no migration"?
- Air pathway minimized
- Volume increase

RE-USE / RECLAMATION ISSUES

- Will there be direct contact?
- Will there be rain contact, ocean contact, or groundwater contact?
- What will applied load be?

CHEMICAL REACTION MECHANISMS

- Precipitation as:
  - hydroxides (OH)
  - silicates (Si)
  - sulfides (S)
- Complexation
- Organic binding

BINDING AGENTS

- Inorganic
- Organic
- Combination

BINDER MATERIAL UTILIZED FOR S/S

Inorganic

- Cement
- Lime
- Kiln dust
- Fly ash
- Silicates
- Clay
- Zeolite

BINDER MATERIAL UTILIZED FOR S/S

Organic

- Asphalt
- Surfactant
- Modified clay
- Activated carbon
- Polyesters

BINDER MATERIAL UTILIZED FOR S/S
Organic
(continued)

- Polyethylene
- Resin
- Epoxide
- Urea formaldehyde

SUMMARY OF 1986 TSDF SURVEY

Process Type	No. Units	Quantity (tons)	Capacity (tons)
Cement	50	291,696	3,880,905
Pozzolanic	37	351,635	2,111,111
Asphaltic	1	0	100,000
Thermoplastic	0	0	0
Organic Polymer	1	157	157
Macroencapsulation	3	306	2,926
Other	15	4,110	11,247
Total	107	656,104	6,106,346

REGULATIONS

RCRA/HSWA
CERCLA/SARA

GOALS OF S/S TREATMENT

Delist → Subtitle D

BDAT → Subtitle C

HAZARDOUS WASTE MANAGEMENT REGULATIONS

- RCRA
  - no free liquids
  - 50 p.s.i.
  - liquids – Release Test
- HSWA
  - BDAT

UNCONTROLLED HAZARDOUS WASTE SITE REGULATIONS

- CERCLA
  - cost effective remedy
- SARA
  - treatment preference

S/S CONSIDERATION AS A REMEDIAL TECHNOLOGY

- Treatment by S/S can achieve substantial reduction of mobility
- S/S treatment may represent the best balancing of the selection criteria (including low level organics)
- Management considerations

COST ESTIMATION

Mobilization	$100,000-200,000/site
Excavation	$10-50/cy
Processing	$50-100/cy
Disposal	$100-250/cy

2. Descriptions of Stabilization and Solidification (S/S) Technologies

Dr. Leo Weitzman
LVW
Durham, North Carolina

Mr. Jesse Conner
Chemical Waste Management
Riverdale, Illinois

1.0 INTRODUCTION

As a result of the 1984 RCRA amendments, no liquids and only limited amounts of chemical wastes may be placed in hazardous waste landfills without being chemically altered prior to disposal. This ruling has resulted in increased quantities of waste being solidified or stabilized. This paper gives a brief description of the S/S industry and the processes that are used.

"Solidification/Stabilization" (S/S), also referred to as waste fixation, is a relatively simple process. The waste is mixed with a binder or mixture of binders. The mass is then cured to form a solid matrix that can be safely disposed in a landfill or other containment.

While solidification and stabilization are mechanically very similar, they are really two different industries. Solidification is the traditional industry which takes a waste that contains free water and solidifies it by reacting it with a binder such as cement or lime. No effort is made to reduce the leachability of any hazardous constituents that may be present in the waste. The goal is only to react all free liquids in the waste with the binder. The stabilization industry is emerging in response to the recent "Land Ban" regulations which are restricting specific categories of waste from hazardous waste landfills unless they are pretreated to a minimum leachability standard. Wastes are stabilized by mixing them with specific types and, usually larger quantities, of binder to fix the hazardous constituents in the solid matrix. The intent in this case is to reduce the leachability of hazardous constituents as measured by the Toxicity Characteristic Leaching Procedure (TCLP).

It is important to differentiate between these two unique industries when evaluating the industries and assessing the impact that regulations will have on them. The initial restrictions on the landfill disposal wastes containing free liquids created the solidification industry. Until this ban, very few wastes were treated in this way prior to disposal. Such treatment can be achieved by the addition of relatively small amounts of binder which converts the waste into a soft granular solid. Such a solid can be readily loaded onto bulk carriers and shipped to a landfill for disposal.

Now, the "Land Ban" restrictions are requiring that more types of waste be treated beyond the point of just chemically binding with free liquids. The treatment must also immobilize contaminants. Such treatment usually requires the addition of more binder and often (but not always) produces a hard, monolithic solid mass that is analogous to soft concrete. Handling such a mass is much like handling large, irregularly shaped concrete chunks which cost more to transport than does the original waste. As a result, facilities that are close to hazardous waste landfills are in a better position to stabilize wastes than are those that have to ship the treated waste.

For the purpose of determining and minimizing organic air emissions, the S/S process can be broken down into three distinct steps:

1. mixing
2. curing
3. storage and landfilling

The mixing step is the basic operation where the waste is placed into a mixer and combined with the binder. The mixer is normally designed for ease of loading of the waste and binder and removal of the mixed material. Once mixed, the material is allowed to cure, at which time chemical reactions between the waste and binder harden the mixture. Curing can take place in the mixing vessel (as when mixing occurs in a drum or other disposable vessel) a temporary storage area, or directly in the landfill where the waste is ultimately placed. It can take as little as a few hours to as 30 days or more.

The final step, storage and disposal, is the goal of the S/S process. The material has hardened and if the binders are appropriate for the application, the waste has been stabilized. At this point, the material may be placed in a landfill and covered.

Until relatively recently, the mixing of wastes and binder has been done in open pits, trenches and bunkers. These do not lend themselves to proper collection and control of air emissions. S/S processes can and do produce both particulate and organic air emissions. Particulate control is being required with increasing frequency by states and by federal regulations. Organic emissions from these processes will soon be regulated on a national level by EPA.

Generally speaking, the type of mixing equipment used will influence how well the organic air emissions are collected and controlled, and the type of binder used will influence the point in the process where they will be released. The mixing equipment's effect on emissions is illustrated by the two extremes of a completely open and a completely closed mixing system. Emissions from a completely open mixing system, an open-pit mixer for example, can be complex, requiring the installation of large enclosures. The enclosures need to be designed so that the waste and binder can readily be added to the mixing vessel, or pit, and the mixture can be removed as well. By comparison, it is relatively easy to duct an enclosed mixer to an air pollution control device. In either case, once the Organic Air Emissions are collected, their removal or destruction can be achieved using readily available equipment.

The type of binder used, determines the temperature of the system during mixing or curing. For example, a binder, based on quicklime (CaO), will get very hot when mixed with an aqueous waste. The high temperature will cause a rapid release of the organic constituents. Because the solidified/stabilized product will then have lost most of the volatile constituents during mixing, it will release fewer organic air emissions during curing, storage, and disposal. If the mixing must occur in open equipment, then a binder with a low heat release is desirable. This will result in minimizing the organic air release until the waste can be placed in a sealed and capped landfill.

Clearly, in order to control air emissions during mixing it is necessary to capture them while mixing is going on and during the handling of the waste before and after mixing. If a greater degree of control is required it may be necessary to capture the emissions during the curing phase as well. Control of organic air emissions, once captured, is straightforward. Standard technologies such as condensation, adsorption, thermal incineration, or catalytic incineration can be used to collect or destroy the organic air emissions captured. Particulate emissions control, while straightforward, does require special consideration since the particulate is wet and includes cement or other natural or synthetic pozzolans which can clog and damage the collection equipment.

The following sections discuss the types of mixing equipment and binders commonly used in commercial S/S processes.

2.0 STABILIZATION PROCESS DESCRIPTION

The various stabilization processes commonly used are described in detail by Cullinane and Jones (5). The following discussion is designed merely to offer an overview of the steps involved. S/S can be broken down into two components. The first is the type of process used to mix and handle the wastes and binders, and the second is the type of binder used. With very few exceptions, each mixing method can be used for each type of binder. The exceptions are discussed in Section 4.

2.1 PROCESS TYPES

2.1.1 Open Pit or Trench Mixing

This is the simplest and most commonly used S/S process. The waste is placed in a lined trench, lagoon, or pit, and the binder is mixed with it. Most frequently, a backhoe or a similar device is used to mix the waste and binder. This process is frequently used to S/S waste or soil at field remediation as well as at permanent locations.

S/S is often used in field remediation to solidify a pond or lagoon in place. The binders are then simply dumped into the pond or lagoon and a backhoe is used to blend them in. The solidified material may then be either left in place and capped or excavated and landfilled elsewhere.

When open trench mixing is used at a fixed location, as in a treatment facility, it is typically performed in a lined trench made of concrete. The trench is sometimes placed in a large, open building to protect it from the elements and can be above or below grade. Truckloads or drumloads of the waste are emptied into the trench and the binder is added, either with a backhoe, front-end loader or through a bulk solids handling system. The mass is then mixed with either a backhoe arm or with the blade of the front-end loader. When the front-end loader is used, the mixing is performed by kneading the waste and binder against the wall of the trench. After the mass is mixed, the backhoe or front-end loader is used to load the pasty mass into a truck or roll-off container for transport. The mixture is then, typically taken to a storage area for curing or to the landfill itself.

After the mass hardens, usually within one day, it is tested (if required by regulations) and, if it satisfies the regulations, it is landfilled.

2.1.2 In-Situ S/S--(Egg Beater Mixing)

This process is a recent development in which soil is mixed with the binder in-place, without excavation. A mobile process is used that combines several drills with bits up to six feet in diameter on one truck or trailer. The process literally drills down into the contaminated soil and as the drill penetrates it, binder is injected. The drill mixes the binder and soil. If emission control is needed, a shroud fits over the drilling assembly.

2.1.3 In-Drum Processing

This is a small-scale process in which the binder and waste are mixed in a drum or other disposable container. An open-head drum with a standard drum mixer mounted in it is typically used for this application. After mixing, the mixer is removed and waste-binder matrix allowed to set. The lid is put on the drum and the waste, drum and all, is disposed of in a landfill.

2.1.4 Reactor Processing

This process is a simple scale-up of the in-drum processing system. The bulk waste material and binder are mixed in mechanical mixing vessels. The vessel can be either open or closed top and should have provisions for loading and discharging the solids and waste.

2.1.5 Batch and Continuous Type Closed-Vessel Processing

This process is similar to Reactor Processing with the exception that the reactor is totally enclosed to facilitate collection of the air emissions during mixing. The reactor will generally be purged with a controlled air flow or, if the organic can form an explosive mixture, an inert gas such as nitrogen. The mixing vessel can be a simple closed reactor with facilities for loading and discharging solids or a continuous mixer such as a ribbon mixer or pug mill.

2.1.6 Mobile Plant Processing

Any of the processes can be mounted on a trailer or truck and operated in the field. The vapor collection and control equipment, if needed, would also have to be mobile. The concepts for mobile plant processing are identical to those for similar fixed systems and does not need to be discussed separately.

2.2   SOLIDIFICATION/STABILIZATION AGENTS

In order for S/S to be effective, the binder used must:

a. React with free water in the waste and form a solid.
b. Bind with the metals and organics to reduce their chemical nature and/or their leachability.
c. Bind with the Organic Air Emissions in the waste to reduce their chemical nature and ability to vaporize.

Ideally, the chemical processes used should involve chemical reactions with the hazardous components in the waste; however, practically, this does not happen very often for all components. Frequently, the binder reacts with the free water and metals in the waste and the resultant matrix traps the organic constituents so that they cannot be readily released. Laboratory tests (1,3) have shown that while the typical inorganic processes appear to immobilize the metals well, they do not reduce the emissions of the volatile organic compounds significantly. Many simply absorb or absorb the free liquid without reacting with it.

Sorption is often used as a means of solidifying wastes. It is not, by itself, considered to be a stabilization process since it does not meet the criterion that free water must be chemically combined into a solid matrix. Sorption is frequently used to clean up hazardous materials spills. The inorganic sorbents are sometimes used when the waste is a free flowing liquid to make it easier to handle. In this case, the absorbent and waste would then be stabilized by mixing it with other agents such as cement to make a solid. Common absorbents are expanded clay and treated organic materials such as corn-cobs, or simple sawdust. Organic sorbents are generally used to collect liquids when the product will be sent to an incinerator for disposal.

Binders fall into two categories, inorganic and organic. They are also identified by their chemical mechanism. By far, the most commonly used binders are inorganics such as cement kiln dust, flyash, and other waste materials that can chemically react with water. Because of the large amounts of binder that are often required to solidify/stabilize a waste, cost of the raw materials dominates the selection process. With development of new regulations, leaching and organic air emissions, the binder's performance will, probably, assume greater significance.

The vast majority of commercial S/S is performed with inorganic binders. While a number of organic binders have been proposed, and marketed, their performance has not been fully demonstrated to date. This coupled with their high cost have kept them from making significant inroads into the market. Within both categories, a variety of proprietary binders have been proposed or are marketed by vendors. The majority of the wastes solidified at present uses generic binders discussed below.

Table 2-1 lists some of the proprietary binders presently available. As can be seen most are sufficiently similar to generic ones that their behavior regarding organic air emissions can be readily determined from the discussions below which restrict themselves to generic binders.

2.2.1 <u>Inorganic Binders</u>

As mentioned earlier, binder is typically selected on the basis of cost. As a result, whenever, possible, a waste material that reacts with water will be used. Commonly used inorganic binders are:

- cement kiln dust
- lime kiln dust--typically contains significant amounts of quicklime
- coal fly/bottom ash
- mixtures of the above

When these are unavailable or unsuitable, commercial products are used. These include:

- natural pozzolans
- lime (usually a grade of agricultural lime) mixed with flyash
- portland cement, usually mixed with an inert flyash

The solidification of wastes is analogous to the manufacture of concrete, which is a mixture of coarse aggregate (gravel), fine aggregate (sand), and a binder (cement). Water chemically reacts with the binder to form a solid matrix of the components. In solidification, the waste supplies the water and (depending on its composition) a greater or lesser fraction of the aggregate. Clearly, the two mixtures can be formulated differently. When manufacturing concrete, strength is essential. As a result, the binder, water, aggregate and cement are mixed in proportions that optimize this property. The purpose of S/S is to chemically react the water to form a solid, and to immobilize the contaminants. The product only has to achieve a minimal load bearing strength. The hazardous waste regulations only require an unconfined compressibility strength of 50 psi.

It is, clearly, impossible to discuss all combinations of binders herein; however, for the purpose of ORGANIC AIR emissions, this is not necessary. Regardless of the binder formulations used, the concepts are the same and can be illustrated by the following examples:

1. aqueous waste solidified/stabilized with portland cement/flyash
2. aqueous waste solidified/stabilized with lime kiln dust/flyash
3. aqueous waste solidified/stabilized with agricultural lime/flyash

Flyash from some coal-fired power plants are very reactive and will set-up, much like cement, when mixed with water. Such flyashes are sold in commerce and commonly used as a concrete additive. They can be considered to behave in a similar manner to portland cement with regards to organic air emissions. They type of flyash discussed here, is the less valuable variety. When mixed with the lime or lime kiln dust, it participates in the chemical reaction. When mixed with portland cement, however, it is relatively inert. It serves as a source of aggregate and as a bulking agent to absorb the free water.

TABLE 2-1.   COMMERCIAL WASTE STABILIZATION PROCESSES

Vendor	Process Name	Ingredients	Comments
Chemfix, In	Chemfix	Cement + Soluble Silicates	Probably does not fix most volatile organics
IU Conversion	Sealosafe Stablex	Silicates	Probably does not fix oils solvents, grease volatile organics
Dravo Lime	Calcilox	Glassy Slag and Scrubber Sludge	Designed to fix scrubber sludge. probably does not fix most volatile organics
Envirotech (Subsid. of Chemfix)	Envirotech	Cement and Silicates	U.S. Patent 3,837,872
Velsicol	Velsicol	Fly Ash, Scrubber Sludge and Cement	Claims to stabilize organics; not specific
Stabitrol Corp	Terra-Tite	Cement	Probably does not fix most volatile organics
TRW Systems	--	1. Cement, Plaster and Lime	Does not fix volatile organics
		2. Polybutadiene Resin	May Work for organics; very costly
U.S. Gypsum	Envirostone	Gypsum	Does not fix volatile organics

2.2.2 Organics

The organic binders are rarely used commercially because of their high cost compared to inorganics. They have been proposed for use on wastes containing organic constituents. Organic binders can be broken into two broad categories--(1) bitumen and (2) polymers.

As much as S/S with portland cement is analogous to the manufacture of concrete, the use of asphalt/bitumen is analogous to the manufacture of asphalt concrete--common paving asphalt. In the latter case, the bitumen or tar replaces the cement to bind the aggregate. Bitumen is not a suitable binder for wastes containing water. In fact, the solid material has to be dried prior to mixing with the hot tar. This process has only been used to a limited extent for solidifying soils containing low level radioactive materials and not for hazardous wastes. As a result, this binder will not be discussed separately here.

Polymeric binders have been proposed for the S/S of hazardous wastes. In this application, the wastes are mixed with a material such as expanded clay or flyash to absorb free liquids. The mass is then mixed with a polymer or polymer precursor and allowed to harden. Polymers that can be used include epoxy, polyesters, polyolefins, and urea-formaldehyde. Polymeric binders do not react chemically with most types of wastes. Rather, they encapsulate the hazardous constituents and prevent them from being released to the environment. Polymers typically cost on the order of 25 cents to several dollars/lb. This could translate to a cost of $500 and up per ton of waste stabilized just for the binder. Adding the cost of labor, capital, and operating costs, could readily bring the cost of stabilization with polymeric materials up to that for incineration. It is, therefore, unlikely that polymers will be used as binders to a significant extent in the foreseeable future. They will not be discussed further herein.

2.2.3 Organic/Inorganic Mixtures

Combinations of inorganic and organic binders have been proposed in the past to S/S hazardous waste. These include diatomaceous earth with cement and polystyrene; polyurethane and cement, and polymer gels with silicate and lime cement. As with organic binders, these are relatively expensive for the same reason as the polymeric binders. They will not, likely be used in large volumes in the foreseeable future and will not be discussed further herein.

2.3 INDUSTRY TREND

S/S appears to be a growing industry. At present, it is used to eliminate free water from wastes and to immobilize heavy metals to make them suitable for land disposal. As land ban rules take effect, it will be necessary to pretreat more types of waste to reduce their leachability. EPA has specified several types of pretreatment technologies such as incineration, S/S, and chemical treatment as a "Best Demonstrated Control Technology" (BDAT). Of those proposed, industry has had the most experience with incineration and S/S. Incineration has shown itself to be useful for the destruction of organic contaminants but with limited applicability for inorganics, especially heavy metals. S/S has had good success in immobilizing heavy metal and other inorganic contaminants. As the BDAT regulations come into effect for increasingly more categories of waste, the amount of waste being treated in this way will continue to grow.

It is not likely that new types of binders will be used in the foreseeable future. By and large, inorganic binders based on waste products such as cement kiln dust, lime kiln dust or flyash will continue to be used. In applications where waste materials are unavailable or unsuitable, high volume, relatively low cost and readily available inorganic binders such as portland cement or lime will continue to be the binder of choice. These have been shown to satisfactorily immobilize metals (based on current testing procedures) and there appears to be little incentive to use other, more costly binders.

Organic binders have been available for over a decade. They have generally been proposed to stabilize wastes that have a high organic content, the very wastes that are addressed by this document. It does not, appear, however, that such binders will be used to a significant extent in the foreseeable future.

As discussed above, asphaltic binders are not suitable for hazardous waste application (they cannot tolerate water). There does not appear to be any reason why they should be used in this application in the foreseeable future.

Polymeric binders are not likely to be used for stabilizing hazardous waste because of their very high cost. The raw materials of even the least expensive polymer could cost more than $0.25/pound or $500/ton. Many binders could cost even more. By the time the handling and other costs are added, S/S with this type of binder could easily cost $1,000 or more per ton. At these costs, it would be more economical to pretreat a waste prior to S/S.

For example, an aqueous waste containing organics and heavy metals could be pretreated by a technique such as physical separation and air stripping to remove the volatile organics and then solidified/stabilized with cement kiln dust and flyash. If the organic content of the waste is higher, the waste could be incinerated and then the ash, containing the heavy metals could be stabilized with an inorganic, if necessary. Because of the existence of such alternate waste treatment techniques, it is unlikely that the use of high-cost binders, such as polymerics, will increase.

It is likely, however, that the types of equipment used for S/S will change. At present, the bulk of the processing is conducted in open equipment. Restrictions on organic emissions from these processes will probably shift the economics for many applications in favor of closed mixing equipment. The open systems would require large enclosures or hoods to capture organic emissions. The air flow through these systems would be relatively large with the inherently high costs associated in controlling these. Closed systems could be controlled by using much lower gas flow rates and, as a result, lower capital and operating costs for the portion of the S/S system. While the economics of each individual application will govern the processing equipment selected, it appears that regulations on organic air emissions will result in an increase in the use of enclosed systems.

REFERENCES

1.   Weitzman, L., Hamel, L., and Cadmus, S. "Volatile Emissions from Stabilized Waste." Final Report, Contract 69-02-3993, WA 32 and 37, Risk Reduction Engineering Laboratory, U.S. Environmental Protection Agency, Cincinnati, Ohio, 1988.

2.   Weitzman, L. and Hamel, L. "Evaluation of Solidification/Stabilization as a BDAT for Superfund Soils." Final Report, Contract 68-03-3241, WA 2-18, Risk Reduction Engineering Laboratory, U.S. Environmental Protection Agency, Cincinnati, Ohio, September 1988.

3.   Weitzman, L. "Air Emissions From Hazardous Waste Landfills." Final Report, Contract 68-02-3993, WA 22, Risk Reduction Engineering Laboratory, U.S. Environmental Protection Agency, Cincinnati, Ohio, September 1988.

4.   Balfour, W.D., Wetherold, R.G., and Lewis, D.L. "Evaluation of Air Emissions from Hazardous Waste Treatment, Storage and Disposal Facilities." Final Report Contract, 68-02-3171, Hazardous Waste Engineering Research Laboratory, U.S. Environmental Protection Agency, Cincinnati, Ohio, June 1984.

5.   Cullinane, J., and Jones, L., "Handbook for Stabilization/Solidification of Hazardous Waste." Final Report Interagency Agreement, ********, Risk Reduction Engineering Laboratory, U.S. Environmental Protection Agency, Cincinnati, Ohio, 1988.

6.   U.S. Environmental Protection Agency, Draft EIS. Hazardous Waste TSDF. "Background Information for Proposed RCRA Air Emissions Standards", Volume 1 - Chapters, Volume 2 - Appendices, March 1988.

7.   U.S. Environmental Protection Agency, "Air Emissions from Municipal Solid Waste Landfills - Background Information for Proposed Standards and Guidelines", Preliminary Draft. March 1988.

DISCUSS

- Waste characteristics
- Binders
- Mixing techniques
  (later talk)

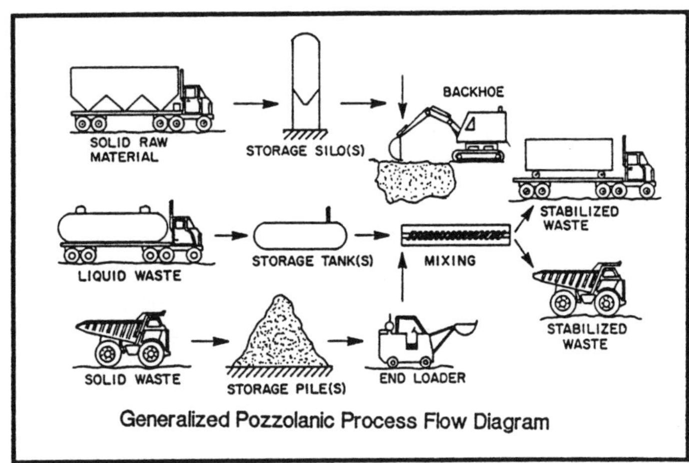

Generalized Pozzolanic Process Flow Diagram

BASIC APPROACH TO STABILIZATION OF INORGANIC CONSTITUENTS

- Mix the waste with materials which convert the target constituents into relatively insoluble compounds
- Encapsulate the insoluble compounds in a matrix which reduces access by leachate and water
  - macroencapsulate
  - microencapsulate

MICRO- VS MACRO-ENCAPSULATION

- Blurred line
- Function of test method

KEY FACTORS

- Waste properties and composition
- Binders
- Mixing techniques

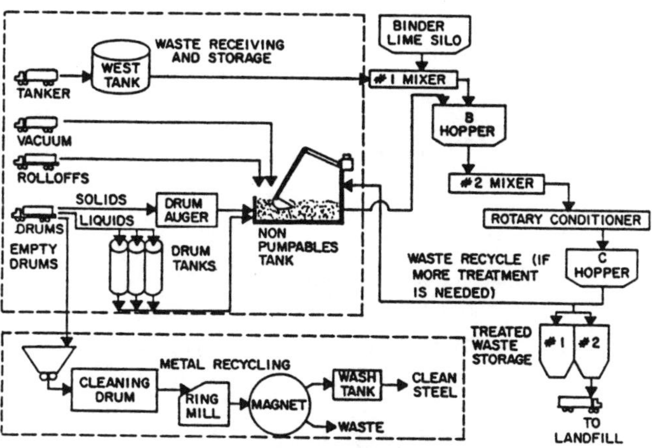

Flow Diagram of S/S Operation of Chem Met Services

IMPORTANT WASTE PROPERTIES

- Water content
  - physical state
- Hazardous constituents
  - inorganic
  - organic

WATER CONTENT

- Binder must chemically react with the water (solidification)
- Commonly done by hydrate formation

WHAT TAKES PLACE DURING S/S

- Water Chemically Reacts
- Hazardous constituents are made less soluble
- Hazardous Constituents are Encapsulated
  - Reduce contact with the environment

FORTUNATELY!

S/S BINDER TYPES
- Cement based binders
  - Portland cement
  - Cement kiln dust
  - Cement/flyash
  - Natural and artificial pozzolans
- Lime/limestone/quicklime
  - Lime kiln dust
  - Lime/flyash
- Absorbents
  - Hydro- and organo- philic clays
  - Wood chips, etc.
    generally not acceptable

(1 of 2)

S/S BINDERS
- Thermoplastic materials
  - asphalt/bitumen
  - thermoplastic polymers
- Thermosetting polymers
- Vitrification

(2 of 2)

POZZOLAN

**Naturally occuring material
that reacts to form a solid
on addition of water or lime
with water**

**VAST MAJORITY OF S/S WITH
CEMENT BASED BINDERS**

WHY?

Consider Portland Cement

CEMENT BASED BINDERS

- Most common type
- Analogous to the manufacture of concrete
    - aggregate (coarse and fine)
    - water
    - cement
- Basic concepts apply

**NORMAL APPROACH
TO S/S**

1. Chemically react all water

2. Insolubilize hazardous constituents

3. Encapsulate products

TYPES OF PORTLAND CEMENTS
(WT %)

Compound	I	II	III	IV
Tricalcium Silicate	53	47	58	25
Dicalcium Silicate	24	32	16	54
Tricalcium Aluminate	8	3	8	2
Tetracalcium Aluminate	8	12	8	12
Total	93	94	90	94

Source: Portland Cement Association

$2(3Ca \cdot SiO_2)$ + $6H_2O$ → $3CaO \cdot 2SiO_2 \cdot 3H_2O$ + $3Ca(OH)_2$

Tricalcium Silicicate   Water   Tobermorite Gel   Calcium Hydroxide

$2(2CaO \cdot SiO_2)$ + $4H_2O$ → $3CaO \cdot 2SiO_2 \cdot 3H_2O$ + $Ca(OH)_2$

Dicalcium Silicate   Water   Tombermorite Gel   Calcium Hydroxide

$4CaO \cdot Al_2O_3 \cdot Fe_2O_3$ + $10H_2O$ + $2Ca(OH)_2$ → $6CaO \cdot Al_2O_3 \cdot Fe_2O_3 \cdot 12H_2O$

Tetracalcium Aluminate   Water   Calcium Hydroxide   Calcium Aluminate Ferrite

$3CaO \cdot Al_2O_3$ + $12H_2O$ + $Ca(OH)_2$ → $3CaO \cdot Al_2O_3 \cdot Ca(OH)_2 \cdot 12H_2O$

Tricalcium Aluminate   Water   Calcium Hydroxide   Tetracalcium Aluminate Hydrate

Source: Portland Cement Association

PORTLAND CEMENT

Dicalcium silicate
Tricalcium silicate $\rbrace$ + H_2O → Tobermorite gel $3CaO \cdot 2SiO_2 \cdot 3H_2O$
+
Lime $Ca(OH)_2$

**The cement ties up
free water four ways**

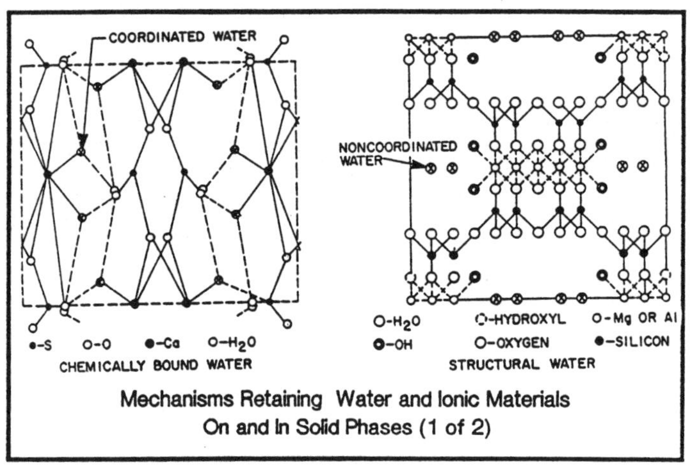

Mechanisms Retaining Water and Ionic Materials
On and In Solid Phases (1 of 2)

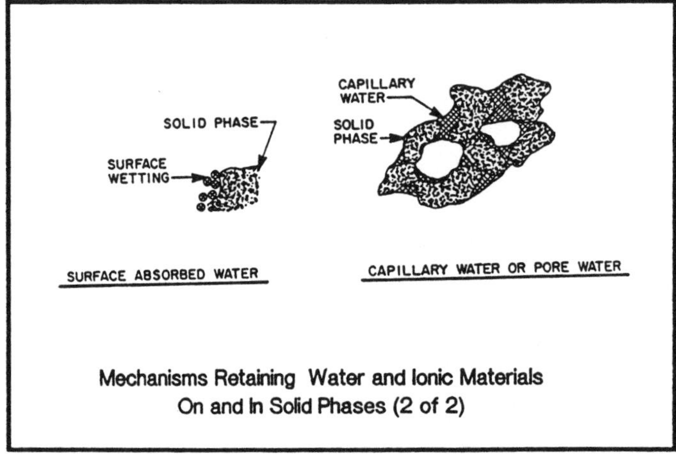

Mechanisms Retaining Water and Ionic Materials
On and In Solid Phases (2 of 2)

**Cement also forms less
soluble compounds**

SOLUBILITY OF SOME COMPOUNDS
OF TCLP CATIONS

Solubility of anionic salt in water

Cation	Cl^-	NO_3^-	OH^-	$CO_3^=$	$SO_4^=$	Oxide
Pb^{++}	S	S	SS	I	SS	I
Zn^{++}	S	S	S	I	S	SS
Cd^{++}	S	S	I	I	S	I
Cu^+	SS	S	-	I	D	I

S - soluble, D - decomposes, R - reacts chemically
SS - slightly soluble (<0.02g/100cc)
I - insoluble (<0.002g/100cc)

(1 of 3)

SOLUBILITY OF SOME COMPOUNDS
OF TCLP CATIONS

Solubility of anionic salt in water

Cation	Cl^-	NO_3^-	OH^-	$CO_3^=$	$SO_4^=$	Oxide
Cu^{++}	S	S	I	I	S	I
Cr^{+3}	I	S	--	--	I	I
Cr^{+6}	S	S	--	I	I	I

S - soluble, D - decomposes, R - reacts chemically
SS - slightly soluble (<0.02g/100cc)
I - insoluble (<0.002g/100cc)

(2 of 3)

SOLUBILITY OF SOME COMPOUNDS OF TCLP CATIONS

Solubility of anionic salt in water

Cation	Cl^-	NO_3^-	OH^-	$CO_3^=$	$SO_4^=$	Oxide
Ni^{++}	S	S	--	I	S	I
Hg^+	I	D	--	I	D	I
Hg^{++}	S	S	--	I	S	SS
Ca^{++}	S	S	S	I	S	R

S - soluble, D - decomposes, R - reacts chemically
SS - slightly soluble ($<$0.02g/100cc)
I - insoluble ($<$0.002g/100cc)

(3 of 3)

CAVEATS

- Mixtures of compounds not always the sum of the components
- Components interfere with chemical reactions, set-up, and S/S
- Theory is a useful start

Solution is a Mixture of Ions

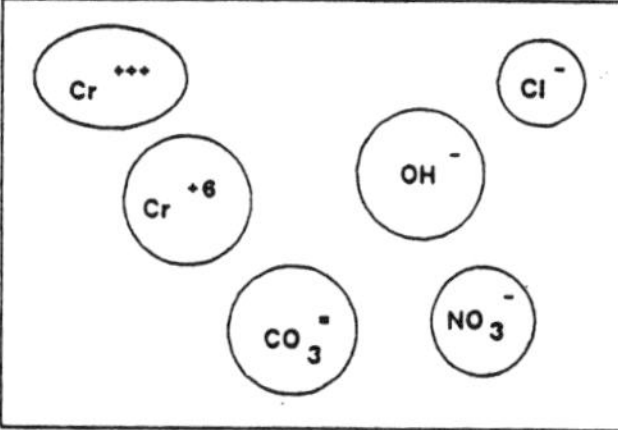

SOLUBILITY PRODUCT

$$An_i \times Cat_j = K_{ij}$$

$$Ca^{++} \times OH^- = \text{Constant}$$

$$Cr^{+3} \times Cl^- = \text{Constant}$$

$$Cr^{+6} \times Cl^- = \text{Constant}$$

Mixtures form complex equilibrium

CONVERT TO INSOLUBLE SALTS
Summary

- Normal approach is to saturate
  with insoluble anion
  i.e. $CO_3^=$, OH^-
- Then combine with other, less
  soluble cations
  i.e. Ca^{++}, Si^{++}
  to form complex insoluble compounds
- Cement, lime, limestone, quicklime,
  and flyash – all sources of these
  anions and cations

NATURAL POZZOLANS

- Some forms of lava and coral
- Rarely used for S/S
- Does not appear to offer any
  operational or cost advantages
  in most situations

FLYASH WITH AND WITHOUT LIME
(Artificial Pozzolans)

- Some coal power plant flyashes will react with water and form hydrates, similar to cement
- Many more flyashes will do so with the addition of lime (or lime kiln dust)

FLYASH

- Contains
    - Silica
    - Alumina
    - Calcium oxide
- Most flyashes do not harden by themselves
- Many flyashes will harden when mixed with $Ca(OH)_2$ or cement
- A few contain enough $Ca(OH)_2$ to harden by themselves

Note: Flyashes can contain metals

LIME/FLYASH MIXTURES
(Artificial Pozzolans)

- Flyash supplies silicon
- Lime supplies $Ca(OH)_2$
- Results in properties chemically similar to cement

LIME BASED BINDERS

- Lime (slaked lime) $Ca(OH)_2$
- Quicklime CaO
- Limestone $CaCO_3$

$$CaO + H_2O \rightarrow Ca(OH)_2$$
quicklime   lime

$$Ca(OH)_2 + CO_2 \rightarrow CaCO_3$$
lime   limestone

LIME KILN DUST

- Contains large amounts of quicklime (lime) – CaO
- Large temperature rise as it is mixed with water to form "slaked lime" $Ca(OH)_2$

WHEN USED WITH MATERIALS LIKE COAL FLYASH

- Calcium compounds react with the iron, silicon and other metals in an analogous manner as in portland cement
- Artificial pozzolans

AGRICULTURAL LIME

- Used to modify pH
- Modifies physical properties
- Generally limited ability to stabilize most metals

NOTE ON CYANIDES

- Alkalinity of cement and lime based binders is important in their stabilization
- If properly done, possible to fix them
- Chemistry is crucial

ADDITIVES / MODIFIERS

- Sodium silicate, $NaSiO_2$
  (water glass)
  - additional silicon
  - claimed to form NaSiAl gels
  - fills pores in product
  - does reduce leaching of
    contaminants at times

TYPES OF SORBENTS

- Flyash
- Limestone screenings
- Clays
- Zeolites
  (see book)

HYDRO- AND ORGANOPHILIC CLAYS

- Commonly used to absorb liquids prior to S/S
- Lab tests have indicated that some organophillic
  clays chemically bond to organic liquids
  - concern with bond strength
- EPA's policy is uncertain now
- In most cases the mechanism is merely
  physical absorption
- May have promise in combination with other binders
  i.e. absorb then bind with cement

NATURAL SORBENTS AND THEIR CAPACITY FOR REMOVAL OF SPECIFIC CONTAMINANTS
FROM LIQUID PHASES OF NEUTRAL, BASIC, AND ACIDIC WASTES

Contaminant	Neutral Waste (calcium flouride)		Basic Waste (metal finishing sludge)		Acidic Waste (petroleum sludge)	
Ca	Zeolite	(5054)[*]	Illite	(1280)	Zeolite	(1390)
	Kaolinite	(857)	Zeolite	(1240)	Illite	(721)
			Kaolinite	(733)	Kaolinite	(10.5)
Cu	Zeolite	(8.2)	Zeolite	(85)	Zeolite	(5.2)
	Kaolinite	(6.7)	Kaolinite	(24)	Acidic F.A.	(2.4)
	Acidic F.A.[†]	(2.1)	Acidic F.A.	(13)	Kaolinite	(0)
Mg	Basic F.A.	(155)	Zeolite	(1328)	Zeolite	(746)
			Illite	(1122)	Illite	(110)
			Basic F.A.	(176)	Basic F.A.	(1.7)
Zn					Zeolite	(10.8)
			Vermiculite	(4.5)		
			Basic F.A.	(1.7)		
Ni			Zeolite	(13.5)		
			Illite	(5.1)		
			Acidic F.A.	(3.8)		
F	Illite	(175)	Kaolinite	(2.6)	Illite	(9.3)
	Kaolinite	(132)	Illite	(2.2)	Acidic F.A.	(8.7)
	Acidic F.A.	(102)			Kaolinite	(3.5)
Total CN–					Illite	(12.1)
					Vermiculite	(7.6)
					Acidic F.A.	(2.7)
COD	Acidic F.A.	(690)	Illite	(1744)	Vermiculite	(6654)
	Illite	(108)	Acidic F.A.	(1080)	Illite	(4807)
			Vermiculite	(244)	Acidic F.A.	(3818)

[*] Bracket represents sorbent capacity in micrograms of contaminant removed per gram of sorbent used. After Sheih (1979) and Chan et al. (1979).
[†] F.A. = fly ash

TYPICAL PHYSICAL AND CHEMICAL PROPERTIES OF COMMONLY USED NATURAL SORBENTS

Sorbent	Bulk density (kg/m^3)	Cation-exchange capacity (meq/100 g)	Anion-exchange (meq/100 g)	Slurry pH	Major mineral species present
Fly ash, acidic	1187	—	—	4–5	Amorphous silicates, hematite, quartz, mullite, free carbon.
Fly ash, basic	1187	—	—	9–10	Calcite, amorphous silicates, quartz, hematite, mullite, free carbon.
Kiln dust	641–890	—	—	9–11	Calcite, quartz, lime (CaO) anhydrite.
Limestone screenings	—	—	—	6–7	Calcite, dolomite.
Clay minerals (soils)	1519			—	Various, e.g., illite.
Kaolinite		5–15	6–20		Can be relatively pure kaolonite.
Vermiculite		100–500	4	—	Can be relatively pure.
Bentonite		100–120	—	—	Smectite, quartz, illite, gypsum, feldspar, kaolinite, calcite.
Zeolite	1543	100–300	—	—	Zeolite (e.g., heulondite, laumonite, stilbite, chabazite, etc.)

From: Sheih (1979), Haynes and Kramer (1982), Grim and Guven (1978).

CONCERNS WITH BINDERS

- May themselves contain
  metals or organics of concern
    - Cement, flyash, etc. can
      contain mercury, cadmium, etc.
      in trace quantities
    - Asphalts can contain naphthalenes,
      other polycyclic organics
- Composition can vary by source

COMPARISON OF REAGENT LEACHING WITH RCRA LIMITS

SAMPLE	TEST TYPE	CONCENTRATION (mg/l) IN WASTE OR LEACHATE			
		CADMIUM	CHROMIUM	LEAD	NICKEL
PORTLAND	TOTAL	1.11	33.70	57.50	24.90
CEMENT	EPT	< 0.01	0.26	0.26	0.02
CEMENT KILN	TOTAL	3.95	29.00	191.00	11.00
DUST (MBI)	EPT	< 0.01	0.07	0.49	< 0.01
TYPE C	TOTAL	2.97	75.40	221.00	52.50
FLYASH	EPT	< 0.01	0.14	0.35	0.02
RCRA EPT LIMITS		1.00	5.00	5.00	
DELISTING LIMITS		0.063	0.315	0.315	
DRINKING WATER STANDARDS		0.01	0.05	0.05	

ASPHALT / BITUMEN

- To understand, consider the manufacture
  of asphaltic concrete

ASPHALT/BITUMEN

- Analogous to asphaltic concrete
- Potential problem with the binder containing organic hazardous constituents
- Organics in the waste tend to dissolve the asphalt/bitumen
- The waste must be "dry"
    - water reduces the cohesion of the solids and asphaltic binder
- The blend is generally temperature sensitive

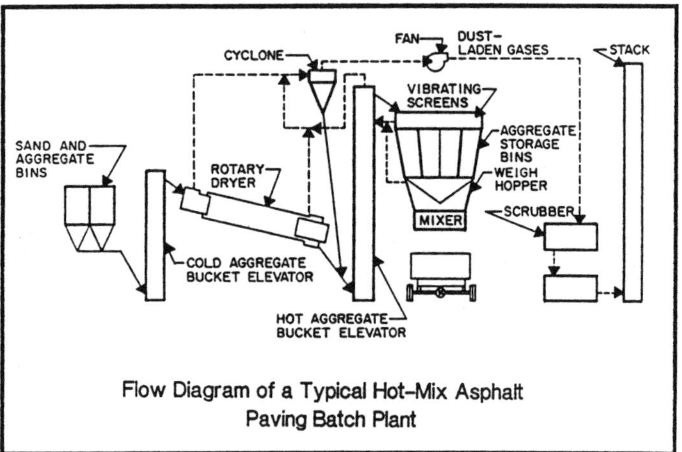

Flow Diagram of a Typical Hot-Mix Asphalt
Paving Batch Plant

POLYMERIC BINDERS
(Thermoplastic, Thermosetting)

- Used to a limited extent with radioactive wastes
    - blocks of vitrified or otherwise fixed waste coated with a polymer
- Used to treat monomer wastes from polymer production
- Has some potential in treating compatible wastes from Superfund sites - usually as a final treatment
- Very highly waste selective

THERMOPLASTICS

- Melt at higher temperatures
- Usually melted into the waste much as asphaltic binders
- Examples:
    - polyethylene
    - polypropylene

THERMOSETTING PLASTICS

- Monomers react at all temperatures
- The polymer does not melt well – often decomposes at higher temperatures
  Examples:
    - bakelite
    - epoxies
- Produce water on polymerization

COMMERCIAL PROPRIETARY PROCESSES

- Typically use variations of basic concepts

COMMERCIAL WASTE STABILIZATION PROCESSES

Vendor	Process Name	Ingredients	Comments
Chemfix, In	Chemfix	Cement + Soluble Silicates	Probably does not fix most volatile organics
IU Conversion	Sealosafe Stablex	Silicates	Probably does not fix oils, solvents, grease, volatile organics
Dravo Lime	Calcilox	Glassy Slag & Scrubber Sludge	Designed to fix scrubber sludge. probably does not fix most volatile organics
Envirotech (Subsid. of Chemfix)	Envirotech	Cement & Silicates	U.S. Patent 3,837,872
Velsicol	Velsicol	Fly Ash, Scrubber Sludge & Cement	Claims to stabilize organics; not specific
Staborol Corp	Terra-Tite	Cement	Probably does not fix most volatile organics
TRW Systems	--	1. Cement, Plaster & Lime	Does not fix volatile organics may work for organics; very costly

Source: Chemical Waste Management

AIR EMISSIONS FROM S/S PROCESSES

- Particulate
  - usually regulated locally
- Organic vapors
  - regulations in final stages of
    of development by EPA/OAQPS

S/S OF ORGANIC CONSTITUENTS

- Rarely react with inorganic binders
- Often interfere with binder setting and inorganic reactions
- Frequently vaporize during S/S

ORGANIC CONSTITUENTS

- May volatilize
- Lack of presence in leachate not necessarily indicative of stabilization
- Air pollution problem

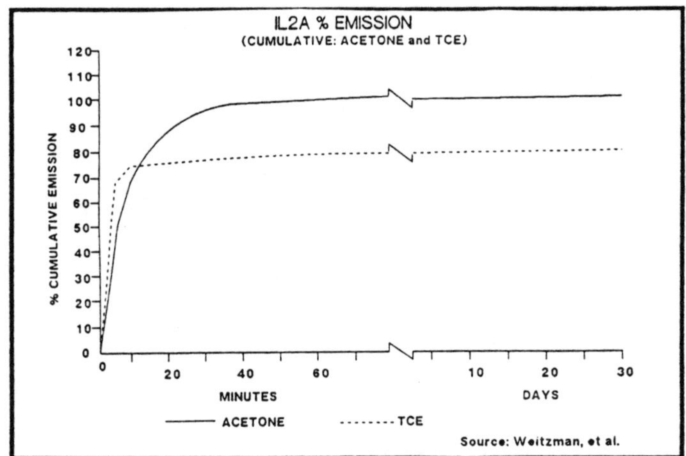
IL2A % EMISSION
(CUMULATIVE: ACETONE and TCE)
% CUMULATIVE EMISSION
120
110
100
90
80
70
60
50
40
30
20
10
0
20
40
60
MINUTES
10
20
30
DAYS
ACETONE
TCE
Source: Weitzman, et al.

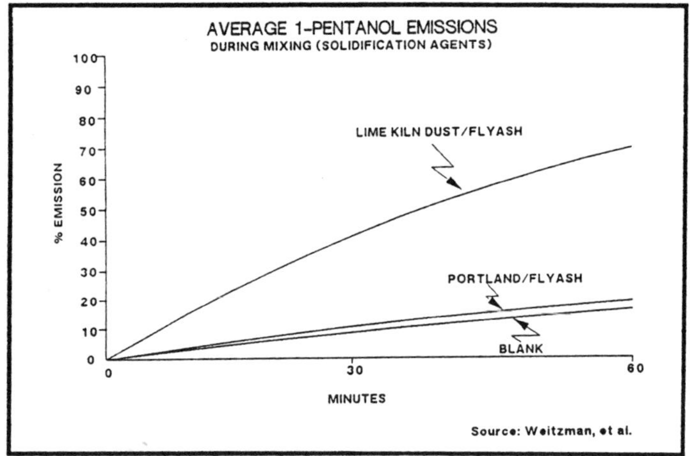
AVERAGE 1-PENTANOL EMISSIONS
DURING MIXING (SOLIDIFICATION AGENTS)
% EMISSION
100
90
80
70
60
50
40
30
20
10
0
LIME KILN DUST/FLYASH
PORTLAND/FLYASH
BLANK
0
30
60
MINUTES
Source: Weitzman, et al.

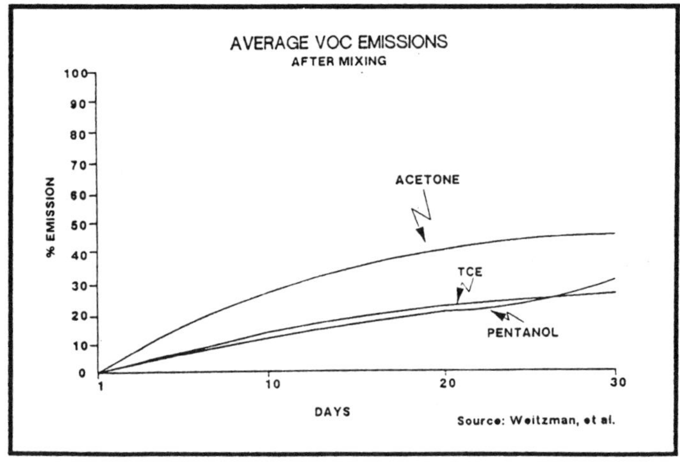
AVERAGE VOC EMISSIONS
AFTER MIXING
% EMISSION
100
90
80
70
60
50
40
30
20
10
0
ACETONE
TCE
PENTANOL
1
10
20
30
DAYS
Source: Weitzman, et al.

CONCLUSION
Organic Contaminants

- Often volatilize

- Interfere with S/S process

- Dissolve or weaken polymers
  and bitumen - compatibility critical

- Weaken structural integrity of
  S/S products

- Increases porosity of product
  - reduces effectiveness of
    encapsulation

HOW TO DEAL WITH
ORGANIC CONSTITUENTS

- Pretreat to remove
  - air strip
  - thermal strip
  - steam strip
  - wet oxidation

- Chemically destroy
  - few organics form stable,
    nonmobile compounds under
    mild conditions
  Exception: monomers $\longrightarrow$ polymers

(1 of 2)

HOW TO DEAL WITH
ORGANIC CONSTITUENTS

- Increase ability of binder to
  encapsulate organics
  - sodium silicate
  - surfactants
  Does not chemically bind or destroy
  organic compound

- Mix in enclosed system with
  collection/control

(2 of 2)

<u>ECONOMICS</u>

<u>100 TPD Operation, All Costs are in $/Ton</u>

Waste	Volume Increase	Chemicals	Processing	Testing	Total Cost Airspace=$20/ton	Airspace=$80/ton
F006-High Solids (35%)						
PC-Based System	7	14	7	2	44	109
KD-Based System	35	17	9	2	55	136
F006 - Low Solids (10%)						
PC-Based System	10	32	6	2	61	127
KD-Based System	65	32	9	2	76	175

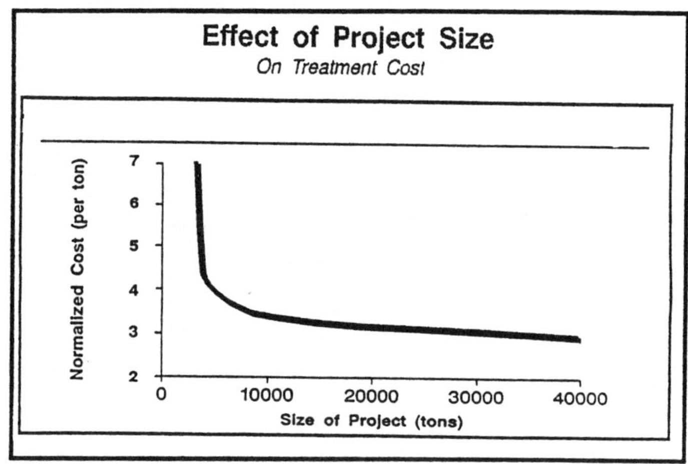

SUMMARY

- Many S/S techniques to choose from
- Need to understand the chemical and physical processes
- Theory is only the first step
- Must then verify and validate theoretical estimates with laboratory and pilot-scale screening tests before selecting alternatives

3. Description of Vitrification Technology

Mr. Jim Hansen
Geosafe Corporation
Kirkland, Washington

The Superfund Amendments and Reauthorization Act (SARA) mandated that the Environmental Protection Agency (EPA) give priority to treatment technologies that are: (1) permanent, (2) capable of reducing the toxicity, mobility, and/or volume of hazardous materials, and (3) capable of being performed on-site and in situ. These criteria have proven to be particularly challenging in the difficult area of remediating contaminated soils, sludges, sediments, and process tailings. A group of vitrification technologies hold the potential to satisfy these objectives for many contaminated solids applications. The accompanying figures and following discussion provide a general description of this developing technology area.

Vitrification technologies are those that involve exposure of hazardous materials to molten glass and related process conditions to affect the destruction, removal, and/or permanent immobilization of hazardous contaminants. Vitrification is defined as conversion of such solids into a glass residual form through the application of heat to the point of fusion. The technologies are applicable to use on solids that are capable of forming a molten, vitreous mass, and of producing a glass-like residual product upon cooling. Typically, the residual product is a solid (super-cooled liquid) containing an amorphous mixture of oxides (primarily silica and alumina) with little or no crystallization prosent.

Exposure of contaminants to vitrification processing results in several desirable results: (1) destruction of hazardous organics by pyrolytic decomposition and/or oxidation, (2) removal (partial or full) of low-solubility, high-volatility, high-solubility inorganics in the residual glass product through chemical incorporation and/or encapsulation. Thus, the vitrification processes may be considered as both thermal treatment (destruction) and immobilization processes.

The various vitrification processes similarly produce a glassy residual product resembling natural obsidian in physical and chemical characteristics. The residual product may be made in granular form, cast into containers, or in multi-thousand ton monoliths. Typically the product has excellent structural, weathering, and biotoxicity characteristics, making it suitable for long-term environmental exposure. The residual typically is able to surpass EPA leach testing requirements (e.g., EP-Tox and TCLP), making it a candidate for delisting as a hazardous waste.

There are several different processes that produce a vitrified product. Most of these are performed ex situ, however one is performed in situ. These basic processes are discussed below.

Ex-Situ Vitrification

Ex-situ vitrification technologies include: (1) electric furnace/melter, (2) plasma centrifugal reactor, and (3) slagging kiln/incinerator.

The <u>electric furnace/melter</u> class includes processes that utilize a ceramic-lined, steel-shelled melter, to contain the molten glass and waste materials to be melted. Some of these processes utilize equipment quite similar to electric glass furnaces that have widespread use for the manufacture of glass products (e.g., bottles, plate products). Such melters involve placement of waste materials and glass batch chemicals directly on the surface of a molten glass bath. The majority of melting occurs at the waste/molten glass interface as heat is transferred from the molten glass. As such waste is heated, organics and inorganic volatiles are evolved and either pyrolyzed or oxidized prior to off-gas treatment to ensure safe air emissions.

Another class of melters involve feeding mechanisms that introduce the waste materials below the molten glass surface. Such method of introduction results in the pyrolysis of organic contaminants within the molten glass, followed by evolution of pyrolyzed off-gases to the space above the glass surface and thence to the off-gas treatment system. Both classes of melters result in the incorporation of nonvaporizable inorganics into the molten glass.

Periodically, the electric melters must be tapped to remove the accumulated glass product. The molten glass may be cast directly into containers. Another alternate utilizes a water bath to produce a granular residual product. The containerized or loose residual product must then be disposed.

The <u>plasma centrifugal reactor</u> varies significantly from the melter/furnace concept. In the plasma reactor, prepared waste materials are fed into a rotating reactor well in which a transferred-arc plasma torch is operating. The plasma torch, which is capable of temperatures exceeding 10,000°C, heats the waste material beyond the point of melting to about (typically) 1,600°C. The melted material is allowed to fall into a slag chamber where it is collected in a container.

Organics and other volatiles emitted during the plasma heating are passed to a secondary combustion chamber into which an oxidizing gas is added. The resulting off-gases are then transferred to an off-gas treatment system to ensure safe air emissions. The containerized slag must eventually be disposed.

Last, solids may be vitrified in rotary kilns and incineration equipment that is operated in a slagging mode. Large kilns are able to accommodate whole drums and mixed solids with little or no pretreatment being required.

Such large facilities are typically stationary as opposed to mobile, on-site facilities. In these facilities, the vitrified product is removed from the exit end of the kiln; it may be cast into containers or granulated as done for the other technologies mentioned above.

Being ex-situ processes, it is necessary that application of each of the above three classes of processes involve excavation, and possible pretreatment of waste materials as required to allowing feeding and proper treatment. In each case the residual product must be disposed; however, the difficulties of such disposal may be minimized if the residual product is of sufficient quality to be delisted.

In-Situ Vitrification

There is one vitrification technology, designated ISV for in-situ vitrification, which brings the benefits of the molten glass exposure and high quality glass residual product to the in-situ application area. The ISV process system has been developed and demonstrated through large-scale; wastes treated include a variety of hazardous chemical, radioactive, and mixed (hazardous chemical and radioactive) wastes.

The ISV process electrically melts inorganic materials (e.g., soil) for the purpose of thermochemical treating free and/or containerized contaminants present within the treatment volume. Most ISV applications involve melting of natural soils; however, other naturally occurring or process residual inorganics (e.g., sludge, tailings, sediments), or process chemicals may be utilized. When used to treat contaminated soil, the process simultaneously destroys and/or removes organic contaminants while chemically incorporating (immobilizing) inorganic contaminants into a chemically inert, stable glass and crystalline residual product.

An array (usually square) of four electrodes is placed to the desired treatment depth in the volume to be treated. As electric potential is applied between the electrodes, current flows through the starter path, heating it and the adjacent soil to temperatures above 1600°C, which is well above typical soil fusion temperatures. Upon melting, typical soils become quite electrically conductive; thus the molten mass becomes the primary conductor and heat transfer medium allowing the process to continue beyond startup. Continued application of electric energy causes the molten volume to grow downward and outward encompassing the desired treatment volume. The rate of melt advance is in the 1 to 2 in/hr range. Individual settings (i.e., the melt involved with a single placement of electrodes) may grow to encompass a total melt mass of up to 1000 ton and a maximum width of about 30 ft. Single setting depths as great as 30 ft are considered possible with the existing large-scale ISV equipment. Adjacent settings are positioned to fuse to each other and to completely process the desired volume at a site.

The molten soil mass is typically in the 1600 to 2000°C temperature range; specific temperatures are dependent on the overall chemistry of the melt. Within the melt, a vigorous, chemically reducing environment is typical. Because soil typically has low thermal conductivity, a very steep thermal gradient (i.e., 150–250°C/in) precedes the advancing melt surfaces. Typically, the 100°C isotherm is less than 1 ft away from the molten mass itself. The soil volume between the 100°C isotherm and the melt is termed the dry zone; this zone has maximum vapor permeability as it exists without water present in the liquid state.

The large-scale ISV system melts soil at a rate of 4 to 6 ton/hr. Since the void volume present in particulate materials (e.g., 20–40% for typical soils) is removed during processing, a corresponding volume reduction occurs. Also, since some of the materials present in the soil are removed as gases and vapors during processing (e.g., humus, organic contaminants), further volume reduction occurs. The volume reduction creates a subsidence volume above the melt and an angle of repose in the soil adjacent to the melt. Upon cooling, an obsidian-like vitrified monolith results (silicate glass and microcrystalline structure); this product possesses excellent structural and environmental properties.

The process utilizes a mobile, on-site equipment system. Electric power is usually taken from a utility distribution system at transmission voltages of 12,500 or 13,800 volts; alternatively the power may be generated on site by a diesel generator. The 3-phase power is supplied to a special multiple-tap transformer (Scott Tee) that converts the power to 2-phase and transforms it to the voltage levels needed throughout the processing. The electrical supply system utilizes an isolated ground circuit which provides appropriate operational safety.

Flow of air through the hood is controlled to maintain a negative pressure (0.5 to 1.0 in H_2O). An ample supply of air provides excess oxygen for combustion of pyrolysis products and organic vapors, if any evolve from the treatment volume. The off-gases, combustion products and air are drawn from the hood (by induced draft blower) into the off-gas treatment system which utilizes the following unit processes to ensure compliant air emissions: (1) quenching, (2) pH controlled scrubbing, (3) dewatering (mist elimination), (4) heating (for dewpoint control), (5) particulate filtration, and (6) activated carbon adsorption. A self-contained glycol cooling system is utilized to cool the quenching/scrubbing solution; this avoids the need for a constant on-site water supply. The amount of moisture present in the exhaust air stream is controlled to accommodate the moisture that is removed form the treatment volume during processing.

A backup off-gas treatment system is provided to accommodate the possibility of power failure. The backup system employs a diesel-powered generator, blower, mist cooler, filter and activated carbon column. The backup system is capable of removing and treating off-gases during a power outage or during the initial cooling time at completion of a setting.

The disposition of contaminants during ISV processing is dependent upon many variables, the most important of which may be grouped into the categories of: (1) pre-melt soil properties, (2) contaminant quantities and properties, and (3) molten zone properties and conditions. Disposition of contaminants depends on the response of the contaminants to the process conditions; individual contaminants may respond to the system of processing variables in several ways, including: (1) change of state, (2) physical movement, and (3) physical/chemical reaction.

There are five (5) basic types of mechanisms that relate to movement responses of contaminants during processing; including: (1) capillary action, (2) concentration-based diffusion, (3) thermally induced vapor transport, (4) melt convective flow, and (5) liquid/vapor adsorption on soil. The combination of processing conditions and responses of contaminants may result in five (5) basic dispositions of the contaminants, including: (1) destruction of compounds, (2) physical removal from the treatment volume to the off-gas treatment system, (3) chemical incorporation into the vitrified product, (4) physical encapsulation within the vitrified product, and (5) continued existence as a minor residual within the treatment zone.

Certain response types and ultimate dispositions are dominant for various classes of contaminants. As organic vapors distribute themselves throughout the void space adjacent to the melt, they increase in temperature to their pyrolysis temperature, where they break down into successively smaller chains of molecules and eventually reach the state of elemental or diatomic gases. Upon pyrolysis, the concentration of the original compound vapor is thereby diminished, resulting in a continued concentration gradient of the original vapor toward the pyrolysis isotherm (i.e., toward the melt). For hazardous organic contaminants, pyrolytic destruction is the dominant disposition; destruction efficiencies in the soil column (prior to off-gas treatment) on the order of 99.9 to 99.995 wt% have resulted from ISV tests involving hazardous organics. The level of destruction can be increased by recycling the organics recovered in the off-gas system to a subsequent ISV setting. Chemical incorporation is the dominant disposition of hazardous inorganic elements (e.g., heavy metals). The ISV silicate glass is very durable relative to environmental exposure and will hold a wide variety of materials in nonleachable form. In addition to chemical incorporation, some amount of volatile or semi-volatile inorganics (e.g., Pb, Hg) may be removed from the treatment volume and recovered in the off-gas treatment system; they may then be disposed by an alternative means or be recycled to a subsequent ISV melt for further disposition by chemical incorporation.

The contaminated soil subject to ISV treatment may be configured in numerous ways for processing. The material may be processed in situ where is presently is, or it may be staged above or below grade for treatment. Special cases also include containerized treatment, stacked settings for deep contamination, continuous feeding arrangements for high volume reduction material, and special conditions for vitrifying underground tanks.

The ISV process can also accommodate significant quantities of inclusions in the treatment volume. Inclusions are defined as highly concentrated contaminant layers, void volumes, containers, metal scrap, general refuse, demolition debris, rock, or other non-homogenous materials or conditions within the waste volume.

REFERENCES

Eschenbach, R.C., et al. "Process Description and Initial Test Results with the Plasma Centrifugal Reactor". Presented at the Forum on Innovative Hazardous Waste Treatment Technologies: Domestic and International, June 19-22, 1989. Retech, Inc., Ukiah, California.

Penberthy Electromelt International, Inc. 1984. "Penberthy PYRO-CONVERTER for Hazardous Wastes Liquid and Sludge, Organic and Inorganic". PC-1. Penberthy Electromelt Int'l., Inc. Seattle, Washington.

Schiegel, Ronald. "Residues from High Temperature Rotary Kilns and Their Leachability". Presented at the Forum on Innovative Hazardous Waste Treatment Technologies: Domestic and International, June 19-22, 1989. W+E Environmental Systems, Germany.

Koegler, S.S., et al. 1988. _Vitrification Technologies for Weldon Spring Raffinate Sludges and Contaminated Soils Phase I Report: Development of Alternatives_. PNL-6704; UC-510. Pacific Northwest Laboratory, Richland, Washington.

Buelt, J.L., et al. 1987. _In Situ Vitrification of Transuranic Waste: An Updated Systems Evaluation and Applications Assessment_. PNL-4800, Supplement 1, Pacific Northwest Laboratory, Richland, Washington.

Geosafe Corporation, 1989. _Application and Evaluation Considerations for In Situ Vitrification Technology: a Treatment Process for Destruction and/or Permanent Immobilization of Hazardous Materials_. GSC 1901, Geosafe Corporation, Kirkland, Washington.

Geosafe Corporation, 1989. "In Situ Vitrification for Permanent Treatment of Hazardous Wastes". Presented at Advances In Separations: A focus on Electrotechnologies for Products and Wastes, April 11-12, 1989. GSC 1903, Geosafe Corporation, Kirkland, Washington.

Reimus, M.A.H. 1988. _Feasibility Testing of In Situ Vitrification on New Bedford Harbor Sediments_. Battelle, Pacific Northwest Laboratories, Richland, Washington.

Timmerman, C.L. 1986. _In Situ Vitrification of PCB-Contaminated Soils_. EPRI CS-4839. Electric Power Research Insitute, Palo Alto, California.

Mitchell, S.J. 1987. _In Situ Vitrification of Dioxin-Contaminated Soils_. Battelle, Pacific Northwest Laboratories.

VITRIFICATION TECHNOLOGIES

DEFINITIONS

GLASS - A solid (super-cooled liquid) containing an amorphous mixture of oxides (primarily silica) with little or no crystallization present

VITRIFY - To convert solids into glass through heat fusion

TREATMENT MECHANISMS

1) Destruction of Organics

   ● Pyrolytic decomposition

   ● Oxidation

2) Removal of Inorganics (partial to complete for low-solubility, high-volatility materials)

3) Immobilization of Inorganics in Residual Glass Product

   ● Chemical incorporation

   ● Encapsulation

TYPICAL RESIDUAL PROPERTIES

Composition:	Analogous to natural obsidian
Strength:	10X concrete
Volume Reduction:	20-40 %
Toxicity Reduction:	Organics are removed/destroyed Inorganic-bearing residual found to have acceptable biotoxicity (EPA)
Mobility Reduction:	Surpasses EP-Tox, TCLP
Wet/Dry Cycling:	Unaffected
Freeze/Thaw Cycling:	Unaffected
Life Expectancy:	DOE glass waste form research indicates geologic time period

VITRIFICATION TECHNOLOGIES

ALTERNATIVE VITRIFICATION TECHNOLOGIES

- **Electric Furnace/Melter**

- **Plasma Reactor**

- **Slagging Kiln/Incinerator**

- **In Situ Vitrification (ISV)**

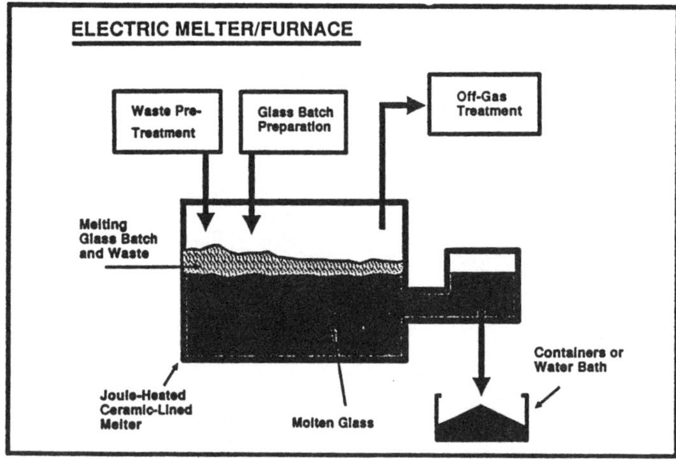

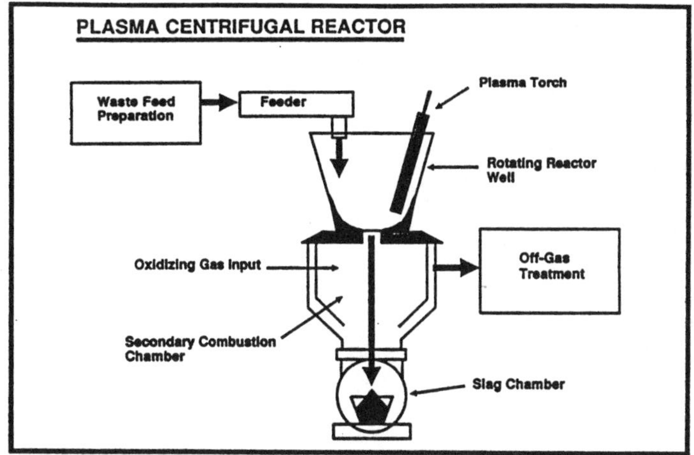
PLASMA CENTRIFUGAL REACTOR
Plasma Torch
Waste Feed Preparation
Feeder
Rotating Reactor Well
Oxidizing Gas Input
Off-Gas Treatment
Secondary Combustion Chamber
Slag Chamber

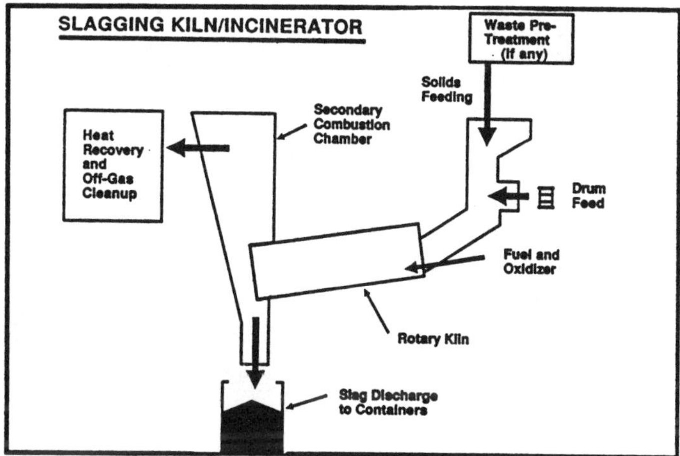
SLAGGING KILN/INCINERATOR
Waste Pre-Treatment (If any)
Solids Feeding
Secondary Combustion Chamber
Heat Recovery and Off-Gas Cleanup
Drum Feed
Fuel and Oxidizer
Rotary Kiln
Slag Discharge to Containers

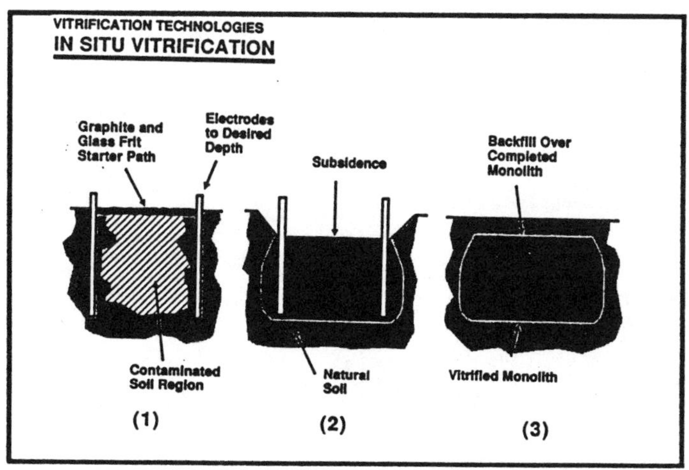
VITRIFICATION TECHNOLOGIES
IN SITU VITRIFICATION
Graphite and Glass Frit Starter Path
Electrodes to Desired Depth
Subsidence
Backfill Over Completed Monolith
Contaminated Soil Region
Natural Soil
Vitrified Monolith
(1)
(2)
(3)

VITRIFICATION TECHNOLOGIES

BASIC COMPARISON OF TYPES   (Page 1 of 2)

	APPLICABLE MATERIALS	WHERE PERFORMED	TYPICAL OPERATING TEMPERATURE	ORGANIC DESTRUCTION MECHANISM
ELECTRIC FURNACE/ MELTER	Dry/Wet Solids	Mobile, On-Site, Ex-Situ	11-1,200 C	Pyrolysis or Oxidation
PLASMA REACTOR	Dry/Wet Solids	Mobile, On-Site, Ex-Situ	1,600 C (glass) 10,000 C (plasma)	Oxidation
SLAGGING KILN/ INCINERATOR	Dry/Wet Solids, Containers (some only)	Transportable, On-Site, Ex-Situ	11-1,500 C	Oxidation
IN SITU VITRIFICATION	Dry/Wet Solids, Containers, Rubble	Mobile, On-Site, In Situ	16-2,000 C	Pyrolysis

VITRIFICATION TECHNOLOGIES

BASIC COMPARISON OF TYPES   (Page 2 of 2)

	TYPICAL THROUGHPUT RATE	PRE-TREAT REQUIRE-MENTS	NATURE OF RESIDUAL	POST-TREAT REQUIRE-MENTS
ELECTRIC FURNACE/ MELTER	0.5-3 tph	Excavate, Sort, Particle Size Screening	Solid-filled Containers or Granular	Disposal of Residual
PLASMA REACTOR	0.1-0.3 tph	Excavate, Sort, Particle Size Screening	Slag in Container	Disposal of Residual
SLAGGING KILN/ INCINERATOR	6 tph 30 drums/h	Excavate (some sort and particle size)	Slag in Container	Disposal of Residual
IN SITU VITRIFICATION	4-6 tph	None	Contiguous Monolith	Backfill Subsidence

IN SITU VITRIFICATION

TECHNOLOGY DEFINITION

In situ electric melting of staged or as-deposited contaminated solids (e.g., soil, sediment, sludge, tailings, various inclusions) for purposes of:

1) organic destruction/removal

2) inorganic immobilization/removal

3) barrier wall formation

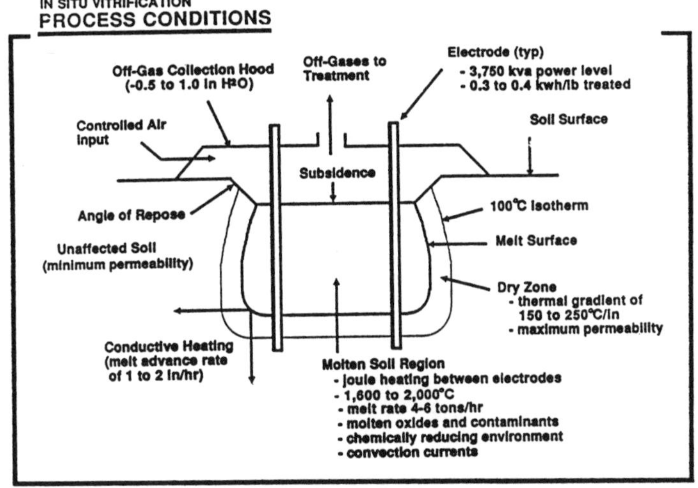
IN SITU VITRIFICATION
PROCESS CONDITIONS
Off-Gas Collection Hood
(-0.5 to 1.0 in H2O)
Off-Gases to
Treatment
Electrode (typ)
- 3,750 kva power level
- 0.3 to 0.4 kwh/lb treated
Controlled Air
Input
Soil Surface
Subsidence
Angle of Repose
100°C Isotherm
Melt Surface
Unaffected Soil
(minimum permeability)
Dry Zone
- thermal gradient of
  150 to 250°C/in
- maximum permeability
Conductive Heating
(melt advance rate
of 1 to 2 in/hr)
Molten Soil Region
- joule heating between electrodes
- 1,600 to 2,000°C
- melt rate 4-6 tons/hr
- molten oxides and contaminants
- chemically reducing environment
- convection currents

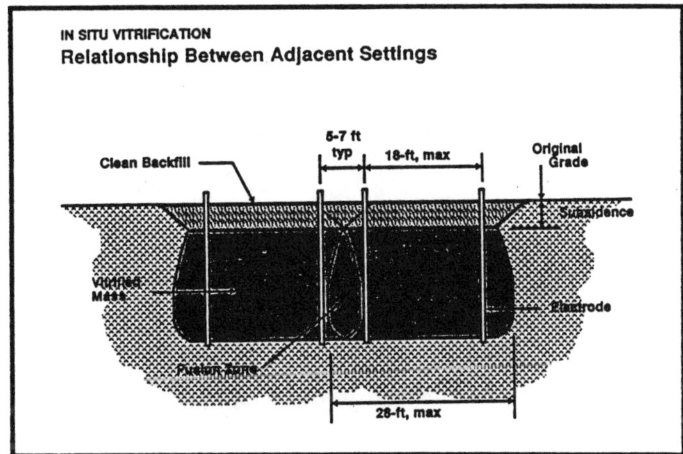
IN SITU VITRIFICATION
Relationship Between Adjacent Settings
5-7 ft
typ
18-ft, max
Clean Backfill
Original
Grade
Subsidence
Vitrified
Mass
Electrode
Fusion Zone
28-ft, max

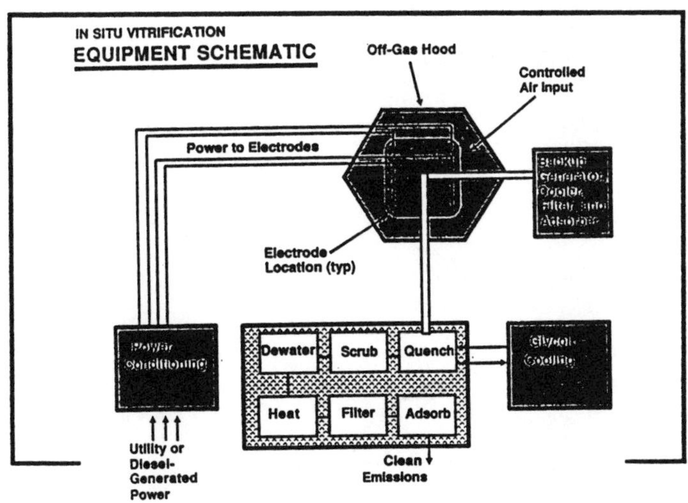
IN SITU VITRIFICATION
EQUIPMENT SCHEMATIC
Off-Gas Hood
Controlled
Air Input
Power to Electrodes
Backup
Generator,
Cooler,
Filter and
Adsorber
Electrode
Location (typ)
Power
Conditioning
Dewater
Scrub
Quench
Glycol
Cooling
Heat
Filter
Adsorb
Utility or
Diesel-
Generated
Power
Clean
Emissions

IN SITU VITRIFICATION

CONTAMINANT MOVEMENT MECHANISMS DURING ISV PROCESSING

- **Capillary Action** - movement of liquids toward dry zone

- **Molecular Diffusion** - concentration-based movement of vapors toward melt

- **Carrier-Gas Transport** - stripping of vapors by water vapor

- **Density Differential** - movement of vapors into melt and toward surface

- **Pressure Differential** - vapor flow toward negative surface pressure

- **Convection** - thermally induced circulation within melt

IN SITU VITRIFICATION

MOVEMENT OF CONTAMINANTS

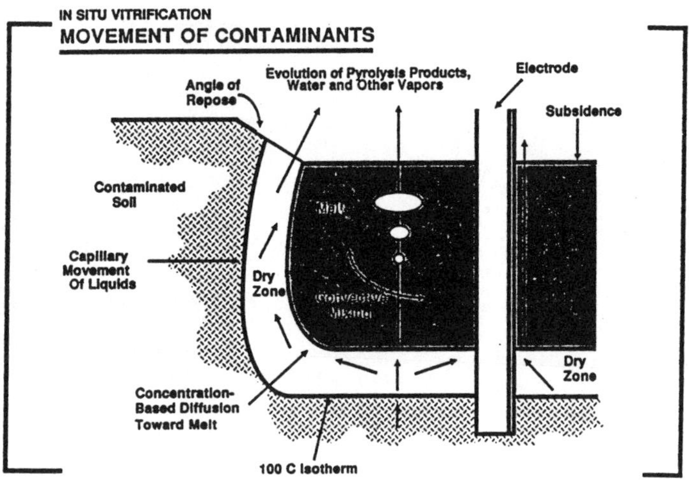

IN SITU VITRIFICATION

POSSIBLE CONTAMINANT DISPOSITIONS

- **Destruction (chemical/thermal)**

- **Removal to off-gas treatment**

- **Chemical incorporation in glass residual**

- **Physical encapsulation in glass residual**

- **Continued existence as adjacent residual**

IN SITU VITRIFICATION
PRIMARY FACTORS AFFECTING
CONTAMINANT DISPOSITION

- Contaminant physical/chemical properties

- Melt chemistry

- Melt temperature

- Dwell time (viscosity, depth)

- Adjacent soil properties

- Soil moisture content

- Degree of overmelting

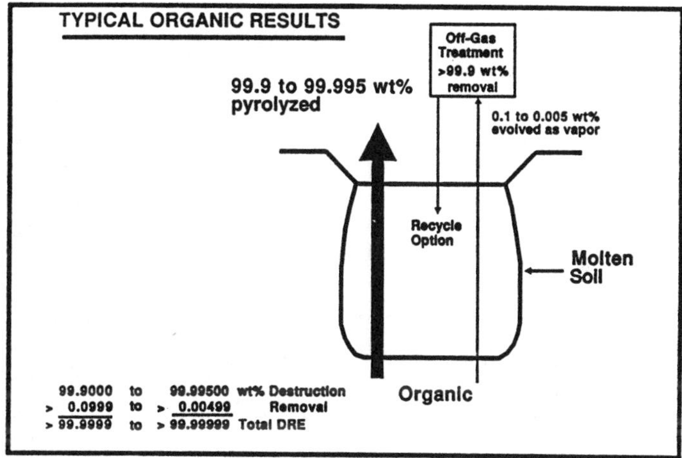

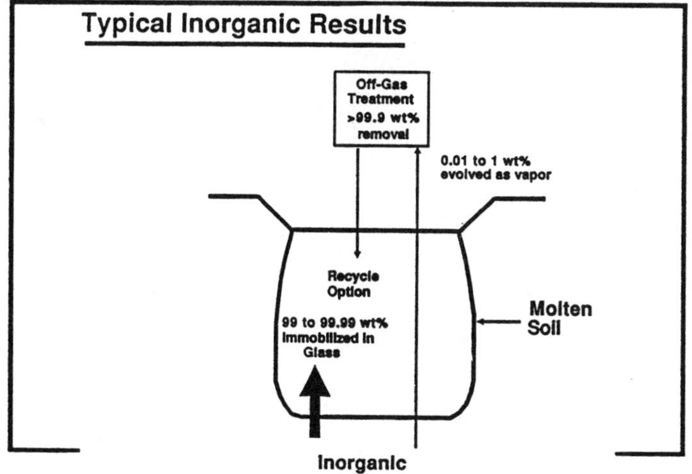

IN SITU VITRIFICATION
TYPICAL CONTAMINANT DISPOSITION

Weight Percent

Destruction	99.995	99.9	95	---	---	---
Removal (Recycle Option)	0.005	0.1	5	0.01	1	5*
Chemical Incorporation	---	---	---	99.99	99	95*
	LV	MV	HV	HS-LV	HV-HS LV-LS	HV-LS
	Organic			Inorganic		

* Mercury is notable exception; nearly all is removed

Note:
L = Low
M = Medium
H = High
V = Volatiles
S = Solubility

IN SITU VITRIFICATION
Residual Product Properties

Composition:	Analogous to Natural Obsidian
Structural Strength:	10X Concrete
Wet/Dry Cycling:	Unaffected
Freeze/Thaw Cycling:	Unaffected
Chemical Leaching:	Zero Organics Present Passes EP-Tox, TCLP
Biotoxicity:	Acceptable (Near Surface Life)
Life Expectancy:	Geologic Time Period

IN SITU VITRIFICATION
APPLICATION OPTIONS

- In Situ - where now existing
- Staged - where placed for treatment
- Barrier Wall - partial treatment plus containment
- In-Container - high-volume reduction materials
- Stacked Settings - very deep treatment

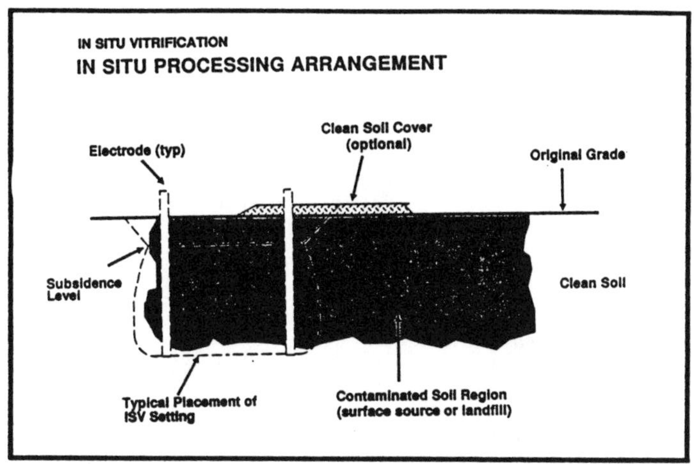
IN SITU VITRIFICATION
IN SITU PROCESSING ARRANGEMENT
Electrode (typ)
Clean Soil Cover (optional)
Original Grade
Subsidence Level
Clean Soil
Typical Placement of ISV Setting
Contaminated Soil Region (surface source or landfill)

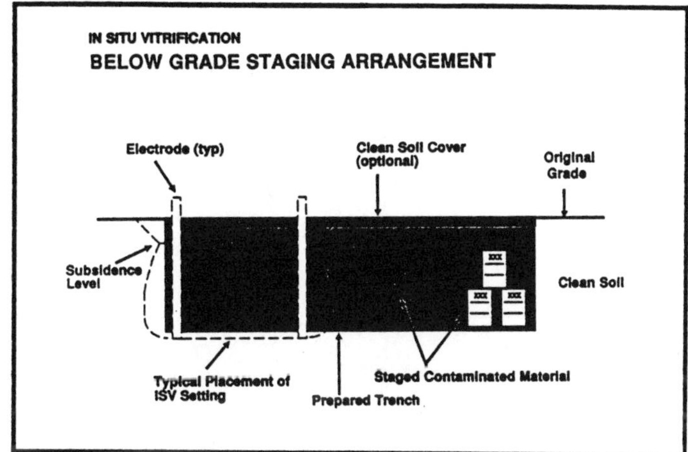
IN SITU VITRIFICATION
BELOW GRADE STAGING ARRANGEMENT
Electrode (typ)
Clean Soil Cover (optional)
Original Grade
Subsidence Level
Clean Soil
Typical Placement of ISV Setting
Prepared Trench
Staged Contaminated Material

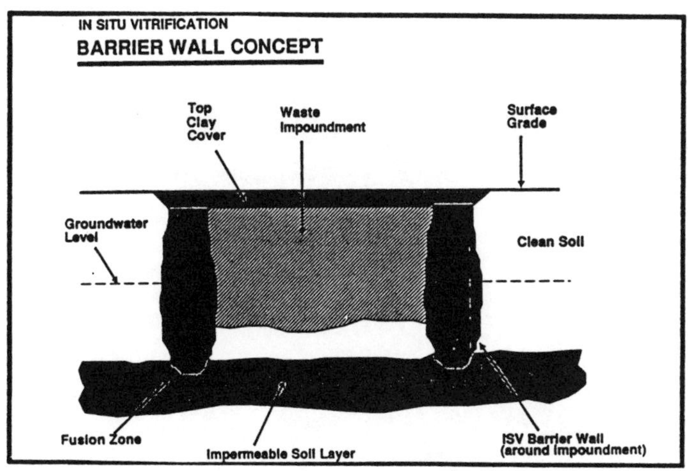
IN SITU VITRIFICATION
BARRIER WALL CONCEPT
Top Clay Cover
Waste Impoundment
Surface Grade
Groundwater Level
Clean Soil
Fusion Zone
Impermeable Soil Layer
ISV Barrier Wall (around impoundment)

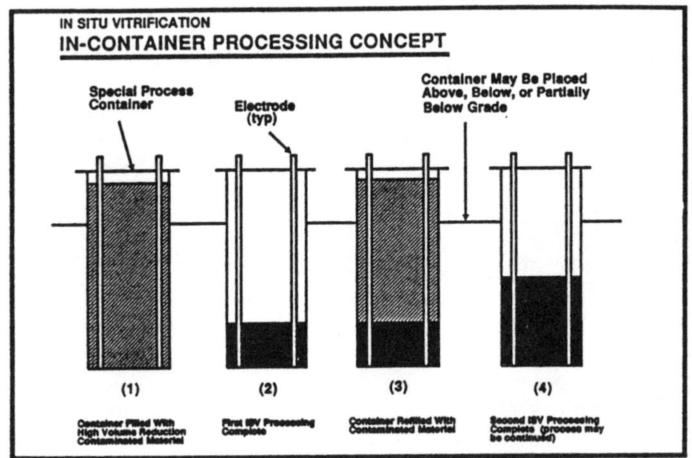

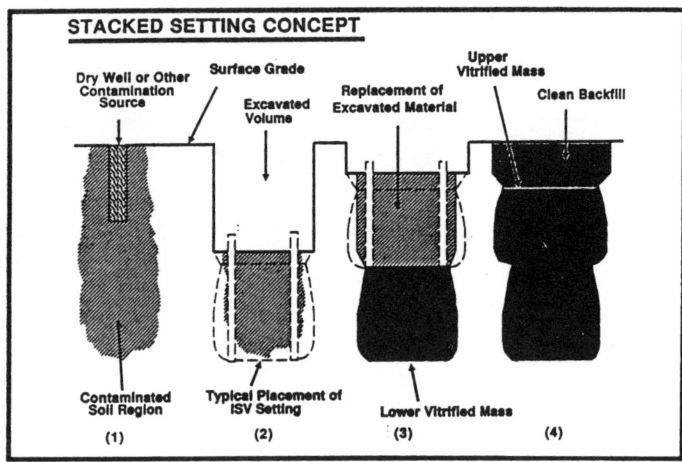

IN SITU VITRIFICATION

BASIC APPLICATION CONSIDERATIONS

- Contaminants (type, concentration, depth, required cleanup levels)

- Solids (type, properties, moisture content, stratigraphy)

- Ground water (location, recharge rate)

- Electricity (availability, price)

- Inclusions (physical/chemical definition)

- Structures (above/below surface)

- Volume (depth, area arrangement)

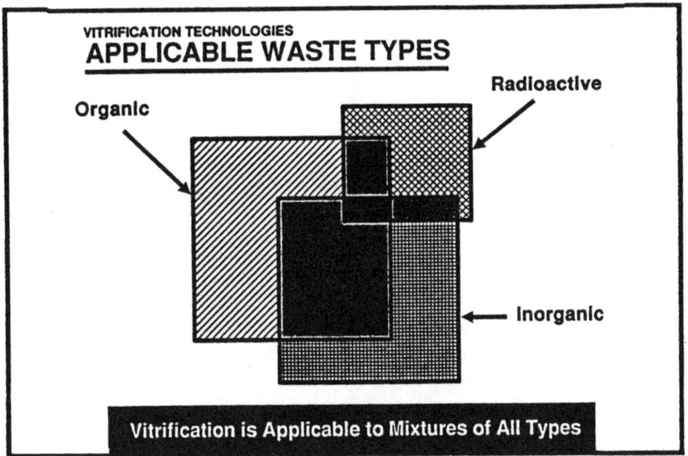

IN SITU VITRIFICATION

WATER CONSIDERATIONS

- Process removes water

- Fully saturated soils, slurries may be processed

- Energy for water removal increases cost

- In-aquifer processing dependent on recharge rate

- In-aquifer options:
    1) Lower water table
    2) Reduce recharge rate

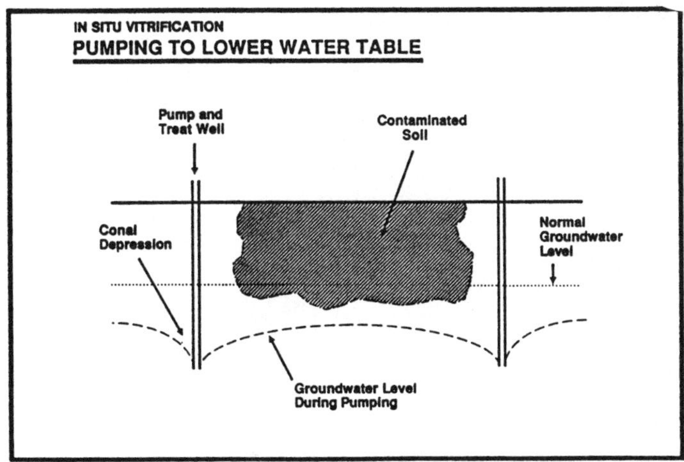

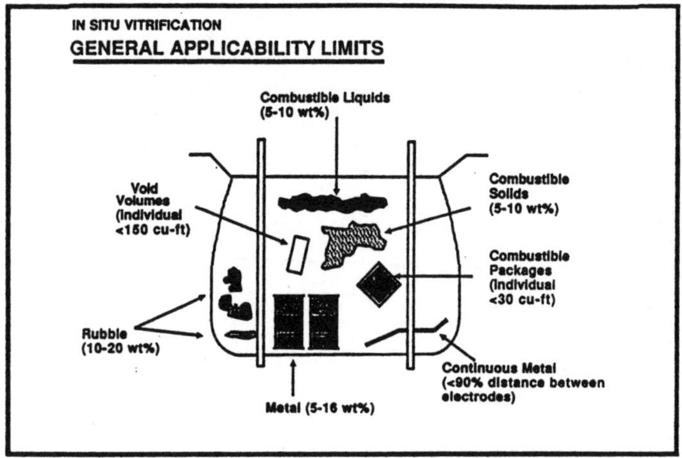

IN SITU VITRIFICATION
REMEDIAL ACTION ELEMENTS

- Site Preparation
- Equipment Mobilization
- Electrode/Starter Path Placement
- Hood Placement
- Vitrification Operations
- Hood Removal
- Clean Backfill Placement
- Equipment Demobilization
- Site Restoration

Note: waste pretreatment and long-term monitoring may be minimized

IN SITU VITRIFICATION
Cost Elements

- Treatability Testing
- Technical Support (RD, Site Prep)
- Equipment Mobilization
- Vitrification Operations
- Equipment Demobilization
- Technical Support (RA, Restoration, O&M)

IN SITU VITRIFICATION

Major Variables Affecting Cost

Variable	Typical Case
Unit Price of Electricity	$0.03 - .06/kwh
Soil/Waste Moisture Content	10 - 20 wt%
Depth of Processing	10 - 20 ft
Size (Volume of Site)	3 - 50,000 cy
Soil Type	Sand - Clay
Staging Requirements	Minimal

Typical ISV Cost: $250 - 350/ton
(excludes treatability testing, technical support, mobilization/demobilization)

Source: Geosafe Corp.

IN SITU VITRIFICATION

POWER REQUIREMENTS

- 12.5 or 13.8 kva supply
- Total of 4 Mw available
- Consumes 0.3 to 0.4 kwh/lb treated
- May be diesel-generated if not readily available from local utility

---

IN SITU VITRIFICATION

TECHNOLOGY BACKGROUND/STATUS

- Originally DOE technology (transuranic oriented)
- Developed/demonstrated through large-scale ($15m since 1980)
- Commercial capability established 1989
- SITE Program evaluation at site selection stage
- Numerous Superfund treatability tests completed
- Numerous RI/FS and RCRA CA studies evaluating
- Superfund procurements in process

4. Physical Testing Methods for Determining the Effectiveness of S/S Processes

Mr. Peter Hannak   Mr. Richard McCandless
CH2M Hill   UC/Center Hill Lab
Waterloo, Ontario   Cincinnati, Ohio

Physical and Engineering testing methods are applicable to:

* untreated S/S wastes
* site characterization (excavation and disposal)
* treatability studies
* S/S process control
* S/S product evaluation

Results of tests may be used directly for the assessment of the waste. However, properties such as porosity and volume increases are calculated from the results of appropriate tests.

The physical properties of S/S wastes provide the means to:

* establish treatment objectives and to measure the effectiveness of contaminant immobilization
* compare a variety of treatment processes and enable the selection of suitable processes
* evaluate economics of the S/S process and minimize associated cost
* assure safety of operations and product
* minimize any air or liquid emissions during the processing phase

The physical/engineering test results are to be used to supplement the results of chemical testing, therefore interpretation of S/S process evaluation should include both groups of tests.

The physical/engineering test may be grouped into two categories as to the type of information which can be obtained.

* characterization of intrinsic waste properties
* site-specific information

The full understanding of the behavior of the waste at a site, can only be described if both types of information are available. The testing methods that are applicable to the untreated wastes includes methods related to:

* liquid solid classification
* particle size analysis
* moisture content
* bulk density
* permeability and
* strength

The inter-relationship of these tests and purpose of their application to S/S process evaluation are discussed in the presentation.

Further discussion of specific methods include tests of regulatory importance (Paint Filter/Liquid Release tests), tests used as guidance (Unconfined Compressive Strength Tests) and tests required for site/waste characterization, including field permeability.

Approach and equipment are described to provide application guidance for the audience. The signification and limitations of the individual methods are presented along with the applicable literature and standard references. The scientific background of complex tests, such as permeability measurement, is presented in the format of charts and equations where necessary.

The function of physical/engineering tests for process characterization are also discussed. These functions include the use of tests for process control and treatability studies on bench and pilot scale.

Mechanisms of interferences affecting the final physical/engineering properties of the S/S wastes are detailed as follows:

* adsorption
* complexion
* precipitation
* nucleation

The inter-relationships of physical/engineering tests performed on the final product are also presented. The applicable tests include:

* bulk density
* specific gravity
* moisture content
* permeability
* weathering resistance
* unconfined compression

These tests are presented in detail including the modifications of procedures to make these tests suitable to S/S wastes.

The application of specific tests under development will expand the understanding of the process and assist in the evaluation of the quality of the S/S waste products. Specific examples include; microcharacterization and dissolution tests.

A discussion concerning the range and inter-relationship/trend of properties is provided. The support for the discussion is based on a series of unpublished experiment results from an international cooperative study

carried out with the participation of the industry. This study
characterized waste streams from a variety of sources including:

- Synthetic wastes
- Wood preservative wastes
- Dredge spoil
- Electroplating wastes

The characterization of waste stabilized by commercial processes provides a
cross-section of S/S waste properties outlined in this presentation.

REFERENCES

Hannak, P., Liem, A., "Development of a Method for Measuring the
Freeze/-Thaw Resistance of Solidified/Stabilized Wastes", Proceedings of the
International COnference on New Frontiers for Hazardous Waste Management,
Sept. 15-18, Pittshurgh, PA EPA/600/0-85/025.

Hannak, P., Liem, A., :Development of New Methods for Solid Waste
Characterization, Part 2: Measurement of Porosity of Solidified/Stabilized
Wastes", Presented to the International Seminar on Solidification and
Stabilization of Hazardous Waste '86, Hazardous Waste Research Centre,
Louisiana State University, Baton Rouge, LA.

Jones, J.N., Bricka, M.R., Myers, T.E. and Thompson, D.W. Factors Affecting
Stabilization/Solidification of Hazardous Wastes. Proceedings of the
International Conference on New Frontiers for Hazardous Waste Management.
EPA 600/9-85-025.

Kingsbury, G., Hoffman, P., Lesnik, B. The Liquid Release Test (LRT).
Proceedings of the Fifth Annual Waste Testing and Quality Assurance
Symposium, July 24-28, 1989. American Chemical Society.

P.E.I. Associates and the Earth Technology Corporation.
Stabilization/Solidification of CERCLA and RCRA Wastes. Physical Tests,
Chemical Testing Procedures, Technology Screening and Field Activities.
EPA, 1989. Draft.

Physical and Engineering Properties of Hazardous Wastes and Sludges. EPA
600/2-77-139, NTIS PB, 272 266/AS.

U.S. Army Engineer Waterways Experiment Station. Physical Properties and
Leach Testing of Solidified/Stabilized Industrial Wastes. NTIS PB 83-147983.

Shively, W. and Dethloff, S. Hazardous Waste Treatment Demonstration Guide,
Test Methods for Solidification/Stabilization of Hazardous Wastes. CH2M
Hill, Second Draft, 1987.

SCOPE

- **Untreated waste/site**
- **Treatability studies**
- **(Process)**
- **S/S wastes**

Does not include soil testing

SIGNIFICANCE OF WASTE CHARACTERIZATION

- Establish treatment objectives
- Select the best treatment option
- Minimize treatment cost
- Assure safety of operations
- Minimize by-products and emissions

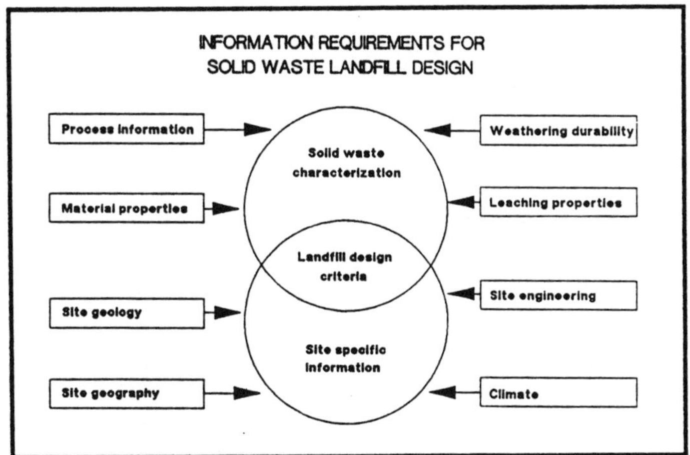

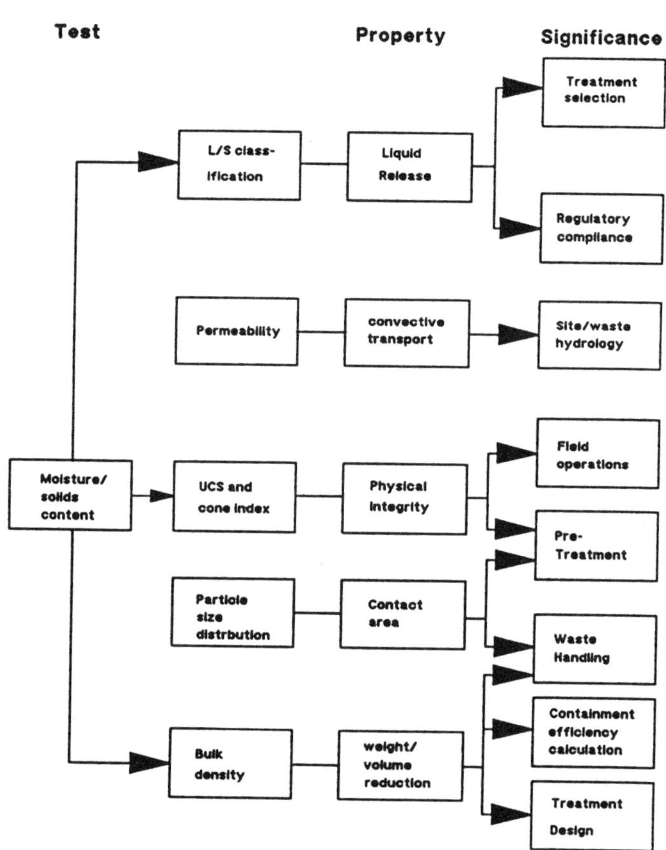

INTERRELATION OF SITE/UNTREATED WASTE TESTING

LIQUID/SOLID CLASSIFICATION

Regulation: Land ban, HSWA prohibits disposal of bulk liquid hazardous waste and hazardous waste containing free liquid

Tests:
- **Paint Filter Test (SW 846-9095)**
- **Liquid Release Test**

MEANS OF LIQUID RELEASE
- Gravitation
- Capillary forces
- Pressure
- Microstructural dehydration
- Wash out
- Degradation
  - physical
  - biological

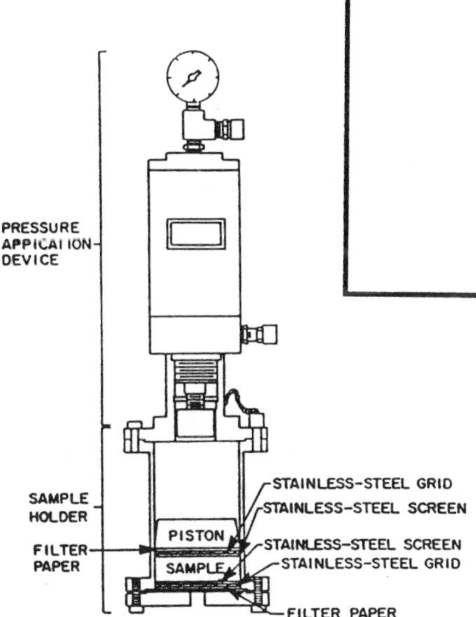

Liquid Release Test Device

PARTICLE SIZE ANALYSIS

- Objectives
    - range measurement
    - distribution measurement
- Significance
    - treatability
    - mixing

PARTICLE SIZE ANALYSIS
Continued

- Method
    - sieve analysis
      measures weight retained,
      range: mm - 0.075 mm

MOISTURE CONTENT

- Objectives
    - determine the amount of free fluid
    - determine solids content
- Significance
    - process selection/compliance
    - need of pretreatment
    - mixing ratio selection
    - field operations
- Method
    - ASTM D2216-80
    - drying at 110 C

BULK DENSITY

- Objective
    - determine weight to volume ratio
- Significance
    - weight to volume conversion (excavation, treatment, shipping) and porosity calculations
- Methods
    - drive cylinder (ASTM D2937-83)
    - sand cone (ASTM D1556-82)
    - nuclear (ASTM D2922-81)
    - screening (ASTM D3402.025)

STRENGTH TESTING

- Objective
    - determine waste/soil stability
- Significance
    - "trafficability" to construction equipment
    - "workability" - ease of processing
    - S/S baseline
- Methods
    - Unconfined Compressive Strength (UCS) (ASTM D2166-85)
    - cone index (ASTM 3441-79)

PERMEABILITY
(Hydraulic Conductivity)

- Objectives
    - measure advective transport of water through the waste
- Significance
    - affects migration of contaminants
    - directly - flow rates
    - indirectly - mechanisms of release
- Methods
    - ASTM 18.04 and D2434
    - Army Corps of Engineers EM 1110-2-1906

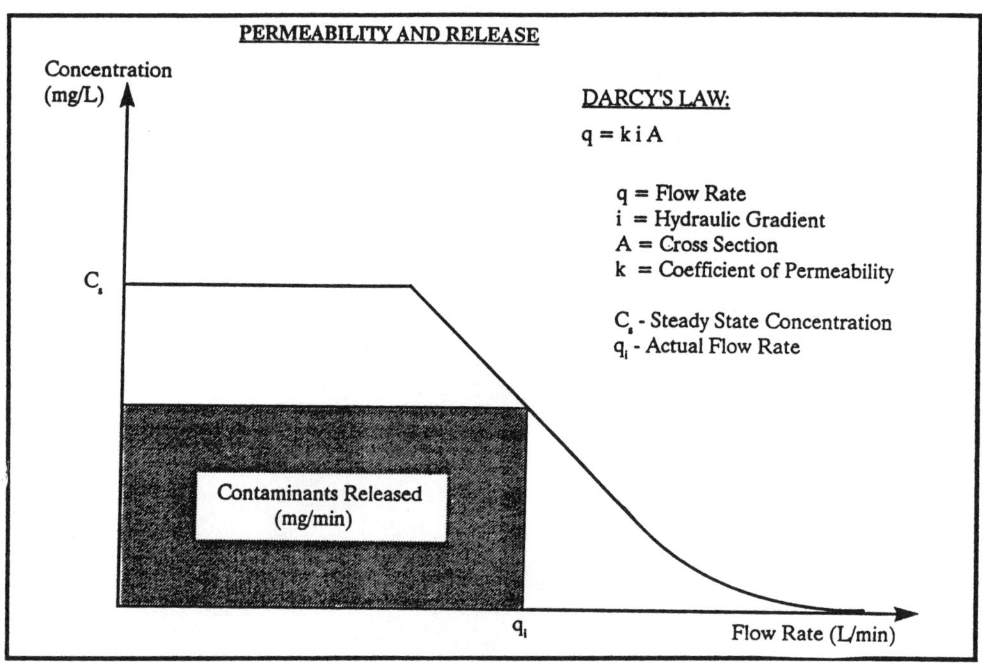

PERMEABILITY FIELD TESTS

- Falling head
- Rising head
- Constant head
- Single well response

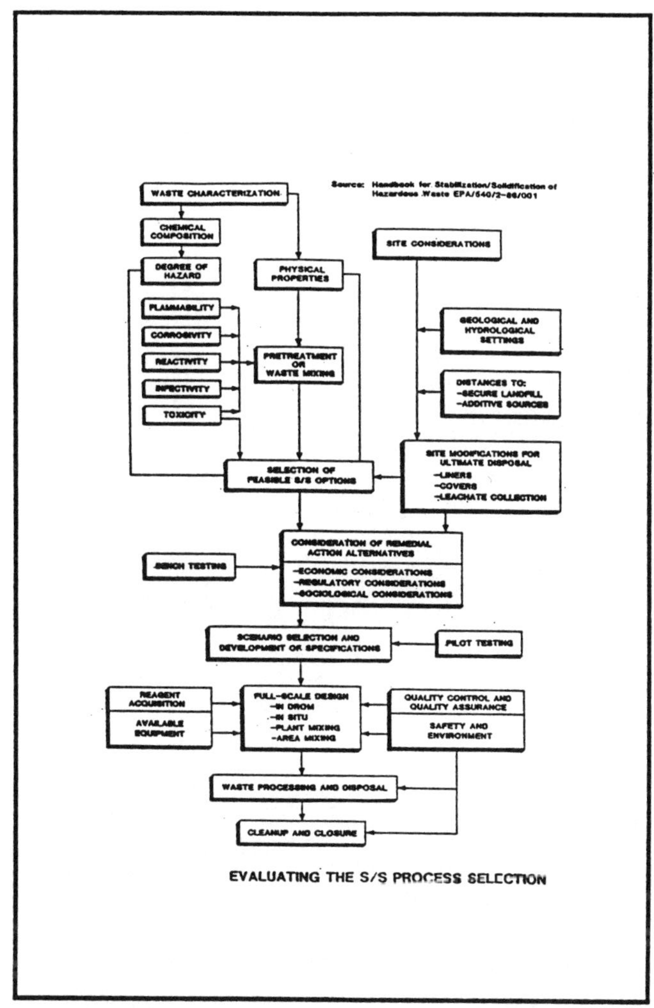

TREATABILITY STUDY STEPS

- Mixing   (ASTM C-305-82)
- Compaction
    - tamping
    - shaker table
- Molding
    - cylinders
    - cubes
- Curing
    - pozzolans
    - cement

Parameters: temperature, humidity, time

INTERFERENCES

- Interfering mechanism:
  - adsorption (dicarboxylic acid)
  - complexation (sucrose)
  - precipitation
  - nucleation (silica gel)
- Accelerators:
  - triethanolamine < 0.06 %
  - calcium formate

INTERFERENCES
Continued

- Retarders:
  - formaldehyde
  - oil and grease
  - phenol
  - sulfates
  - metals (Pb, Cu, Zn)

PROCESS TESTING

- Mixing properties
  - heat of reaction
  - alkalinity
  - spiking
- Strength development
- Additive volume/weight control

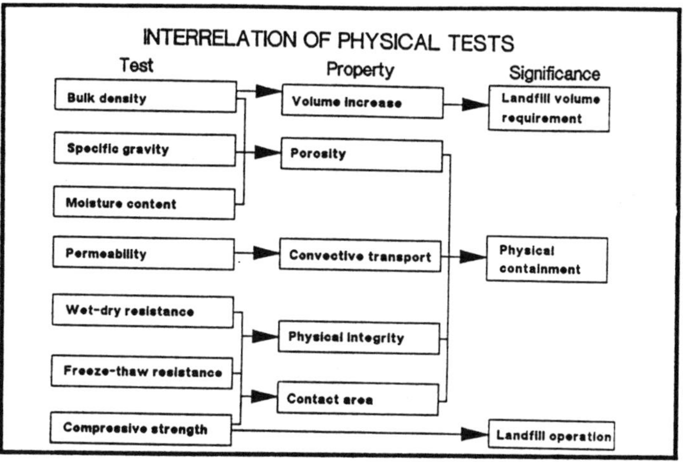

METHOD MODIFICATIONS FOR TREATED WASTES

- UCS - cylindrical vs cube samples (D4633-84 vs C109-86) method of pre-soaking
- Permeability - flexible wall permeameter

ADDITIONAL TESTS FOR TREATED WASTES

- True density/specific gravity
- Solubility
- Weathering properties
    - wet dry (ASTM D-4843)
    - freeze thaw (ASTM D-4842)
- Biodegradation
    - fungus (ASTM G21-70)
    - bacteria (ASTM G22-76)
- Micromorphology (LSU-WES-AEC)

CALCULATED PROPERTIES

$$\text{Weight change factor} = \frac{\text{weight of treated waste}}{\text{weight of untreated waste}}$$

$$\text{Volume change factor} = \frac{\text{bulk density of untreated waste}}{\text{bulk density of treated waste}} \times \text{weight change factor}$$

$$\text{Porosity} = 1 - \frac{\text{bulk density of treated waste}}{\text{true density of treated waste}} (1 - \text{moisture content})$$

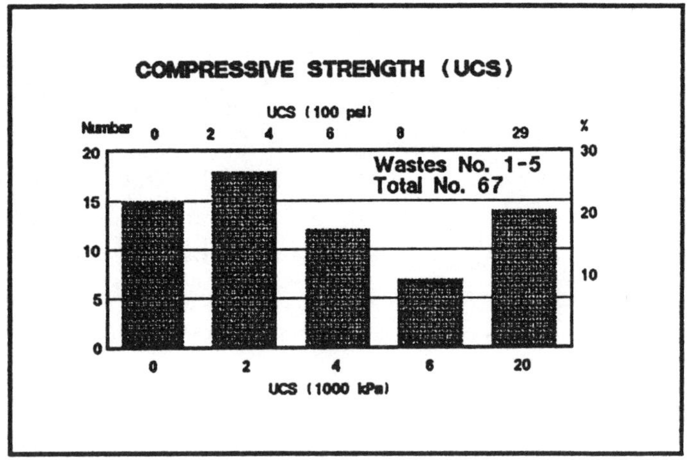

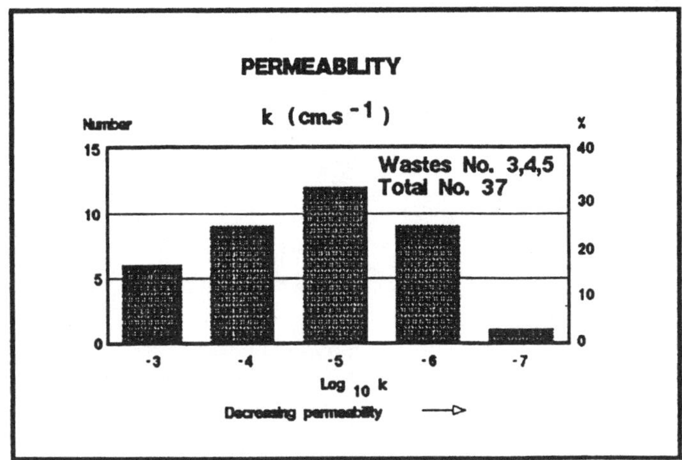

- **Properties are waste dependent**

- **High UCS ⟶▷**
  - **good weathering durability**

- **Low UCS ⟶▷**
  - **not conclusive**

- **Freeze - Thaw Test**
  - **more severe than Wet - Dry Test**

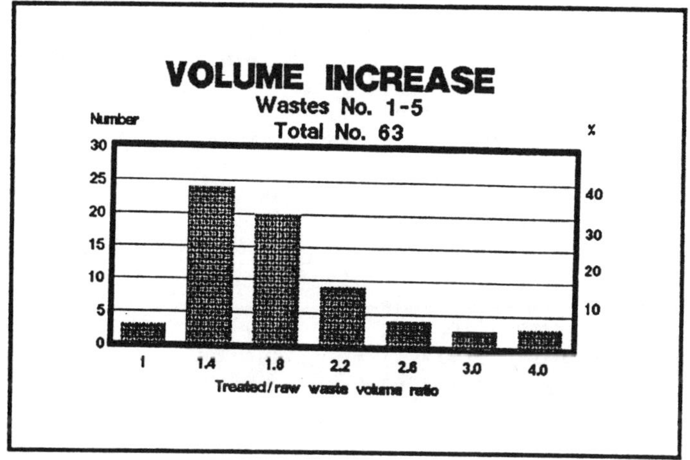

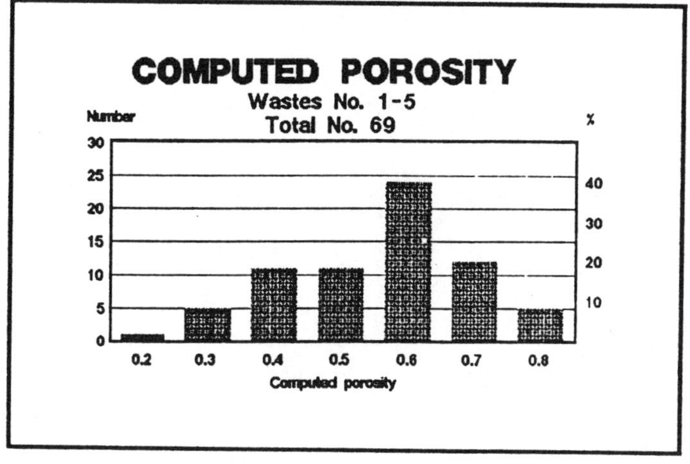

5. Chemical Testing Methods for Determining Effectiveness of S/S Processes

Mr. Carlton Wiles
USEPA/RREL
Cincinnati, Ohio

Mr. Edwin Barth
USEPA/RREL
Cincinnati, Ohio

Dr. John Nobis
PEI Associates, Inc.
Cincinnati, Ohio

Chemical testing procedures applicable to raw or untreated hazardous wastes may also have application to solidified/stabilized (S/S) wastes. This presentation is devoted largely to a discussion of leaching tests since these tests are most often used to evaluate the performance of S/S wastes as a treatment process for hazardous wastes. Emphasis is on the appropriate selection of the leaching tests and the interpretation of laboratory data. The experimental conditions affecting reproducibility of laboratory data and the limitations in extrapolating results to the field are discussed. At best, laboratory leaching data can simulate the behavior of waste forms under "ideal," static, or "worst-case" field conditions. Currently, leach tests are used to compare the effectiveness of various S/S processes, but are not verified to determine long-term leachability of a waste.

The chemistry of the waste and the leaching solution defines the types and kinetics of the chemical reactions that mobilize or demobilize contaminants in the S/S waste. Reactions that can mobilize contaminants adsorbed or precipitated within the waste form include dissolution and desorption. Under nonequilibrium conditions, these reactions compete with demobilizing reactions such as precipitation and adsorption. Nonequilibrium conditions generally develop when a S/S waste is contacted by a leaching solution and can result in a net transfer, or leaching, of contaminants into the leaching solution.

Extraction (or batch extraction) tests refer to a leaching test that generally involves agitation of ground or pulverized waste forms in a leaching solution. The leaching solution may be acidic or neutral. Also, it may vary throughout the extraction tests. Extraction tests may involve one-time or multiple extractions. In either case, leaching is assumed to reach equilibrium by the end of one extraction period; therefore, extraction tests are generally used to determine the maximum, or saturated, leachate concentrations under a given set of test conditions.

Other types of leach tests involve no agitation. The leaching of monolithic (instead of crushed) waste forms is evaluated in these tests. Leaching may occur under static or dynamic conditions, depending on the frequency of the leaching solution renewal. In static leach tests, the leaching solution is not replaced by a fresh solution; therefore, leaching takes place under static hydraulic conditions (low leaching velocities and maximum leachate concentrations for monlithic waste forms). In dynamic leach tests, the leaching solution is periodically replaced with new solution; therefore, this test simulates the leaching of a monolithic waste form under nonequilibrium conditions in which maximum saturation limits are not obtained and leaching rates are high. "Static" and "dynamic," therefore, refer to the velocity, not the chemistry of the leaching solution.

Another key difference between these two leaching tests is that extraction tests are short-term tests lasting from hours to days, whereas leach tests generally take from weeks to months. Because of the crushed nature of the waste and the larger amount of surface area available for leaching, extraction tests (although short-term) are used to simulate "worst-case" leaching conditions.

The following tests are discussed:

- Toxicity Characteristic Leaching Procedure (TCLP)
- Extraction Procedure Toxicity Test (EP Tox)
- California Waste Extraction Test (Cal WET)
- Multiple Extraction Procedure (MEP)
- Monofilled Waste Extraction Procedure (MWEP)
- Equilibrium Leach Test (ELT)
- Acid Neutralization Capacity (ANC)
- Sequential Extraction Test (SET)
- Sequential Chemical Extraction (SCE)
- Materials Characterization Center Static Leach Test (MCC-1P)
- American Nuclear Society Leach Test (ANS-16.1)
- Dynamic Leach Test (DLT)

There are various experimental factors that effect laboratory results and their interpretations. These included:

- Sample nonhomogeneity
- Curing time
- Chemical type of leachant/exractant
- Liquid to solid ratio
- Extraction time
- Particle size
- Amount of oil and grease

The impact of these and other experimental parameters must be considered in an evaluation of the results of several leaching tests on S/S wastes.

As mentioned previously, leaching tests produce results that are not directly applicable to leaching behavior in the field. Nevertheless, the results of several leaching tests or of leaching tests combined with physical tests or microscopic techniques can be used as indicators of field performance and environmental impact.

When used for comparative purposes, results from several leaching tests can help to identify field conditions that result in high concentrations of waste constituents. Therefore, these data may be used to site or design waste facilities that will minimize the leaching of hazardous constituents from the wastes. The data also may be used to predict the leaching of S/S wastes at different stages in time. For example, leaching conditions of a well-managed operational facility where the monolithic S/S waste receives maximum precipitation infiltration may be simulated by the use of the DLT, as this test involves constant renewal of the leaching solution and a monolithic waste form. For a closed facility that has a cover which is maintained and minimizes precipitation infiltration, leaching conditions may be similar to those of the MCC-1P test (i.e., static hydraulic conditions). In the long run, the liners, cover, and waste form may degrade, and leaching conditions may be similar to those found in multiple extraction tests (such as MEP and MWEP).

REFERENCES

1. U.S. Environmental Protection Agency. 1986. Test Methods for Evaluating Solid Waste. Volumes 1A-1C: Laboratory Manual Physical/Chemical Methods: and Volume II: Field Manuals, Physical/Chemical Methods, SwW846, Third Edition, Office of Solid Waste. Document Control No. 955-001-00000-1.

2. ASTM D1498-76, Standard Practice for Oxidation-Reduction Potential of Water.

3. American Society for Testing and materials. 1981. Standard Methods for Chemical Analysis of Hydraulic Cement, ASTM Committee C-1 on Cement. Philadelphia, Pennsylvania. July 1981.

4. U.S. Environmental Protection Agency. 1979. Methods for the Chemical Analysis of Water and Wastes. Office of Research and Development. EPA-600 4-79-020, March 1979.

5. Standard Methods for the Examination of Water and Wastewater, 16th Edition, APH, AWWA, WPCF, 2985 Washington, D.C.

6. ASTM C186-86. Standard Test Method for Heat of Hydration of Hydraulic Cement.

7. Bishop, P.L. 1986. Prediction of Heavy Metal Leaching Rates from Stabilized/Solidified Hazardous Wastes. In Toxic and Hazardous Wastes Proceedings of the 18th Mid-Atlantic Industrial Waste Conference.

8. American Nuclear Society (ANS). 1986. ANSI/ANS-16.1-1986 American National Standard Measurement of the Leachability of Solidified Low-Level Radioactive Wastes by a Short-Term Test Procedure. Prepared by the American nuclear Society Standards Committee Working Group ANS-16.1. Published by the American Nuclear Society, La Grange Park, Illinois. Approved April 14, 1986, by the American National Standard Institute, Inc.

9. Bishop, P.L. 1988. Leaching of Inorganic Hazardous Constituents From Stabilized/Solidified Hazardous Wastes. Hazardous Waste and Hazardous Materials, 5(2):129-144.

10. California Code, Title 22, Article 11. Criteria for Identification of Hazardous and Extremely Hazardous Wastes, pp. 1800.75-1800.82.

11. Cote, P., and D.P. Hamilton. 1984. Leachability Comparison of Four Hazardous Waste Solidification Processes. In: Proceedings of the 38th Industrial Waste Conference, May 1983, Purdue University, West Lafayett, IN.

12. Cote, P.L., and T.R. Bridle. 1987a. Long-Term Leaching Scenarios for Cement-Based Waste Forms. Waste Management and Research, Vol. 5, pp. 55-66.

13. Cote, P.L., T.W. Constable, and A. Moreira. 1987b. An Evaluation of Cement-Based Waste Forms Using the Results of Approximately Two Years of Dynamic Leaching. Nuclear and Chemical Waste Management, Vol. 1, pp. 129-139.

14. Environment Canada and Alberta Environmental Center. 1986. Test Methods for Solidified Waste Characterization.

15. Jones, L.W. 1986. Interference Mechanisms in Waste Stabilization/ Solidification Processes: Literature Review. Unpublished Report. U.S. Environmental Protection Agency, Hazardous Waste Engineering Research Laboratory, Cincinnati, Ohio.

16. U.S. Environmental Protection Agency, Generic Treatability Protocol for Solidification/Stabilization Treatment for Contaminated Soils (1989)

17. U.S. Environmental Protection Agency. 1986b. Test Methods for Evaluating Solid Waste, Method 9095, SW-846, Third Edition. November 1986.

18. U.S. Environmental Protection Agency. 1986. A Procedure for Estimating Monofilled Solid Waste Leachate Composition. Technical Resource document SW-924, 2nd Edition. Hazardous Waste Engineering Research Laboratory, Office of Research and Development, Cincinnati, Ohio, and Office of Solid Waste and Emergency Response, Washington, D.C.

19. U.S. Environmental Protection Agency. 1986g. SW-846 Test Methods for Evaluating Solid Waste, Vol. 1C: Laboratory Manual Physical/Chemical Methods, Third Edition. Office of Solid Waste and Emergency Response, Washington, D.C.

20. U.S. EPA. 1988d. Evaluation of Test Protocols for Stabilization/ Solidification Technology Demonstrations. Revised Draft Report. U.S. Environmental Protection Agency, Office of Research and Development, Cincinnati, Ohio. Prepared by PRC Environmental Management, Inc. Contract No. 68-03-3484, April 25, 1988.

21. Weitzman, L., L.E. Hamel, and E. Barth. 1988. Evaluation of Solidification/Stabilization as a Best Demonstrated Available Technology. Paper presented at 14th Annual Hazardous Waste Engineering Laboratory Conference, Cincinnati, Ohio, May 1988.

OVERVIEW OF CHEMICAL TESTING SECTION

- Theory of contaminant migration
- Review of chemical testing procedures
- Review leach data

LEACH TESTING

VS

EXTRACTION TESTING

LEACH TEST PURPOSE

- Predict field performance
- Accelerated leaching
  - steady state vs rate
- Comparison

LEACH TEST VARIABLES

- Monolith vs particles
- Surface area
- Liquid/solid ratio
- Agitation vs static
- Flow rate
- Leachate composition
- Temperature

LEACH TEST SCALE

- Beaker
- Column
- Lysimeter
- Groundwater collection system

LEACH TEST DEFINITIONS

- Diffusion
- Advection
- Flux
- Porosity
- Tortuosity

COMPARE LEACH RESULTS TO

- Health based levels
- Technology based levels
- Dilution levels
- Regulatory levels
- % reduction
- Different processes

LEACHING DATA

Leach Rate $\qquad L = \dfrac{mass}{(area)(time)}$

MODELLING S/S PERFORMANCE

Use data from multiple extract leach test
to estimate plume generation over time

STEP 1   Obtain leach data (ANSI 16.1)

STEP 2   Fick's law model

STEP 3   Plot cumulative released vs. time
              Determine release rate

STEP 4   Incorporate release rate into
              saturated zone model (Wilson & Miller)
                    -velocity
                    -aquifer thickness
                    -waste volume
                    -porosity

EXTRACTION TESTS

- Refers to leaching tests involving agitation of ground .waste with extractant leaching solution
- May involve one-time or multiple extractions
- Used to determine the maximum or saturated leachate concentrations under a given set of test conditions

EXTRACTION TESTS

Continued

- Applicable to untreated hazardous wastes
- May also be applicable to organic solidified/stabilized wastes

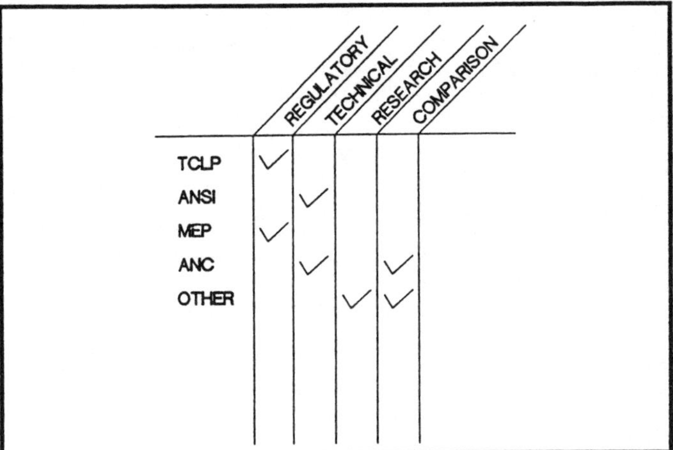

LEACH TEST METHODS AND APPLICATIONS

- TCLP   (Toxic Characteristic Leaching Procedure)
- EP TOX   (Extraction Procedure Toxicity Test)
- Cal WET   (California Waste Extraction Test)
- MEP   (Multiple Extraction Procedure)
- MWEP   (Monofilled Waste Extraction Procedure)

LEACH TEST METHODS AND APPLICATIONS
Continued

- ELT   (Equilibrium Leach Test)
- ANC   (Acid Neutralization Capacity)
- SET   (Sequential Extraction Test)
- SCE   (Sequential Chemical Extraction)

LEACH/EXTRACTION TESTING

TCLP   Grind to 9.5 mm sieve(.375 in)
              Acetic acid(pH=3, 5)

TWA   Hexane or MeCl digestion

ANSI 16.1   Distilled water renewal
              No buffer

TOXIC CHARACTERISTIC LEACHING PROCEDURE (TCLP)

- Purpose – to determine amounts of constituents available for leaching in an acid medium (codisposal with municipal waste)

TCLP CAGE MODIFICATION

- Short-term tests to evaluate stability or instability of S/S wastes

- Requires tumbling a solid sample inside of stainless steel cage that is inside of TCLP extraction jar containing leaching buffer

- Well solidified wastes remain intact and poorly solidified wastes are significantly degraded

SUMMARY PROCEDURAL DIFFERENCES BETWEEN TCLP AND EP TOX

Experimental Parameter	TCLP	EP TOX
Filter size, 1×10^{-6} m	0.6-0.8	0.45
Filter pressure, psi	50	75
Leaching solution	Acetate buffered, pH approx. 3 or 5	Acetic acid, pH approx. 5
Period of extraction, H	18	24
Liquid/solid ratio	20:1	16:1

MULTIPLE EXTRACTION PROCEDURE (MEP)

- Used for delisting
- Multiple extractions - usually nine, but more possible if last three do not decrease leachate concentrations
- Results used to determine maximum leachate concentrations
- Results used with EP Tox to compare leachability of hazardous constituents under mild and acidic conditions

CALIFORNIA WASTE EXTRACTION TEST (Cal WET)

- Differs from the TCLP and EP TOX in the following parameters:
    - different leaching solution (sodium citrate buffered solution at pH of 5; or, for hexavalent chromium, distilled water)
    - smaller liquid to solid ratio (10:1)
    - smaller particle size (< 2.0 mm)
    - longer extraction period (48 hrs.)

ANSI 16.1

- Infinite leachant
  (5 day, 90 day)
- QC
- Indicator of performance
  with model
- Utilized by NRC

ANS-16.1 PROCEDURE

- Monolithic cylinder (length:diameter 0.2-5.0)
- Leached with distilled water at V/S ratio of
  10 cm at ambient temperature
- Rinse sample to zero contamination at
  surface
- Immerse in water which is replaced after
  2, 7, 24, 48, 72 hrs., and then 4, 5, 14,
  28, 43, 90 days

PSA MODIFIED ANSI/ANS-16.1 LEACH METHOD

- Developed to deal with specific sample
  preparation & handling required by soft &
  porous waste forms from hydraulic binders, e.g.
  cements, pozzolans, gypsums, phosphates, etc.
- Changes made in (1) specimen preparation &
  handling, (2) site-specific leachant, (3) vessel
  material & cleaning procedure, (4) leachate
  handling & stabilization, (5) variable length
  leach intervals, & (6) statistical design

DYNAMIC LEACH TEST
(DLT)

- Modified version of ANS-16.1 which renews leaching solutions more frequently and changes V/S ratio
- Results on cement containing heavy metals show that Leachate Index (LX) values vary within 1 unit
- Organic compounds have LX values between 5-10 (higher leach rates) and metals higher (slower rates)

MATERIALS CHARACTERIZATION CENTER STATIC LEACHING TEST (MCC-1P)

- Static leaching test developed for high-level radioactive waste
- Involves leaching monolith with water at ratio of leaching solution to surface area solids (V/S) of 10 - 200 cm
- Period and temperature of extraction may vary
- Used for organic and polymer S/S processes

MONOFILL WASTE EXTRACTION PROCEDURE (MWEP)

- Formerly Solid Waste Leach Test (SWLT)
- Involves multiple extractions of a monolith or crushed waste with water
- Used to derive reasonable leachate compositions in monofilled disposal facilities

ACID NEUTRALIZATION CAPACITY
(ANC)
Continued

- Test used to determine buffering capacity of solidified/stabilized waste form

- The higher the buffering capacity of the waste, the greater the possibility of maintaining alkaline conditions and minimizing the amount of metals leached

NOTE:

PC - Portland Cement
KD - Kiln Dust
LF - Lime/Flyash

PHASE I SARA BDAT PROGRAM
Synthetic Soil III

Metal	Raw TWA	Raw TCLP	Treated TCLP	% Removal	Binder
As	690	6.39	ND	100	PC
Cd	2380	33.1	ND	100	PC,KD
Cr	1260	ND	.07	increase	PC,KD,LF
Cu	9550	80.7	.09	100	PC
Pb	15100	19.9	ND	100	PC
Ni	1540	17.5	ND	100	PC
Zn	34450	359	0.69	100	PC

PHASE I SARA BDAT PROGRAM
Synthetic Soil IV

Metal	Raw TWA	Raw TCLP	Treated TCLP	% Removal	Binder
As	540	9.58	ND	100	PC
Cd	3790	35.3	ND	100	PC,KD,LF
Cr	1400	.06	.07	increase	PC,KD,LF
Cu	11250	10.0	0.17	100	PC
Pb	15680	70.4	0.37	99	PC
Ni	1550	26.8	ND	100	PC,KD,LF
Zn	28660	396	0.74	100	PC

OKLAHOMA SITE
TCLP Results

Volatiles	Untreated	Treated
Benzene	ND-TR	ND
Ethyl Benzene	ND-TR	ND
Toluene	TR-0.027	ND

FLORIDA SITE
Treated Waste ppm

	MCC	ANS 16.1	EP Tox	OILY EP
Lead	0.05	0.03	0.4	61
PCBs	0.001	0.001	N/A	N/A

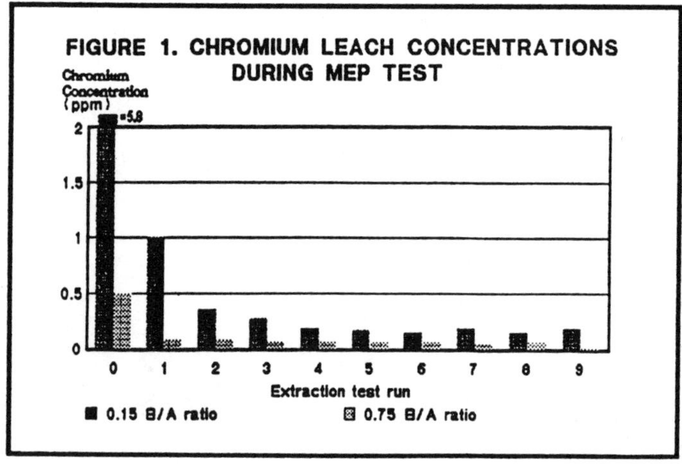

ORGANOPHILIC CLAY RESULTS

	Raw TWA	Treated TWA	Treated TCLP
Bis (2-chloro isopropyl ether)	8528	ND	ND
Naphthalene	18060	1445	ND
Phenanthrene	20184	ND	ND
Benzo (A) anthracene	30460	ND	ND

CONCLUSIONS

- Leaching tests not directly applicable to leaching in the field

- Results of several leaching tests and some physical tests can be used as <u>indicators</u> of field performance

- Data may be used to design waste facilities that will minimize the leaching of hazardous constituents

- Data may also assist in predicting the leaching of S/S wastes at different stages in time

6. Technology Screening Procedures for Determining if S/S Should Be Implemented

Dr. Leo Weitzman
LVW
Durham, North Carolina

Mr. Jesse Conner
Chemical Waste Management
Riverdale, Illinois

The decision on whether to use S/S, another technology or a combination of technologies at a site requires the evaluation of fundamental concepts, experience and the results of laboratory and pilot-scale tests. This section presents a series of steps to be taken in the selection process. It discusses the following factors and their interrelations:

1. Waste Characteristics
2. Technology/Binder Selection
3. Pretreatment Requirements
4. Site Characteristics
5. "Product" Acceptability Tests
6. The Iterative Nature of the Screening Process

This discussion builds on the concepts described in the earlier section, "Description of S/S Technologies." It combines the concepts with empirical results obtained in the past and gives guidelines for how one should set up a screening program appropriate to a specific waste/site problem.

REFERENCES

Chemical Rubber Publishing Company, <u>Handbook of Chemistry and Physics</u> 41st Edition, 1960.

Pauling, L. <u>General Chemistry</u>, second edition, W.H. Freeman & Co. San Francisco, CA, 1953.

Rowe, G. <u>Evaluation of Treatment Technologies for Listed Petroleum Refinery Wastes</u> Report prepared for The American Petroleum Institute Technology Task Force, American Petroleum Institute, Washington, DC, December, 1987.

Cullinane, J.M., Jones, L.W., <u>Handbook for Solidification/Stabilization of Hazardous Wastes</u> Final Report, Interagency agreement AD-96-F-2-A145 between the EPA and the U.S. Army Corp. of Engineers, Waterways Experimental Station, Vicksburg, MI, 1989.

Weitzman, L., Hamel, L.R., Cadmus, S.R. <u>Volatile Emissions from Hazardous Waste</u> Final Report Prepared for the U.S. Environmental Protection Agency, Risk Reduction Engineering Laboratory, Contract 68-02-3993, WA 32, 34.

Weitzman, L., Hamel, L.R., Cadmus, S.R. <u>Evaluation of Solidification/ Stabilization as a BDAT for Superfund Soils</u>. Final Report prepared for Environmental Protection Agency, Risk Reduction Engineering Laboratory, Contract 68-03-3241, WD-18, 1988.

Environmental Protection Agency, <u>Review of In-Place Treatment Techniques for Contaminated Surface Soils</u>. EPA-540/2-84-003a, Sept, 1984.

Kyles, J.H., Malinowski, K.C., Stanczyk, T.F. <u>Solidification/Stabilization of Hazardous Waste, A Comparison of Conventional and Novel Methods</u>, Toxic and Hazardous Waste, Proceedings of the Ninteenth Mid-Atlantic Industrial Waste Conference, Technomic Publishing Co. Lancaster, PA, 1987.

Wiles, C.C. <u>A Review of Solidification/Stabilization Technology</u>, Journal of Hazardous Materials, 14(1987)5-21, Also available as EPA/600/J-87/019

<u>Stabilization/Solidification of CERCLA and RCRA Wastes</u>, PEI Associates and Earth Technology Corporation, EPA/May 1988.

GOALS OF TECHNOLOGY SCREENING

1. To determine suitability and likelihood
   of success of immobilization

2. To select appropriate immobilization
   technologies
   a. for RCRA TSD facilities
   b. for CERCLA remediations
      (and RCRA corrective actions)

WHAT DOES GOAL NO. 1 REALLY MEAN?

**In the broadest sense EVERY WASTE
is potentially suitable for
immobilization**

**BOTH SUITABILITY AND LIKELIHOOD
OF SUCCESS DEPEND UPON**

- Matching the waste to appropriate
  immobilization technologies

- Matching each technology's needs (along
  with the waste's characterization) to
  appropriate pretreatment requirements

SCREENING

**Comparing waste characteristics to technology
capabilities is NOT a "one-time" operation**

**It is reiterated (at greater depth)
throughout the technology selection
process at such stages as**

**- "conceptual" screening
- detailed screening
- feasibility study
- process design**

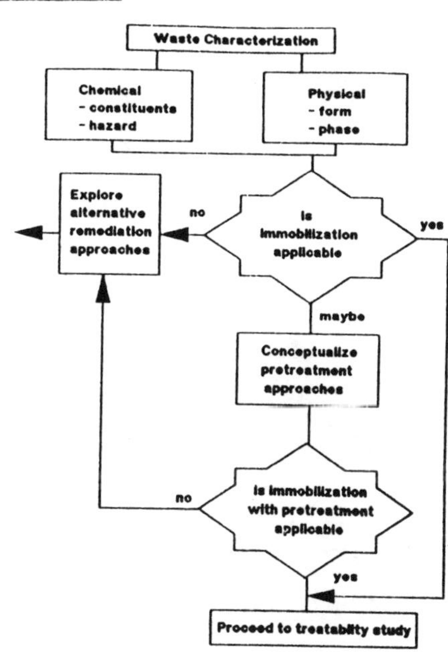

EVALUATION AND SELECTION OF IMMOBILIZATION
FOR SITE REMEDIATION (CERCLA or RCRA C.A.)

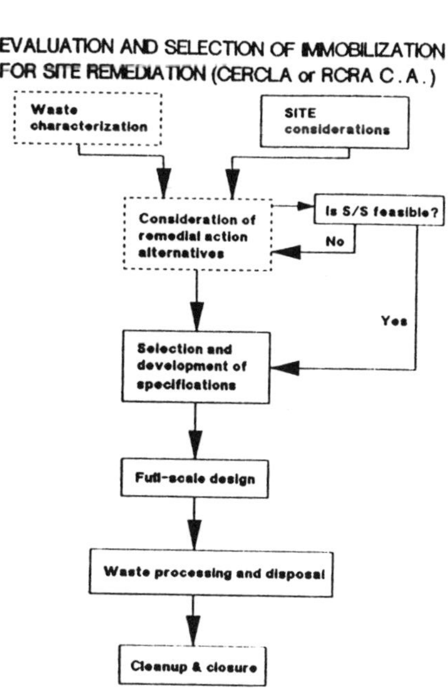

IMPORTANT WASTE CHARACTERISTICS
Chemical Composition
(Concentration & toxicity of each constituent)

- Organic components
    - polar, nonpolar
    - volatile, semivolatile, nonvolatile
- Inorganics
    - acids
    - oxidizers
    - metals
    - soluble salts

IMPORTANT WASTE CHARACTERISTICS
Physical Characteristics

- Phase & form (liquid, solid, sludge, soils)
- Total solids
- Particle size distribution
- Presence of debris

How do we match waste characteristics

to potential immobilization technologies?

IMMOBILIZATION TECHNIQUES

- Cement based binders
    - portland cement
    - cement kiln dust
    - natural and artificial pozzolans
    - cement/flyash
- Lime/limestone/quicklime
    - lime kiln dust
    - lime/flyash
- Absorbents
    - hydro- and organo-philic clays
    - wood chips, etc., generally not acceptable

(1 of 2)

IMMOBILIZATION TECHNIQUES

- Thermoplastic binders
    - asphalt/bitumen
    - thermoplastic polymers
- Thermosetting polymeric binders
- Vitrification

(2 of 2)

COMPATIBILITY OF WASTE TYPES vs. IMMOBILIZATION METHOD

Waste Component	Cement-Based S/S
Inorganics	
- acid wastes	Cement will neutralize acids
- oxidizers	compatible
- sulfates	may retard setting
- halides	easily leached, may retard setting
- heavy metals	compatible
- radioactive materials	compatible

COMPATIBILITY OF WASTE TYPES
vs. IMMOBILIZATION METHOD

Waste Component	Lime-Based S/S
Inorganics	
- acid wastes	compatible
- oxidizers	compatible
- sulfates	compatible
- halides	easily leached, may retard setting
- heavy metals	compatible
- radiaoactive materials	compatible

COMPATIBILITY OF WASTE TYPES
vs. IMMOBILIZATION METHOD

Waste Component	Thermoplastic solidification (little data available)
Inorganics	
- acid wastes	should be neutralized before incorporation
- oxidizers	may cause matrix breakdown; fire
- sulfates	may dehydrate and rehydrate causing splitting
- halides	may dehydrate
- heavy metals	compatible
- radiaoactive materials	compatible

COMPATIBILITY OF WASTE TYPES
vs. IMMOBILIZATION METHOD

Waste Component	Thermosetting Polymers (little data available)
Inorganics	
- acid wastes	compatible
- oxidizers	may cause matrix breakdown
- sulfates	compatible
- halides	compatible
- heavy metals	acid pH solubilizes metal hydroxides
- radiaoactive materials	compatible

COMPATIBILITY OF WASTE TYPES
vs. IMMOBILIZATION METHOD

Waste Component	Vitrification
Inorganics	
- acid wastes	should first be neutralized
- oxidizers	will react in process
- sulfates	compatible
- halides	fluorine-silicate interactions possible (defluorination as pretreatment)
- heavy metals	compatible
- radiaoactive materials	compatible

COMPATIBILITY OF WASTE TYPES
vs. IMMOBILIZATION METHOD

Waste Component	Cement-Based S/S
Organics *	
- nonpolar	impedes setting, volatiles will escape in processing
- polar	may significantly retard setting, decreases durability

* Generally, organics should be removed or destroyed as "pretreatment" before cement-based S/S

COMPATIBILITY OF WASTE TYPES
vs. IMMOBILIZATION METHOD

Waste Component	Lime-Based S/S
Organics *	
- nonpolar	impedes setting, decreases durability, volatiles may escape
- polar	phenols and alcohols retard setting, decreases durability

* Generally, organics should be removed or destroyed as "pretreatment" before lime-based S/S

**COMPATIBILITY OF WASTE TYPES
vs. IMMOBILIZATION METHOD**

Waste Component	Thermoplastic Solidification
Organics	
- nonpolar	compatible, may vaporize upon heating
- polar	compatible, may vaporize upon heating

**COMPATIBILITY OF WASTE TYPES
vs. IMMOBILIZATION METHOD**

Waste Component	Thermosetting polymers
Organics	
- nonpolar	may impede setting
- polar	compatible

**COMPATIBILITY OF WASTE TYPES
vs. IMMOBILIZATION METHOD**

Waste Component	Vitrification
Organics*	
- nonpolar	will volatilize, will combust if oxygen is available
- polar	will volatilize, will combust if oxygen is available

* Vitrification of residues as part of an organics incineration scheme is often proposed

COMPATIBILITY OF PHYSICAL CHARACTERISTICS
vs. IMMOBILIZATION METHOD

Waste Component	Cement-Based Binders
Phase or Form	
- liquid	acceptable
- sludge	acceptable (pumping concerns)
- solid	acceptable (mixing concerns), solids must be wettable
- contaminated soil	acceptable
- total solids	any range
particle size dist.	prefer smaller size and random distribution
Presence of debris	must be removed

COMPATIBILITY OF PHYSICAL CHARACTERISTICS
vs. IMMOBILIZATION METHOD

Waste Component	Lime-Based Binders
Phase or Form	
- liquid	acceptable
- sludge	acceptable (pumping concerns)
- solid	acceptable (mixing & wetting concerns)
- contaminated soil	acceptable
- total solids	any range (0-100%)
- particle size dist.	prefer smaller size and random distribution
- Presence of debris	must be removed

COMPATIBILITY OF PHYSICAL CHARACTERISTICS
vs. IMMOBILIZATION METHOD

Waste Component	Thermoplastic
Phase or Form	
- liquid	aqueous liquids not applicable
- sludge	water evaporated in process
- solid	acceptable
- contaminated soil	acceptable
- total solids	10-100%
- particle size dist.	smaller size preferred
- Presence of debris	must be removed

**COMPATIBILITY OF PHYSICAL CHARACTERISTICS
vs. IMMOBILIZATION METHOD**

Waste Component	Thermosetting polymers
Phase or Form	
- liquid	information lacking
- sludge	
- solid	
- contaminated soil	
- total solids	
- particle size dist.	
- Presence of debris	

**COMPATIBILITY OF PHYSICAL CHARACTERISTICS
vs. IMMOBILIZATION METHOD**

Waste Component	Vitrification
Phase or Form	
- liquid	not applicable
- sludge	dewatering required
- solid	acceptable
- contaminated soil	acceptable
- total solids	40% - 100%
- particle size dist.	in reactor - solids should be crushed in situ - no restrictions
- Presence of debris	in reactor - remove debris in situ - no metallic debris > 0.9 X electrode spacing

---

INTERFERENCES TO CHEMICAL S/S SYSTEMS

- Sulfur
- Calcium chloride
- Sodium arsenate
- Sodium borate
- Sodium phosphate
- Sodium halides
- Soluble metal salts
  - tin, zinc, copper, lead

MANIFESTATIONS OF INTERFERING CONSTITUENTS

- Spalling and cracking
- Set retardation; hardening and waste containment are impeded
- Flash set; mixing incomplete as a result of quick set, equipment can be fouled

(1 of 2)

MANIFESTATIONS OF INTERFERING CONSTITUENTS

- Chelated/complexed toxic constituents may accelerate leaching, even if waste is successfully S/S
- Some waste constituents can react and cause swelling and disintegration of S/S mass long after setting reactions complete
- Oxidizers can cause slow deterioration of the organic binder matrix

(2 of 2)

INTERACTIONS BETWEEN S/S BINDERS AND SPECIFIC ORGANICS

| Organic Compounds | Portland Cement | | Clay-Cement |
	Type 1	Type II & IV	
Alcohols & Glycols	Durability: decrease	Durability: decrease	Durability: decrease
Aliphatic & Aromatic Hydrocarbons	Set time: increase Durability: no effect	Set time: increase	Data unavailable
Chlorinated	Set time: increase Durability: decrease	Set time: increase Durability: decrease	Data unavailable

INTERACTIONS BETWEEN S/S BINDERS AND SPECIFIC INORGANICS

Chemical Group	Portland Cement		Clay-Cement
	Type 1	Type II & IV	
Heavy Metal Salts & Complexes	Set time: increase Durability: decrease	Set time: increase Durability: no change	Set time: increase Durability: decrease
Inorganic Acids	Set time: no effect Durability: decrease	Set no effect Durability: no effect	Durability: decrease
Inorganic Bases	Set time: no effect Durability: no effect	Set time: no effect Durability: no effect Durability (decreases with KOH & NaOH)	Durability: decrease

In general, one or more methods of immobilization will be good candidates as part of an integrated process for ANY waste.

GOALS OF TECHNOLOGY SCREENING

1. To determine suitability and likelihood
    of success of immobilization

2. To select appropriate immobilization
    technologies
    a. for RCRA TSD facilities
    b. for CERCLA remediations
        (and RCRA corrective actions)

DIFFERENCE BETWEEN GOALS 2a AND 2b

- RCRA treatment facilities will be driven by the regulations and by the technology they have available in their facility
- RCRA land disposal facilities will be driven by regulations
- Remediation actions will have the full set of technologies as options
- Remedial actions will be driven by site characteristics and regulatory and institutional restrictions as well as waste characteristics

---

GOALS OF TECHNOLOGY SCREENING

1. To determine suitability and likelihood of success of immobilization

2. To select appropriate immobilization technologies
   a. for RCRA TSD facilities
   b. for CERCLA remediations
      (and RCRA corrective actions)

Any one treatment facility will probably have one (or a few) specific immobilization technologies in place

and

will have a limited menu of pretreatment options available

- AND SO -

"Screening" at a RCRA TSD Facility means determinig whether each proposed waste is treatable by that immobilization technology

TSD

Treatment, Storage, and Disposal of hazardous wastes under the Resource Conservation and Recovery Act (RCRA) and subsequent amendments. A unit or facility which disposes of RCRA wastes at the facility may also be referred to as a "secure landfill" or "minimum technology unit"

SCREENING CONCERNS
Treatment vs. Disposal Facilities

- Treatment question - can "this" waste be treated to be acceptable for disposal
- Disposal question - can "that" material pass all the required tests for acceptance here

SELECTION FACTORS INVOLVED IN THE USE OF S/S AT A TSD FACILITY

- Regulations
- Waste Characteristics
- Operational and Economic Factors
- Site Methodology
- Test Methods

S/S DECISION TREE AT A RCRA TSD FACILITY

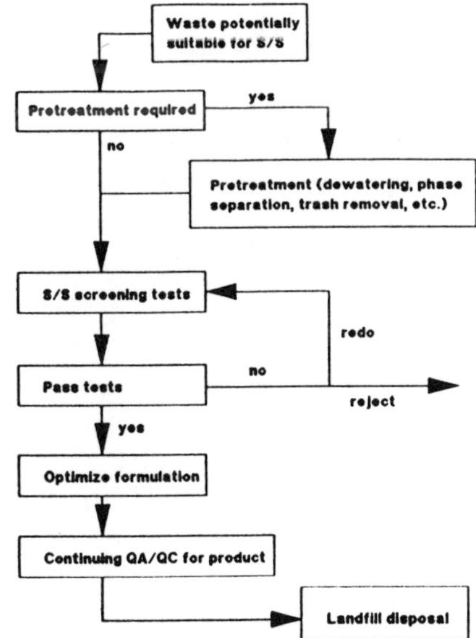

DETERMINING IF S/S IS APPLICABLE AT A RCRA TSD FACILITY

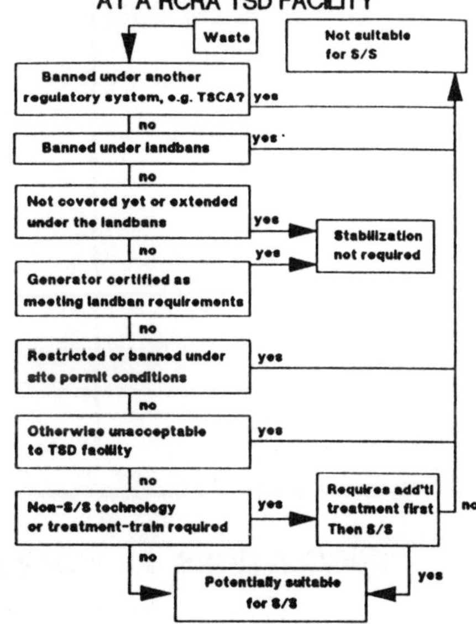

SELECTION FACTORS INVOLVED IN THE USE OF S/S AT A TSD FACILITY

- Regulations
- Waste Characteristics
- Operational and Economic Factors
- Site Methodology
- Test Methods

REGULATIONS AFFECTING DISPOSAL

- Landbans
- Permit Conditions
- TSD Conditions
- Other Regulatory Systems

REGULATIONS AFFECTING DISPOSAL
Landbans

- Liquid-in-Landfill Ban
- Solvent Ban
- California List Ban
- Scheduled Wastes
  - First Third
  - Second Third
  - Third Third
  - Soft-Hammer
  - Extensions

REGULATORY LEVELS IN LEACHATES
AFTER 2nd 3rd LANDBAN
(July 1989)

Concentration of Constituent (mg/l) in Waste or Leachate

Description	As	Ba	Cd	Cr	Pb	Hg	Ni	Se	Ag
Lowest level in any 1st3rd waste code	0.004		0.066	0.094	0.180	0.025	0.048	0.025	0.072
Delisting levels	0.315	6.300	0.063	0.315	0.315	0.013	3.150 2.205	0.063	0.315
RCRA Characteristic levels	5.000	100.0	1.000	5.000	5.000	0.200		1.000	5.000
Drinking water stds.	0.050	1.000	0.010	0.050	0.050	0.002		0.010	0.050

REGULATIONS AFFECTING DISPOSAL
Permit Conditions

- EPA Region
- State

REGULATIONS AFFECTING DISPOSAL
TSD Conditions

- Company Policy
- Protection of Landfill Liners

REGULATIONS AFFECTING DISPOSAL
Other Regulatory Systems

- TSCA: PCBs

SELECTION FACTORS INVOLVED IN THE USE OF S/S AT A TSD FACILITY

- Regulations
- Waste Characteristics
- Operational and Economic Factors
- Site Methodology
- Test Methods

SELECTION FACTORS INVOLVED IN THE USE OF S/S AT A TSD FACILITY

Waste Characteristics

- Chemical Characteristics
  - Metal Content
  - Metal Speciation
  - Reactivity
    - Presence of Cyanide or Sulfide

(1 of 2)

SELECTION FACTORS INVOLVED IN THE USE OF S/S AT A TSD FACILITY

Waste Characteristics

- Chemical Characteristics
  - Corrosivity
  - Ignitability
  - Organic content
    - Non-toxic Organics
    - Toxic organics
    - Volatile organics
  - Radioactivity

(2 of 2)

**SELECTION FACTORS INVOLVED IN
THE USE OF S/S AT A TSD FACILITY**
**Waste Characteristics
(Physical)**

- Total solids
- Viscosity
- Particle size distribution
- Phase separation
- Physical state

(1 of 2)

**SELECTION FACTORS INVOLVED IN
THE USE OF S/S AT A TSD FACILITY**
**Waste Characteristics
(Physical)**

- Dustiness, Odor
- Organic emissions
- Trash content

(2 of 2)

**SELECTION FACTORS INVOLVED IN
THE USE OF S/S AT A TSD FACILITY**

- Regulations
- Waste Characteristics
- Operational and Economic Factors
- Site Methodology
- Test Methods

**SELECTION FACTORS INVOLVED IN
THE USE OF S/S AT A TSD FACILITY**
Operational and Economic Factors

- Availability and cost of binders
- Quality and consistency of binders
- Pretreatment requirements
- Materials handling
- Volume and weight increase

**SELECTION FACTORS INVOLVED IN
THE USE OF S/S AT A TSD FACILITY**

- Regulations
- Waste Characteristics
- Operational and Economic Factors
- Site Methodology
- Test Methods

**SELECTION FACTORS INVOLVED IN
THE USE OF S/S AT A TSD FACILITY**
Site Methodology

- Waste profile
- Acceptance test on sales sample
- Fingerprint of received waste
- QC/QA on continuing basis
    - Sampling procedure
    - Sampling frequency

**SELECTION FACTORS INVOLVED IN
THE USE OF S/S AT A TSD FACILITY**

- Regulations
- Waste Characteristics
- Operational and Economic Factors
- Site Methodology
- Test Methods

**DETERMINING FACTORS IN THE ACCEPTABILITY
OF "IMMOBILIZED" WASTE AT A TSD FACILITY**
Test Methods

- Paint Filter Test (PFT)
- Toxicity Characteristic
  Leaching Procedure (TCLP)
- Site specific tests
  - Screens for:
    - Ignitability
    - Corrosivity
    - Reactivity
    - Radioactivity

GOALS OF TECHNOLOGY SCREENING

1. To determine suitability and likelihood
   of success of immobilization

2. To select appropriate immobilization
   technologies
   a. for RCRA TSD facilities
   b. for CERCLA remediations
      (and RCRA corrective actions)

EVALUATION AND SELECTION OF IMMOBILIZATION
FOR SITE REMEDIATION (CERCLA or RCRA C.A.)

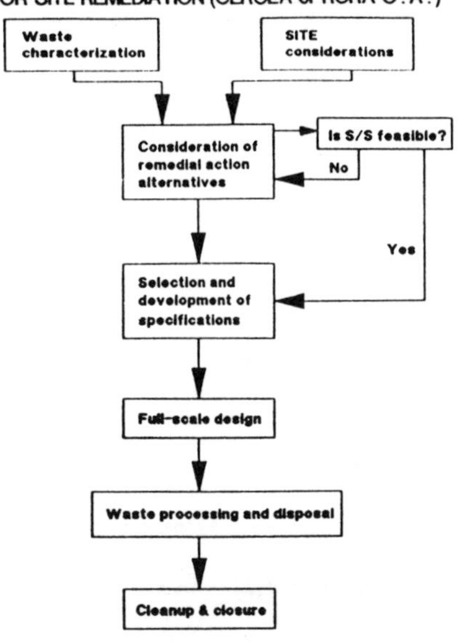

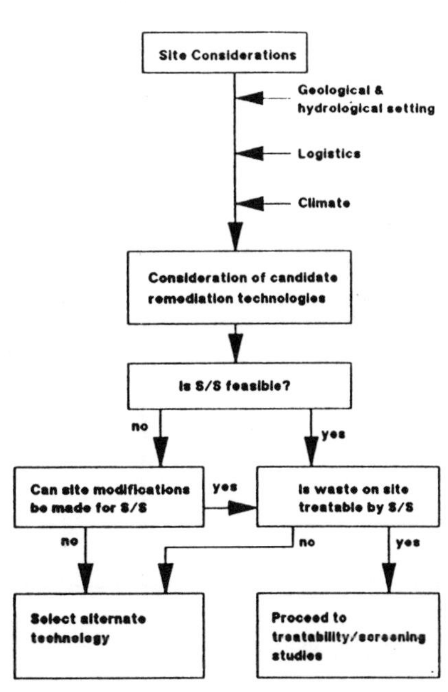

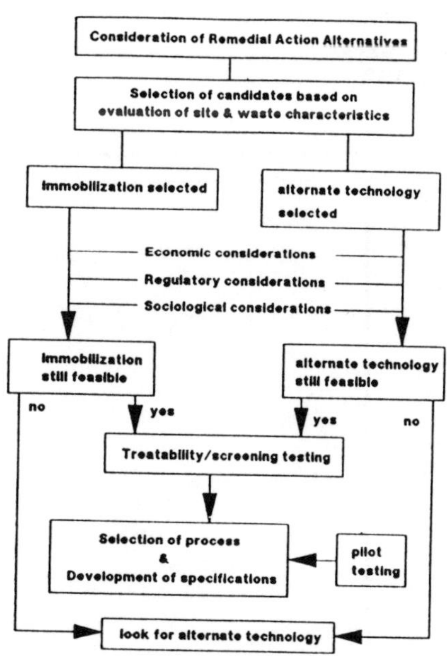

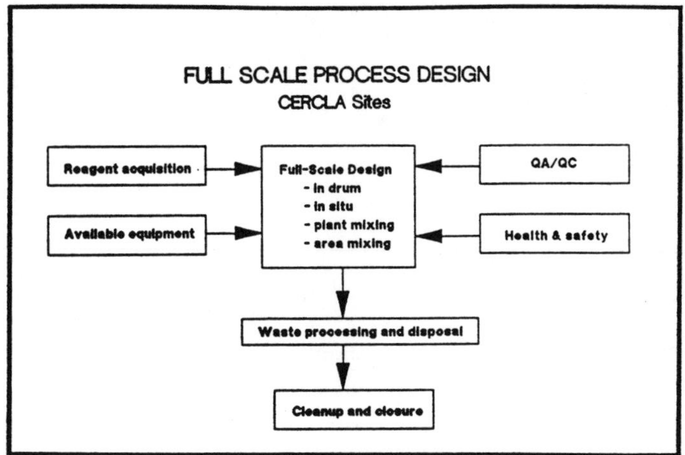

FACTORS IN SELECTING AN IMMOBILIZATION PROCESS/SYSTEM

- Waste characteristics
- Immobilization technology
- Site characteristics
    - water table
    - climate
    - soil characteristics
    - site layout
    - logistics
    - other

SITE CHARACTERISTICS
Will Limit:

- Immobilization technology chosen
- Pretreatments used
  (further limiting technologies)
- The full-scale design process
- Final specifications imposed on the
  immobilized product to be disposed
    - if on site, spec's determined for
      each site
    - if off site, spec's same as TSD

SITE CONSIDERATIONS

- Water table
- Climate
- Soil characteristics
- Site layout
- Logistics
- Other

WATER TABLE & CLIMATE
Questions to be Answered First

- Does immobilization require engineered controls to allow implementation and placement on site
- Will immobilized material be disposed off site

WATER TABLE/RAINFALL

Wetter	Engineered controls needed	Drier

for on-site disposal

Higher	Cost of on-site disposal	Lower

Note: On site means at the CERCLA remedial site
      Off site means final disposal at another location

ENGINEERED CONTROLS

- Dikes
- Berms
- Groundwater diversion
- Liners; grout curtains
- Waste isolation

SITE CONSIDERATIONS

- Water table
- Climate
- Soil characteristics
- Site layout
- Logistics
- Other

SOIL TYPE

| Porous, Sandy | Engineered controls needed for on-site disposal | Clay, Impervious |

←

| Higher | Cost of on-site disposal | Lower |

←

Note: On site means at the CERCLA remedial site
       Off site means final disposal at another location

SITE CONSIDERATIONS

- Water table
- Climate
- Soil characteristics
- Site layout
- Logistics
- Other

SITE LAYOUT CONCERNS

- Available area
- Staging/storage space
- Topography
    - grading
    - drainage concerns
- Proximity of neighbors
    - noise
    - blowing dust
    - odors/volatiles emission

SITE CONSIDERATIONS

- Water table
- Climate
- Soil characteristics
- Site layout
- Logistics
- Other

LOGISTIC CONCERNS

- Available access
    - equipment
    - binder material delivery
- Proximity to suppliers
- Proximity to disposal facility
  (if immobilized product is to be
  disposed off site)

TREATABILITY STUDY TO SELECT S/S BINDER SYSTEM

- Sampling program
- Analytical program
- Binder screening
- Binder quantity determination
    - S/S product performance
    - Criteria for success
- QA/QC for all steps

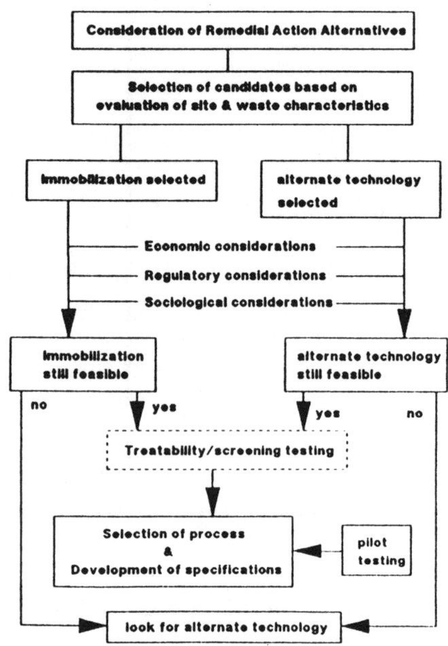

SAMPLING PROGRAM

- Waste characteristics
- Soil characteristics
- Site characteristics
- Specific considerations noted

ANALYTICAL PROGRAM

- What tests/measurements?
- Accuracy; reproducibility
- How many measurements necessary?
  -statistics

TREATABILITY STUDY TO SELECT S/S BINDER SYSTEM

- Sampling program
- Analytical program
- Binder screening
- Binder quantity determination
  - S/S product perfromance
  - Criteria for success
- QA/QC for all steps

SCREENING OF BINDER SYSTEM

- Which candidate systems meet physical requirements
- Which candidate systems meet chemical (leaching) requirements

PHYSICAL SCREENING

- Will the binder/waste matrix harden?
  - penetrometer
  - UCS
  - visual

CHEMICAL (LEACHING) SCREENING

- Will the binder/waste matrix pass applicable tests
  - EP toxicity
  - TCLP
  - ANS 16.1
  - WET
  - etc.

**TREATABILITY STUDY TO SELECT S/S
BINDER SYSTEM**

- Sampling program
- Analytical program
- Binder screening
- Binder quantity determination
    - S/S product perfromance
    - Criteria for success
- QA/QC for all steps

BINDER QUANTITY DETERMINATION

- Set up screening tests at varying
  ratios of:
    - binder to contaminated soil or waste
    - water to binder, or water to solids
- Measure performance at each condition

EXAMPLE

- Screening tests
    - binder:waste (B/W) 0.1, 0.5, 1.0,.....
    - water:solids (W/S) 0.5, 0.7, 1.0,.....
- Will it harden in 24 hours
    - UCS
    - penetrometer
    - visual

Within general ranges of system selected

prepare a large number of samples

for further testing

CHEMICAL SCREENING TEST
**Using the lowest addition that hardened
to selected criteria**

- TCLP
- EP toxicity
- MEP
- ANSI 16.1

WHAT RATIO WILL HARDEN
IN 24 HOURS?

B/W \ W/S	0.5	0.7	1.0
0.1	2*	2*	2*
0.5	2*	2*	2*
0.7	2*	2*	2*

* - No. of samples
- Test for hardness at 24 hours
- Will give a broad range for more specific testing

NARROW RANGE FOR TESTING BASED ON SCREENING RESULTS

B/W \ W/S	0.5	0.6	0.7
0.2	X	X	X
0.3	X	X	X
0.4	X	X	X
0.5	X	X	X

TREATABILITY STUDY TO SELECT S/S BINDER SYSTEM

- Sampling program
- Analytical program
- Binder screening
- Binder quantity determination
  - S/S product perfromance
  - Criteria for success
- QA/QC for all steps

Regulatory guidance on product "specification" (e.g. leaching standards, performance criteria, etc.) are not yet "set in concrete". So we must consider other regulatory guidelines

LEACHING STANDARDS
(mgs/L)

	E.P. TOX Characteristic	(MEP & others) Delisting	Landban F006	(TCLP) K061
Arsenic	5.0	0.315	—	—
Barium	100.0	6.3	—	—
Cadmium	1.0	0.063	0.066	0.14
Chromium	5.0	0.315	5.2	5.2
Lead	5.0	0.315	0.51	0.24

LEACHING STANDARDS FOR ORGANICS
(mgs/L)

HW No.	Contaminant	Regulatory level
D019	Benzene	0.07
D021	Carbon Disulfide	14.4
D022	Carbon Tetrachloride	0.07

Tentative OAQPS tests for air emissions

CONCLUSION

- In general, one or more methods of immobilization will be good candidates as part of an integrated treatment process for ANY waste

- But you must match the waste to the technology & select apptopriate pretreatments

7. Field Implementation Procedures Utilized
for Stabilization/Solidification

Mr. Richard McCandless   Mr. Peter Hannak   Mr. Robert Maxey
UC/Center Hill Lab   CH2M Hill   The Earth Technology Corp.
Cincinnati, Ohio   Waterloo, Ontario   Alexandria, Virginia

There are usually six functions which must be satisfied for the successful implementation of a stabilization/solidification project. They are:

- Waste removal
- Untreated waste transportation
- Chemical reagent storage
- Waste/reagent mixing
- Stabilized/solidified waste transportation
- Stabilized/solidified waste replacement

In addition, waste may have to be stockpiled, moved or further processed in pretreatment operations such as dewatering or neutralization. Some processes may not require all of the steps, such as in-situ stabilization/ solidification where the reagent is brought to the waste and it is stabilized/solidified in place.

<u>Waste Removal</u>

Due to the large quantities of waste which are usually stabilized/ solidified, waste removal has generally been accomplished by the use of traditional earth-moving equipment. This equipment includes tracked backhoes, draglines, bulldozers and front-end loaders. This equipment has been in use for many years in the construction industry. Thus, its application to hazardous waste use is merely one of adaptation. The physical state of the waste may indicate the sequencing of operations necessary for the site. For example, the presence of a liquid above the waste can be managed by removal of this supernatant and treatment as a separate waste. The safety aspects of the waste relate to the equipment operators and other workers. Complete enclosure of the operator space in construction equipment and the provision of breathing air may be necessary.

Tracked backhoes and draglines are used for the remote removal of materials. Typically a backhoe is used in cases where use of equipment in an area is prevented for reasons of soil stability. Backhoe operations can occur in hazardous waste sludges which ordinarily would not support equipment. The rate at which waste is removed needs to be assessed in order that rates of material removal and transportation be as evenly matched as possible.

Draglines have application where reaches longer than 40 feet are necessary. They are delivered to a site in pieces, usually by truck, and assembled. This delivery and assembly needs to be taken into consideration as they can consume many days' time. Draglines cannot provide removals more accurately than a couple of feet. Thus, excess material is usually removed during dragline operation.

Bulldozers and front-end loaders maneuver in and on the material which is being removed. The bulldozer pushes the material by scraping. In a stabilization/solidification operation, this movement of material may be to an area where it is stabilized/solidified or to where it is loaded. Front-end loaders are capable of removal and loading of solid material. This is an advantage where the material must be placed in a dumptruck or a hopper for transportation or processing.

Waste Transportation

Depending upon the nature of the waste and of the site, waste can be transported by dump truck, pump and hose, or a fixed system such a conveyor belt or screw auger system. Dump trucks are commonly used for solids transportation, particularly when the liquid content is low and travel distance is over a quarter-mile. An important consideration is that of spillage, particularly since the waste may be hazardous. Truck beds should be lined with plastic to prevent escape of waste when the waste contains liquid or when the trucks will travel off site. In addition, the trucks should be covered with tarpaulins. Such covering is essential when the material is of a small particle size and subject to wind dispersion.

Conveyor transportation is used at a site where large amounts of waste are to be moved over a fixed distance for long periods of time. Their setup is complicated and costly, but this is offset by their ease of use once set up. Conveyors are incapable of moving liquids and spillage is a problem unless the material to be stabilized/solidified is dewatered or its water is decanted. Since a conveyor is a piece of moving equipment, time must be allotted for maintenance and repair. The use of sidewalls may be necessary to prevent unwanted dispersal of hazardous materials.

Pumping of hazardous waste solids may be feasible when they are sludge-like. In this instance the transportation function is also the removal function. Pumping has the advantage that the waste can be directly sent to the processing equipment.

Chemical Reagent Storage

Material stockpiling is an obvious requirement for the satisfactory operation of a stabilization/solidification process. Primary requirements are: (1) sufficient storage of agent so that operations are not delayed; (2) ability to keep the agent dry, since virtually all stabilization/ solidification agents are dry; and, (3) ease of unloading upon delivery from the supplier and to the process. Bins and hoppers are the primary storage vessels for solid agents. These hoppers are sometimes transportable. Liquid storage is occasionally required. Sometimes solid materials are delivered to the site in bags, which are palletted, or in bulk for storage in piles. Typical chemical storage tanks are routinely used for this function.

Waste/Reagent Mixing

Mixing of the hazardous waste and its stabilization/solidification agent is the heart of the stabilization/solidification process. Most of the stabilization/solidification activities that have occurred in the United States have used area mixing. In area mixing, the waste is usually neither removed nor transported prior to the addition of a stabilization/ solidification agent. Instead, the stabilization/solidification agent is delivered to the area to be stabilized/solidified and mixed with the waste directly. Often, tracked backhoes are used for mixing. Typically, the water has enough waste present for reaction with the stabilizer/solidifier and the waste. The waste is mixed until the operator perceives that mixing is complete. The benefits to this type of mixing are: (1) the waste usually need not be removed from the site; (2) the operation can proceed with traditional earth moving equipment; and (3) it is inexpensive. Offsetting these benefits is the probable incomplete and inadequate mixing of the waste.

A wide range mixing equipment exists for use in industries which process solids. Typical mixing equipment, suitable for use in stabilization/ solidification operations, are as follows: pug mills, ribbon blenders, mueller mixers, extruders, screw conveyors. Waste and reagent can be metered into this equipment upon mixing as a means of quality control.

Some waste stabilization/solidification contractors report the use of pug mills for mixing operations. Pug mills are manufactured by several companies; contractors occasionally modify the pug mill to suit their special needs. Pug mills consist of double screws into which one material is mixed with another on a flow-through basis. The materials are generally mixed prior to their addition to the pug mill, but this is only partial mixing.

The placement of soil-like materials can be performed in accordance with procedures used in the roadbuilding industry. Here, the stabilized/solidified waste is placed in lifts of eight to ten inches using earth-moving equipment such as graders or bulldozers. Following placement, the waste is compacted. To assure proper compaction, the optimum moisture content of the material needs to be determined. The optimum moisture content is that moisture which gives the greatest density for a given material. The addition of moisture is necessary because too little moisture does not provide the required lubrication for the soil particles to slide past one another during compaction. Too much moisture causes the material to approach the density of water, which is presumably lower than that of the waste. Thus, for soil-like stabilized/solidified hazardous waste, the amount of water in the waste must be controlled as the waste is compacted.

A second factor which affects the compaction of the waste is the particle size distribution of the material. Poorly graded wastes, which are predominantly of one particle size only, do not compact well. Well-graded materials compact better since the void volume of the larger particles is occupied by smaller particles giving a better fit. This consideration is usually one that cannot be controlled unless the particle size distribution of the stabilization/solidification agent can be regulated.

REFERENCES

Caterpillar, Inc. 1987a. Caterpillar Performance Handbook, Edition 18. Peoria, Illinois.

Caterpillar, Inc. 1987b. Caterpillar Performance Handbook: Hydraulic Excavators. Peoria, Illinois.

Cullinane, M.J., Jr., L.W. Jones, and P.G. Malone. 1986 Handbook for Stabilization/Solidification of Hazardous Wastes. EPA/540/2-86/001. Hazardous Waste Engineering Research Laboratory, U.S. Environmental Protection Agency, Cincinnati, Ohio.

Curry, M. 1986. Fixation/Solidification of Hazardous Waste at Chemical Waste Management's Vicery, Ohio, Facility. In: Proceedings of Hazardous Materials Control Research Institute Conference, December 1986. pp. 297-302.

Green, D.W., ed. 1984. "Processing Bulk Solids," in Perry's Chemical Engineers Handbook, 6th ed. McGraw-Hill, New York. pp. 21-3-10.

Kyles, J.H., K.C. Malinowski, and T.F. Stanczyk. 1987. Solidification/Stabilization of Hazardous Waste - A Comparison of Conventional and Novel Techniques, Toxic and Hazardous Wastes. In: Proceedings of the 19th Mid-Atlantic Industrial Waste Conference, June 21-23, 1987. Jeffrey C. Evans, ed. pp.554-568.

National Lime Association. 1987. Lime Stabilization Construction Manual. Bulletin 326. Arlington, VA.

National Research Council. 1987. Lime Stabilization-Reactions, Properties, Design and Construction. Transportation Research Board, Washington, D.C.

Rowe, G., and API Waste Treatment Task Force. 1987. Evaluation of Treatment Technologies for Listing Petroleum Refinery Wastes. Final Report, The American Petroleum Institute.

Tittlebaum, M.E., et. al., 1985. "State of the Art on Stabilization of Hazardous Organic Liquid Wastes and Sludges." CRC Critical Reviews in Environmental Control, 15(2):191-193.

U.S. Environmental Protection Agency. 1986h. Mobile Treatment Technologies for Superfund Wastes. Office of Solid Waste and Emergency Response. EPA/540/2-86/003, September 1986.

GENERAL PERSPECTIVES

- Highly site-specific
- Dependent on other project phases
- Expanded QA/QC program

PROJECT DEVELOPMENT

- Site/waste characterization — Define problem(s)
- Technology screening — Identify solution & formulate implementation approach
- Treatability testing — Demonstrate technical feasibility for a practical implementation scheme
- Pilot testing/ design — Develop/specify tailored remedy

▷ IMPLEMENTATION

BASIC IMPLEMENTATION FUNCTIONS

- Pretreatment
- Removal
- Storage
- Mixing
- Replacement

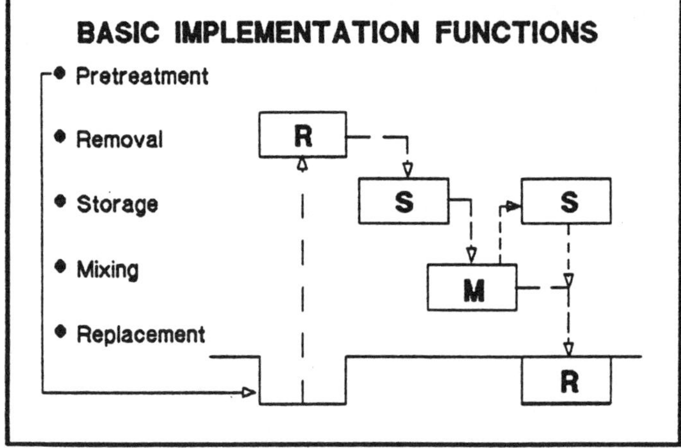

GENERAL PREPARATIONS

- Access

- Site preparation - utilities
    - grading for roads
    - security

- Health & safety - protective clothing
    - monitoring equipment
    - dust control
    - emergency medical
    - communications

- Decontamination - non-expendable equipment
    - workers
    - collection/treatment of rinseates

PRETREATMENT

- Removal of large objects

- Dewatering - water table suppression
    - drainage beds

- Neutralization - safety
    - protection of equipment
    - as part of treatment

- Initial homogenization

WASTE REMOVAL

- Operation in waste
    - bulldozers scrape waste
    - front-end loaders move and
      load waste; they can be tracked
      or use low pressure tires

- Remote operation
    - backhoes
    - draglines

- Consider equipment reach, cycle
  time (make certain enough truck/
  conveyor capacity is present
  to haul waste)

TRANSPORTATION
- Pump & hose to remove liquids
- Dump truck
    - line & cover
    - decontamination necessary
- Conveyor belt
    - not for liquids
    - sidewalls necessary
- Screw auger

UNTREATED WASTE STORAGE
- Sloped to collect liquids
- Lined to keep base clean
- Rocks, twigs, etc. removed to avoid liner puncture
- Drainage pipes, sump & water treatment system necessary to handle liquids
- Covered to prevent waste dispersion
- Bermed to retard waste slumping
- Rubber tire loader to move waste

CHEMICAL REAGENT STORAGE
- Tank for liquids
- Solids storage
    - bins & hoppers
    - palleted bags for lime & cement
    - storage piles
- Requirements
    - keep solid reagents dry
    - ease of unloading is important
    - have sufficient quantity of reagent on-hand to prevent delays in operation

WASTE/REAGENT MIXING

- Mobile plant
- Area
- In situ
- In drum

MOBILE PLANT WASTE/REAGENT MIXING

- Pug mills, ribbon blenders, mueller mixers
- Process control (QC) is good
- S/S waste replacement
    - place in lifts of 6" to 8"
    - compact considering optimum moisture content & particle size

---

AREA WASTE/REAGENT MIXING

- Uses traditional earth moving equipment
- Technique
    - use LPTs or wide tracks if in waste
    - use timber matts to support backhoes
- Lime stabilization techniques
    - soil preparation
    - lime spreading
    - soil/lime mixing
    - compaction to achieve practical density
    - curing

AREA WASTE/REAGENT MIXING
(Continued)

- Advantages
    - inexpensive
    - waste stays on site
      (no transportation &
      disposal costs)
- However, mixing may be poor

IN-SITU MIXING PROCESSES

- Used largely for pits, ponds,
  lagoons, where liquids & sludges
  are present
- Vendors
    - Geo-Con
    - Enreco
    - Harmon
    - others

GEO-CON PROCESS

- S/S agent delivered by auger &
  caisson into waste
- Mixing done by lifting & turning action
- Depths to 100 ft. can be stabilized
- Overlapping bore patterns allow for
  complete coverage
- Advantages
    - waste not excavated
    - water table not lowered
    - fugitive emissions controlled
    - feed of agent can be metered
      for good QC

GEO-CON PROCESS
Continued

- Cannot penetrate masses with boulders or debris
- Has other geotechnical applications
    - dams
    - slurry walls
- SITE Program demonstration project

ENRECO IN-SITU PROCESSES

- S/S agent fed through injector on backhoe
- Mixing done by back & forth action of delivery head
- Depths to 100 ft can be stabilized
- Volume increases of 10% can be expected

HARMON

- HSS system
    - auger on front end of bulldozer
    - good for 8" to 10" lifts
- PF-5 injector
    - 5 injection tubes have impellers & augers to promote mixing
    - these tubes are at the end of a backhoe which reaches into the waste
    - can be used to depths of 10 ft

IN-DRUM WASTE/REAGENT MIXING

- Use drum as mixing vessel &
  waste container

- Reagent will require additional
  volume (more drums)

- Important considerations
    - how to access drum (bung
      hole, top removal)
    - safe operation?

- Required equipment
    - onsite chemical storage
    - chemical batching system
    - mixing system
    - drum handling system

REPLACEMENT

- Pumpable product
    - fluid or "concrete like"
    - time
    - vibration to remove air voids

- Compactable product
    - cohesive or friable "soil-like"
    - moisture conditioning for
      maximum density

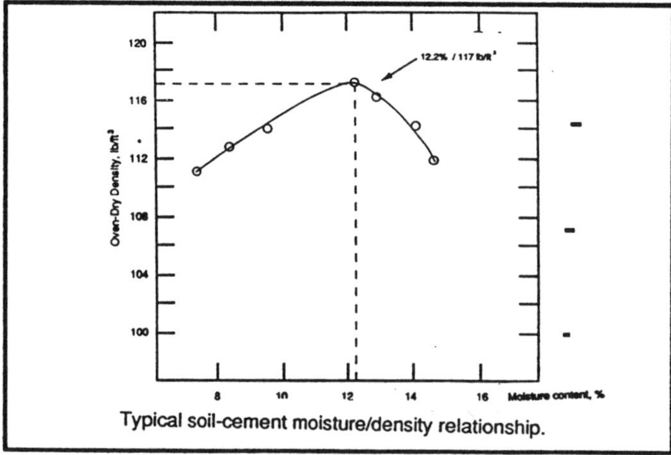

Typical soil-cement moisture/density relationship.

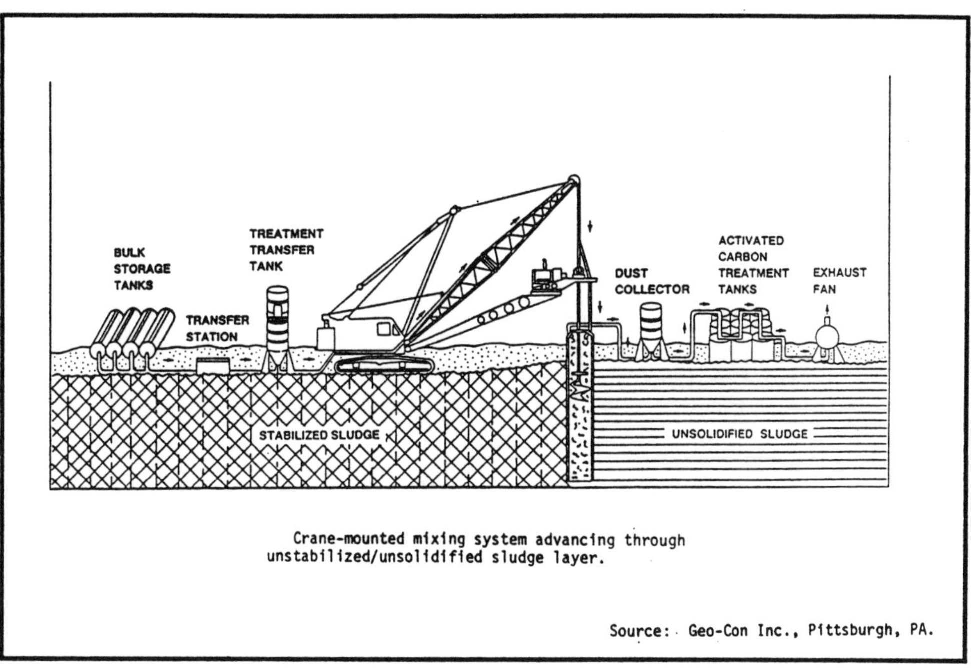

Crane-mounted mixing system advancing through
unstabilized/unsolidified sludge layer.

Source: Geo-Con Inc., Pittsburgh, PA.

OTHER FACTORS

- Segregation
- Thickness
- Exposed surface area
- Cover

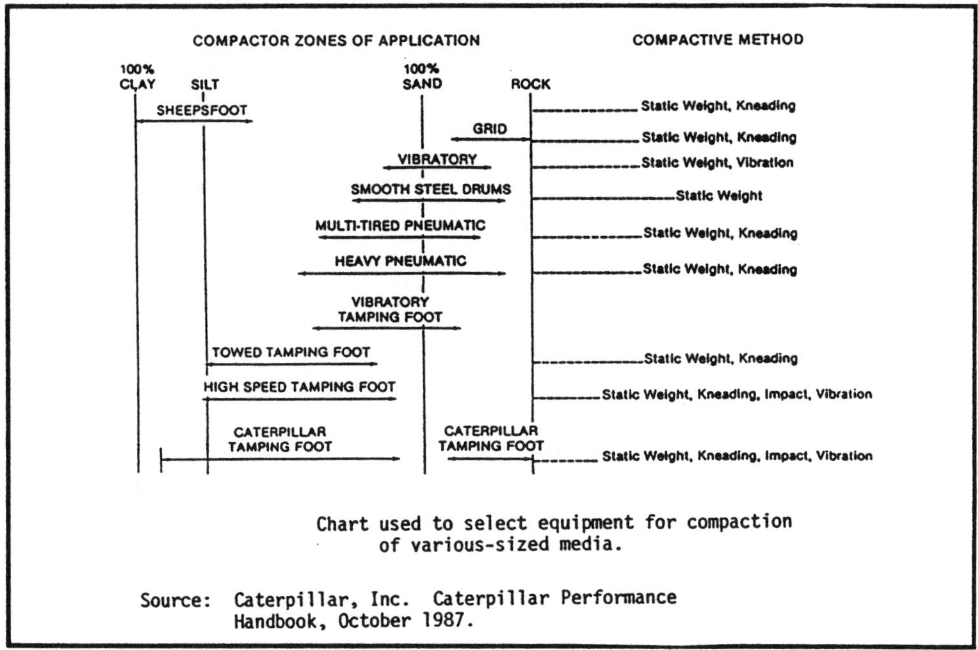

Chart used to select equipment for compaction
of various-sized media.

Source: Caterpillar, Inc. Caterpillar Performance
Handbook, October 1987.

8. Quality Assurance Procedures for Ensuring Long-Term Performance

Mr. Richard McCandless
UC/Center Hill Lab
Cincinnati, Ohio

Mr. Peter Hannak
CH2M Hill
Waterloo, Ontario

Overview

Quality Assurance/Quality Control (QA/QC) is the backbone of all solidification/stabilization (S/S) actions involving Superfund and RCRA wastes. From waste characterization through bench testing, pilot testing and implementation of the remedy, a through and well managed QA/QC program is the only means by which confidence in the finished product can be estimated.

Quality Assurance is defined as all activities designed to ensure that a specified product performance will be achieved with a stated level of confidence. Quality Control is a part of QA and involves all activities intended to control the quality of a product so that it <u>does</u> meet the specified criteria. Whereas QA relates to planning and overall verification (what and how), a QC program involves checks on the measurement systems to demonstrate that they are in calibration and being applied properly at the right time and place. A simplification of the concept might be stated as: "QA IS DOING THE RIGHT THING; QC IS DOING THINGS RIGHT."

Treatments of this subject often tend to focus on QA/QC objectives (accuracy, precision, etc.) and can be less than enlightening since statements describing the level of numerical control anticipated or desired do not help to answer the questions "What should I measure and how?" "How good is good enough and how will I know when I've succeeded?" The usual answer to the latter question is "good statistics", but good judgement and common sense are at least as important. Cookbook methods do not exist and cannot be offered here. What can be offered are suggestions based on limited experience that may help to minimize oversight, avoid latent pitfalls and in general improve our ability to generate reliable information.

A sound engineering strategy and a good understanding of the relationship between project phases is key to making appropriate decisions along the way. Each phase of the project builds upon the observations and findings of the proceeding phase; for better or worse. For practical reasons, however, this dicussion focuses on specific problem areas in specific phases of a typical project. For example, although the topic of mixing is relevant in several phases, it is discussed in connection with pilot testing since this phase represents the "bridge" between idealized bench testing and the real conditions to be encountered in implementation. A general outline of this session follows:

Site Conditions

- Characterization and Sampling Practice for Known and Unknown Conditions

Bench Testing

- Physical Sample Size and "Representativeness"
- Chemical Baseline Needs
- Significant Variability in Analytical Data and other Factors affecting Test Results

Pilot Testing

- Mixing Practice, Parameters and Performance

Implementation

- Process Control Requirements and suggested methods for Monitoring Uniformity
- Prescriptive versus Performance Contracts and Impacts on QA/QC

REFERENCES

AICHE Equipment Testing Procedure, <u>Dry Solids, Paste and Dough Mixing Equipment</u>; Second Edition, American Institute of Chemical Engineers, New York, N.Y., 1979.

Barth, E.F., and McCandless, R.M., et.al., <u>General Guide for Treatability Assessment</u>; Draft S/S Technology Information Bulletin for the USEPA Engineering Forum. USEPA Risk Reduction Engineering Laboratory, Cincinnati, Ohio, July, 1989.

<u>Chemical Quality Management – Toxic and Hazardous Wastes</u>; U.S. Army COE Engineer Regulation 1110-1-263, December 30, 1985.

<u>Contract-Construction Quality Management</u>; US Army Corps of Engineers, ER 1180-1-6, July 31, 1986.

<u>Handbook for Stabilization/Solidification of Hazardous Wastes</u>; EPA/540/2-86/001, June 1986.

Horwitz, W., Kamps, L.R. and Boyer, K.W., <u>Quality Assurance in the Analysis of Foods for Trace Constituents</u>; Journal of the Association of Official Analytical Chemists, Inc., Vol. 63, No. 6, 1980.

Keller, D.J., <u>Construction Quality Assurance for Stabilization/ Solidification</u>; Draft Final Report, W.A.# 2-17. USEPA Risk Reduction Engineering Laboratory, Cincinnati, Ohio, August 31, 1989.

Perry, R.H., _Chemical Engineers' Handbook_; McGraw-Hill Book Co., New York, N.Y., 1984.

Quality Control in Remedial Site Investigation--Hazardous and Industrial Solid Waste Testing; C. L. Perket, Editor. ASTM STP 925, Fifth Volume, 1986. ASTM Publication Number (PCN) 04-925000-16.

Soil Sampling Quality Assurance User's Guide; Second Edition, EPA/600/8-89/046, March, 1989.

Taylor, John Keenan, _Quality Assurance of Chemical Measurements_; Lewis Publishers, Inc., Chelsea, Michigan, 1987.

OBJECTIVES
ACCURATE
PRECISE
COMPLETE
REPRESENTATIVE
COMPARABLE

ANALYTICAL QA
STATISTICS

QA: DOING THE RIGHT THING
(what)

QC: DOING THINGS RIGHT
(how)

PHASED ENGINEERING APPROACH

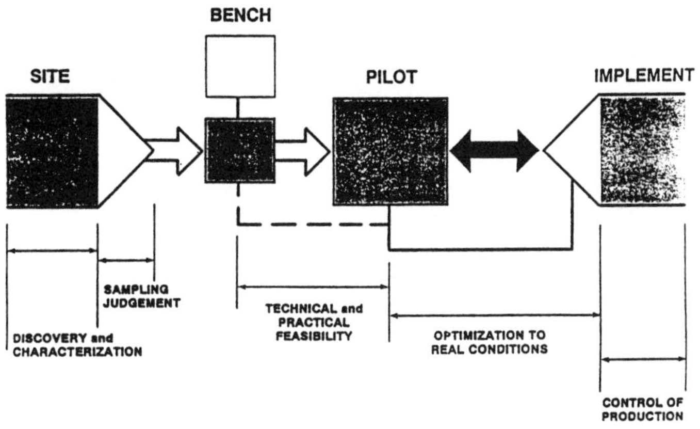

TOPICS for REVIEW

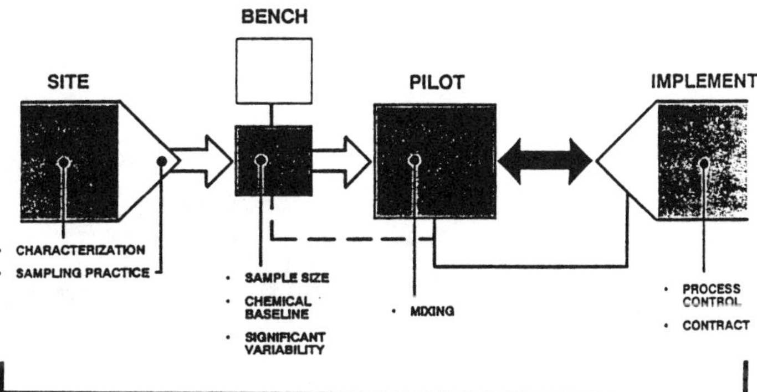

TOPICS for REVIEW

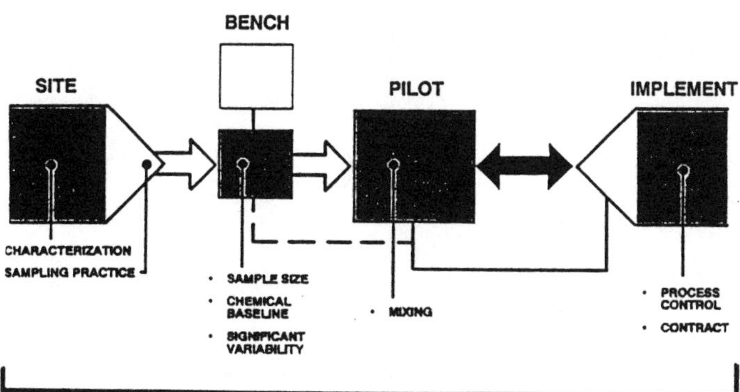

CHARACTERIZATION AND SAMPLING PRACTICE
Critical Information

- Waste types
- Conditions
- Distribution
- Quantities

CHARACTERIZATION AND SAMPLING PRACTICE
Sampling Design

- "ASTM Standard Guide for General Planning of Waste Sampling" (in development)
- Not for unknown or abandoned sites or wastes

CHARACTERIZATION AND SAMPLING PRACTICE
Sampling Schemes

- Judgement: deciding through visual observation or knowledge of the site
- Systematic: statistical basis for entire waste body only where known to be homogeneous; otherwise by waste stratigraphy
- "n" can be calculated based on estimate of variance of NORMAL population
- Lilliefors "goodness of fit" test for normality of population generally required

CHARACTERIZATION AND SAMPLING PRACTICE

- Product
  - Credible "worst case" composite for bench testing
- Level of effort

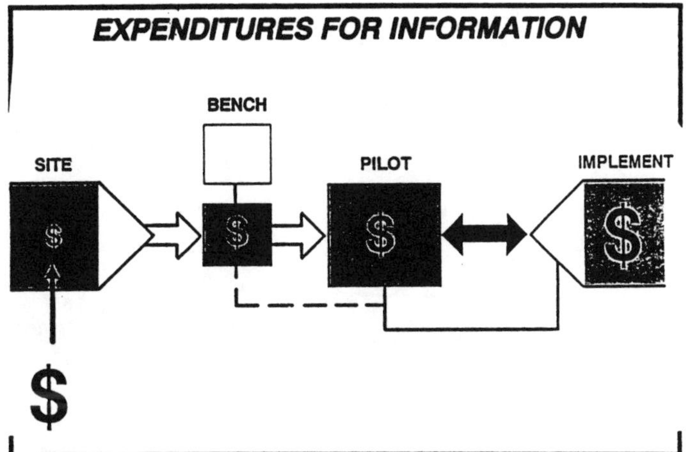

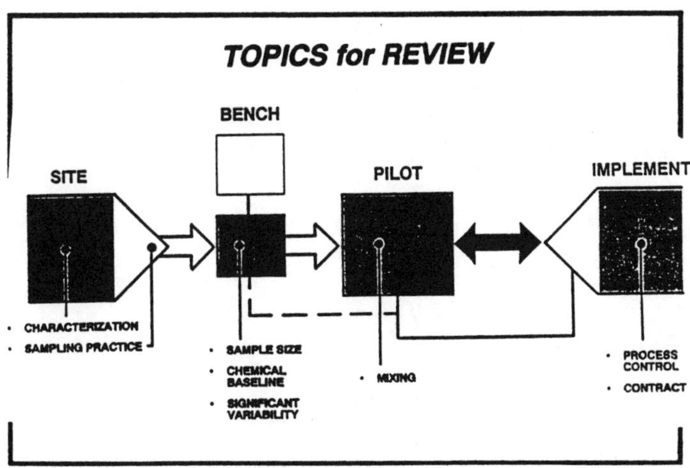

SAMPLE SIZE

- Representative physical size....not "n"

- Some guidance for physical tests (ASTM)
  - concrete specimens for strength:
    minimum dimension of sample = 3 times
    maximum dimension of largest particle

  - grain size distribution of soils:
    mass of sample for test "proportional"
    to nominal diameter of largest particle

SAMPLE SIZE
(continued)

- Some guidance for physical tests (ASTM)
  - recognize limitations of test equipment

- No guidance for chemical tests
  - acid digestion for metals limited
    to a few grams
  - need to test "leach test size" samples

FRACTIONATION/ACID DIGESTION
FOR HEAVY METALS
("FAD"; under development)

- Combines gravimetric separation
  (ASTM D4371) with cold acid
  extraction (MSA 19-3.4)

- Untreated waste or treated waste
  before "set"

- Measure amount soluble in strong
  acid; not "total"

FRACTIONATION/ACID DIGESTION FOR HEAVY METALS
("FAD"; under development)
(continued)

- Test entire "leach test size" specimen
  - 1 L 20% nitric acid for 100g waste (S.G.=1.1)
  - agitate 7 hours, separate (settle)
  - filter and weigh "floaters" and "sinkers"
  - analyze extraction liquid

CHEMICAL BASELINE

What?	Why?
• Waste	• Safety check • Verification of "worst case" • Interference
• Binder	• Unexpected leach test results • Hazardous?

• Waste & binder mix	• Homogeneity a bad assumption
	• Measure actual variability for
	  -Calculations of confidence intervals
	  -performance comparisons

How?

SIGNIFICANT VARIABILITY

- How much variability should I expect in measuring the concentrations of analytes?

- How can I estimate variability in order to calculate "n"?

- Review

  CV: The variability expressed as a percentage of the mean

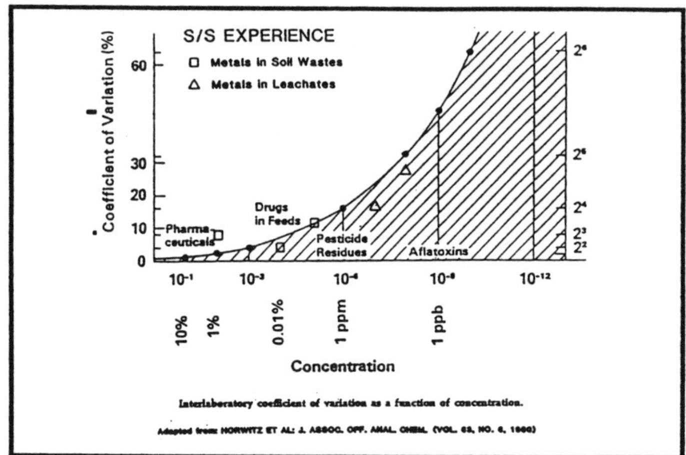
S/S EXPERIENCE
Metals in Soil Wastes
Metals in Leachates
Coefficient of Variation (%)
Pharmaceuticals
Drugs in Feeds
Pesticide Residues
Aflatoxins
10%
1%
0.01%
1 ppm
1 ppb
Concentration
Interlaboratory coefficient of variation as a function of concentration.
Adapted from: HORWITZ ET AL: J. ASSOC. OFF. ANAL. CHEM. (VOL. 63, NO. 6, 1960)

SOURCES OF SIGNIFICANT
VARIABILITY
Sample Handling
Mixing
Molding
Curing
Storage
Shipping

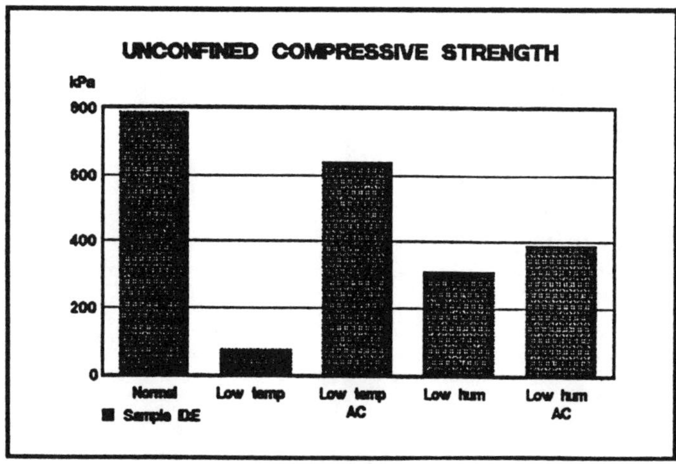
UNCONFINED COMPRESSIVE STRENGTH
kPa
800
600
400
200
0
Normal
Low temp
Low temp AC
Low hum
Low hum AC
Sample I.E

OBSERVED VARIABILITY
Replicates for Physical Testing

Test Method	# Products Tested	Median	No. of Replicates	Precision
Bulk Density (g/cc)	69	1.44	3	+/- 0.14
Specific Gravity Wet Method	69	2.48	4	+/- 0.28
Dry Method	55	2.34	2	+/- 0.20

OBSERVED VARIABILITY
Replicates for Physical Testing
(Continued)

Test Method	# Products Tested	Median	No. of Replicates	Precision
Moisture Content (%,w/ww)	69	25.8	3	+/- 1.6
Hydraulic Conductivity (m/s)	37	3×10^{-8}	4	630%
UCS (kPa)	66	1760	4	56%

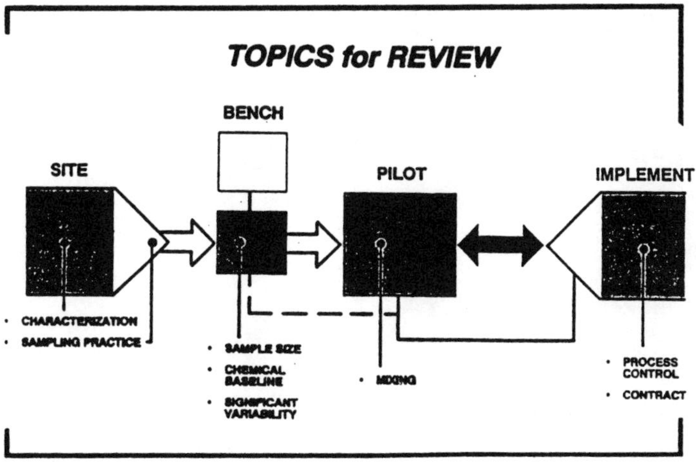

MIXING

- Objective
  - Intimate association of waste and binder
- Practice - a materials science of three domains
  - Liquid-liquid
  - Solid-solid
  - Liquid-solid

MIXING

Bench-Scale Approach

- Small quantities
- Closest approach to ideal mixing
- May not be practical or representative of pilot scale

MIXING

Pilot-Scale Approach

- Large quantities
- Practical optimization to real conditions

MIXING
Approach to Scale-up

- Power requirement proportional to unit process volume so long as equipment geometry unchanged
- Mixing time requirement inversely proportional to mixing speed

IMPORTANT MIXING PARAMETERS
Waste and Binder

- Particle size distribution
- Bulk density
- True density
- Particle shape
- Surface area and charge characteristics

IMPORTANT MIXING PARAMETERS
Waste and binder
(continued)

- Flow characteristics
- Friability
- State of agglomeration
- Moisture content
- Density, viscosity and surface tension of liquids added

IMPORTANT MIXING PARAMETERS
Equipment

- Mechanism design - blades, baffles, cams,
  impellers, rollers, rods, balls,
  screw augers, extruders

- Intensity

- Duration/retention time - proper
  combination of mechanism design and
  intensity should produce desired blend
  in a few minutes; 15 minutes maximum

IMPORTANT MIXING PARAMETERS
Performance

- Mixing trials and evaluation required
- Refer to AIChE procedures for
  conducting and interpreting performance
  tests on mixing equipment

"Dry Solids, Paste and Dough Mixing Equipment
Testing Procedure", American Institute of Chemical
Engineers, Second Edition, New York, 1979

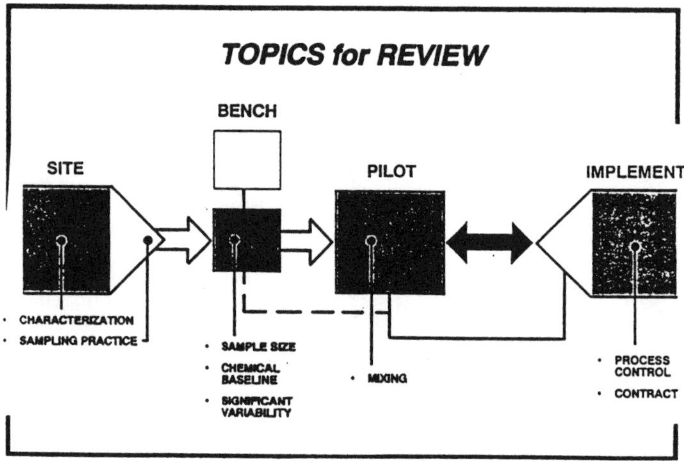

PROCESS CONTROL

- Excavation and pre-processing
- Mixing - quantities, time, energy
- Ratio control
- Documentation
- Experience

"SLOW" QC TOOLS FOR VERIFICATION ON-SITE

PHYSICAL	WASTE IN	PRODUCT OUT	TIME FACTOR
Unconfined Compressive Strength		X	D
Cone Penetrometer		X	D
Cement Content of Freshly Mixed Soil-Cement (ASTM D-2901)		X	H
CHEMICAL			
Acid Neutralization Capacity	X	X	D
Fractionation/Acid Digestion	X	X	D
Inert Spikes	X	X	D
LEACHING			
TCLP	X	X	D,W
EP Toxicity	X	X	D,W

"QUICK" QC TOOLS FOR MONITORING PROCESS UNIFORMITY

PHYSICAL INDICATORS	WASTE IN	PRODUCT OUT	TIME FACTOR
Color - Visual	X	X	M
Texture - Visual	X	X	M
Water Content	X	X	M,H
Grain Size	X	X	H
% Fines (-200 Sieve Size)	X	X	M
Unit Weight	X	X	M
Viscosity - Slump	(X)	X	M
- Shear	(X)	X	M
Temperature	X	X	M
Optical Tracers - Dyes		X	H
- Microtaggants		X	H
CHEMICAL INDICATORS			
pH	X	X	M
X-Ray Fluorescence	X	X	M

CONTRACT OPTIONS

- Prescriptive or "methods" contract
  - low contractor risk; low price
- Performance or "results" contract
  - low government risk so long as criteria clearly defined
  - higher contractor risk and cost may be offset by latitude for innovation

SUMMARY

Construction Contract Quality Management Checklist

- Waste characterization
  - Quantity
  - Condition & distribution
  - Interferants
- Treatment objectives
  - Target analytes
  - Treatment levels
  - Sampling & testing procedures
  - Required physical properties
  - Volume increase restrictions

SUMMARY

Construction Contract Quality Management Checklist

- Process Requirements

"Methods"
  - Materials
  - Quantities
  - Proportions
  - Pre-treatment requirements
  - Premixing needs for processes sensitive to waste variability

"Results"
  - Required physical and chemical performance
  - Methods used to assess performance

SUMMARY

**Construction Contract Quality
Management Checklist**

- Testing requirements
  - Methods for waste & binder
    characterization
  - Calibration of weight & volume
    measuring equipment
  - Mixing performance
  - Quick and simple "on-line"
    production tests

SUMMARY

**Construction Contract Quality
Management Checklist**

- Post-placement tests
  - Strength, leachability
  - In situ or sampled
  - Frequency
  - Duration
  - Chain-of-custody

SUMMARY

**Construction Contract Quality
Management Checklist**

- Contractor QC plan
  - Qualifications of technical and
    supervisory personnel
  - Outline of construction process
    with details on S/S methods,
    equipment & schedule
  - QC sampling & testing methods to be used
  - Action to be taken if criteria are not met
  - Health and safety

SUMMARY
Construction Contract Quality
Management Checklist

- ◆ Government QA plan
  - Identify those responsible for:
    - Review/approval of contractor QC plan
    - Conduct pre-construction and periodic meetings for QM review
    - Evaluation of construction inspections and review of submittals
  - Perform:
    - Job site sampling & testing
    - Inspection of corrective actions

SUMMARY

- ◆ Invest resources to define the problem(s) as well as practical
- ◆ Establish objectives that are sensible and achievable
- ◆ Do not assume - measure
- ◆ Treat and test with implementation scheme in mind
- ◆ Calibrate indirect measurement systems used in QC
- ◆ Maintain a clear division of responsibilities in the field and document...

  "If it's not written down, it didn't happen"

OSHA REGULATED HAZARDOUS SUBSTANCES
Health, Toxicity, Economic and Technological Data

Occupational Safety and Health Administration
U.S. Department of Labor

This two-volume book provides industrial exposure data and control technologies for more than 650 substances currently regulated, or candidates for regulation, by the Occupational Safety and Health Administration (OSHA). The health, toxicity, economic and technological data provided for each substance are intended to serve as a reference for those who are potentially exposed to one or more of these substances in their workplace, or for those who have supervisory or management responsibility for workers potentially exposed. OSHA "permissible exposure limits" (PELs) for these 650 substances reflect all updates and changes as presented in the *Federal Register*.

The information on each substance in the book includes, as available, synonyms, trade names, physical description, health effects, toxicity/exposure limits, industry use data, and NIOSH National Occupational Exposure Survey data, NIOSH National Occupational Hazard Survey data, OSHA/exposure data, engineering controls, personal protective equipment, and storage.

Indexes included in the book provide cross referencing by synonyms, trade name, and Chemical Abstract Service (CAS) number.

Below is an alphabetical listing of the first 61 substances in the book.

Acetaldehyde
Acetic acid
Acetic anhydride
Acetone
Acetonitrile
2-Acetylaminofluorine
Acetylene
Acetylene tetrabromide
Acetylsalicylic acid
Acrolein
Acrylamide
Acrylic acid
Acrylonitrile
Aldrin
Allyl alcohol
Allyl chloride

Allyl glycidyl ether
Allyl propyl disulfide
alpha-Alumina
Aluminum: alkyls
Aluminum: metal and oxide
Aluminum: pyro powders
Aluminum: soluble salts
Aluminum: welding fumes
4-Aminodiphenyl
2-Aminopyridine
Amitrole
Ammonia
Ammonium chloride (fume)
Ammonium perfluorooctanoate
Ammonium sulfamate
Amosite
n-Amyl acetate
sec-Amyl acetate
Aniline and homologues
Anisidine (o-, p- isomers)
Antimony and compounds
Antimony trioxide as Sb
ANTU
Argon
Arsenic and compounds, as As
Arsenic trioxide production
Arsine
Asbestos
Asphalt fumes
Atrazine
Azinphos-methyl
Barium sulfate
Barium, soluble compounds
Benomyl
Benzene
Benzidine
Benzidine (based dyes)
Benzo(a)pyrene
Benzoyl peroxide
Benzyl chloride
Beryllium and compounds
Biphenyl
Bismuth telluride
Borates, tetra, sodium salts
Boron oxide
plus 589 other substances

ISBN 0-8155-1240-6 (1990)   6" x 9"   2 volumes   2294 pages

HOW TO SELECT
HAZARDOUS WASTE TREATMENT TECHNOLOGIES
FOR SOILS AND SLUDGES
Alternative, Innovative and Emerging Technologies

by

Tim Holden, John Newton, Paul Sylvestri, Max Diaz
Versar, Inc.

Colin Baker
Camp Dresser & McKee

Jonathan G. Herrmann
U.S. Environmental Protection Agency

Pollution Technology Review No. 163

This book is a guide for screening feasible alternative, innovative and emerging treatment technologies for contaminated soils and sludges at CERCLA (Superfund) sites. The technology data were selected from individual treatment technology vendors.

Perhaps the most critical decision in the Superfund cleanup process is treatment selection, which can be both difficult and controversial. Determination of the cleanup level appropriate to a particular site is complicated by the fact that each site has different physical conditions and different wastes to be addressed, and many concerned contingents with competing interests.

This guide has been developed to help those responsible for remedy selection to identify potentially applicable treatment systems for the remediation of uncontrolled hazardous waste sites. It contains technical information useful for determining the feasibility and availability of different treatment technologies. This guide is intended for use as a screening tool to facilitate the scoping of site investigations and feasibility studies. It is not a substitute for in-depth engineering analyses.

For each of the treatment technologies, information is presented on (a) the generic system, (b) individual, unique systems, (c) developmental status, (d) process schematics, (e) characteristics affecting treatment performance, and (f) contacts. Some limited information is also presented about pretreatment, materials handling, and residuals management requirements.

An Addendum presents additional technology profiles for innovative and emerging technologies. These technologies have been evaluated and documented as part of the Superfund Innovative Technology Evaluation (SITE) Program. The book also includes a Vendor Index.

A **condensed table of contents** appears below.

ISBN 0-8155-1213-9 (1989)   6"x9"   150 pages

DUST CONTROL HANDBOOK

by

V. Mody and R. Jakhete
Martin Marietta Laboratories

Pollution Technology Review No. 161

This handbook consolidates information developed by industry and government laboratories on dust control engineering techniques for metal and nonmetal mineral processing. Although designed for the minerals processing industry, the technology is pertinent for other industries as well.

The concepts of dust, its prevention, formation and control are examined, including wet and dry control systems, personal protection, and testing methods. Costing methodologies are also examined.

Prospective users of this handbook include maintenance foremen, plant engineers, mill supervisors, and plant safety directors, or the like. The information is described and illustrated in a manner to assist one to have enough information to implement a dust control technique without need to refer to other sources.

CONTENTS

1. DUST AND ITS CONTROL
What Is Dust?
How Is Dust Generated?
Types of Dust
Why Is Dust Control Necessary?
Health Hazard Factors
How Is Dust Controlled?

2. PREVENTING DUST FORMATION
Belt Conveyors/Transfer Chutes
Enclosures/Crushers/Screens
Storage Bins and Hoppers
Bucket Elevators
Feeders/Screw Conveyors
Pneumatic Conveyors
Grinding Mills/Dryers
Stockpiles/Haul Roads
Truck and Railroad Car Dumping
Powder Handling and Packing

3. DUST CONTROL SYSTEMS
Why Dust Control?
Types of Dust Control Systems
Selection of a Dust Control System
Dust Collection Systems

Wet Dust Suppression Systems
Airborne Dust Capture Systems
Design of a Water-Spray System

4. COLLECTING AND DISPOSING OF DUST
What Is a Dust Collector?
Necessity for Dust Collectors
Types of Dust Collectors
Types of Inertial Separators
Fabric Collectors
Wet Scrubbers
Electrostatic Precipitators
Unit Collectors
Selecting a Dust Collector
Fan and Motor
Disposal of Collected Dust

5. SPECIFIC ILLUSTRATIONS

6. ESTIMATING COSTS OF DUST CONTROL SYSTEMS
Necessity for Cost Estimates
Cost Estimating Procedures
Cost Compounds
Cost Justification

7. CONTROLLING SURROUNDING DUST SOURCES
Contributing Sources/Factors
Administrative Measures
Preventive Maintenance Program
Proper Operating Procedures
Managerial Measures
Outside Dust Source Control
Recirculation of Uncontrolled Emissions

8. SAMPLING DUST IN THE WORK ENVIRONMENT
Personal Dust Sampling
ABC's of Personal Dust Sampling
Personal Dust Sampling Equipment
Sampling Procedure
Errors in Respirable Dust Sampling

9. TESTING DUST CONTROL SYSTEMS
Reason for Testing
Dust Collection System Testing
Wet Dust Suppression
Evaluating Dust Control Systems

ISBN 0-8155-1182-5 (1988) 8½"x11" 203 pages